ISLAMIC BIOETHICS: PROBLEMS AND PERSPECTIVES

INTERNATIONAL LIBRARY OF ETHICS, LAW, AND THE NEW MEDICINE

VOLUME 31

For other titles published in this series, go to
www.springer.com/series/6224

Islamic Bioethics:
Problems and Perspectives

DARIUSCH ATIGHETCHI

Second University,
Naples, Italy

 Springer

Library of Congress Control Number: 2008941017

ISBN 978-1-4020-9615-0 (PB)
ISBN 978-1-4020-4961-3 (HB)
ISBN 978-1-4020-4962-0 (e-book)

Published by Springer
P.O. Box 17, 3300 AA Dordrecht, The Netherlands

www.springer.com

Printed on acid-free paper

To BIANCA, my mother

TABLE OF CONTENTS

Note... xi
Acknowledgements xiii

1 INTRODUCTION TO MUSLIM LAW 1
  The Origins 1
  The Present.. 7
  Conclusion 9

2 FEATURES OF ISLAMIC BIOETHICS 13
  The Value of the Different Positions... 13
  The Problems of Ethical Pluralism 14
  Algeria, Tunisia, Pakistan And Iran.. 16
  The Dependence on Muslim Law 18
  The Political Dimension of Islamic Bioethics 19
  The Principles of Bioethics.. 21
  Cultural Sensibilities and Medical Ethics 23
  Bioethics and Society.. 26
  Bioethics and Apologetics 28
  Bioethics and Muslim Countries... 28
  Conclusion 28

3 SOME ASPECTS OF MEDICAL ETHICS 31
  Introduction... 31
  Principles and Characteristics 33
  The Doctor–Patient Relationship.. 37
  Men and Women 39
  Autonomy and Consent of the Patient 47
  Saudi Arabia, Libya, Tunisia, Algeria 51
  The Problem of Penal Mutilation.. 54
  Doctors and Penal Mutilation 57
  Conclusion 63

4   CONTRACEPTION AND POPULATION CONTROL   65
     Some Classic Formulations...   65
     The Contemporary Debate: The Pro-Contraception
     Jurists   71
     Sterilisation   78
     The Opponents of Contraception...   79
     The Socio-Political Context...   82
     The Case of Iran...   86
     Conclusion   89

5   ABORTION   91
     Introduction..   91
     The Sources of Tradition..   92
     Before Ensoulment...   95
     After Ensoulment   98
     The Penal System...   100
     The Duration of Pregnancy   105
     Three Bioethical Problems...   106
     The Debate and Contemporary Opinions   111
     Rape, Adultery and Fornication...   115
     Legislation in Some Countries...   119
     Abortion and the Defence of Honour...   131
     Conclusion   133

6   ASSISTED PROCREATION   135
     Introduction..   136
     Legal Adoption   139
     Juridical-Religious Formulations..   140
     Opinions in Shi'ite Islam   148
     Problems Relative to the Embryo   151
     Society and Legislation..   154
     Conclusion   159

7   THE DEVELOPMENT OF ORGAN TRANSPLANTS........................   161
     Ethical-Juridical Principles   161
     Some Features of the Debate   163
     Transplants from Living Donors...   168
     Transplants from Corpses   170
     The Debate on the Criteria of Death...   174
     The Organ Trade   178
     Uterine Transplantation...   180
     Xenotransplantation   181
     Some National Legislations   183
     Conclusion   196

8   AIDS 199
     Introduction 199
     Particular Aspects 207
     The Countries 210

9   THE OPINIONS ON GENETICS 235
     Principles and Values 235
     The Debate on Genetics 237
     Human Cloning 241
     Positions Tolerating Human Cloning 245
     Research on Stem Cells 248
     The Abortion of Handicapped Foetuses 250
     Consanguineous Marriage 254
     Pre-Natal Diagnosis 259
     Conclusion 265

10 THE END OF LIFE 267
     Suffering and Illness 267
     The Incurably and Terminally Ill Patient 271
     Palliative Care 272
     Information and Consent of the Seriously and
     Terminally Ill Patient 274
     The Living Will 283
     Euthanasia 285
     Suicide and Martyrdom 290
     Death 296
     Post-Mortems 297
     Conclusion 303

11 FEMALE GENITAL MUTILATION IN SPECIFIC
     MUSLIM AREAS 305
     Some Historical-Juridical Elements 305
     The Opinions in Favour of Genital Mutilation 309
     The Opinions Against Genital Mutilation 311
     The Debate in Some Countries 316

12 THE KORAN AND MODERN SCIENCE 327
     Introduction 327
     Scientific Exegesis of the Koran 332
     Moderate Concordism 345
     The Opponents of Scientific Exegesis 347
     Conclusion 349

13 GENERAL CONCLUSIONS 353

     AUTHOR INDEX 355

     SUBJECT INDEX 365

NOTE

For the terms in Arabic and other languages, a simplified system of transcription has been adopted without special signs to distinguish similar consonants or long vowels. The proper names of historic figures appear in their commonest spelling, for modern names they are shown as they appear in the sources. As there is no distinction between first name and surname, the names are shown in such a way as to make them easy to recognise.

To help the reader confirm the data, we have chosen sources translated into Western languages.

ACKNOWLEDGEMENTS

Francesco Castro (Full Professor of Muslim Law, Second University, Rome-Tor Vergata); Renzo Pegoraro (President of the Lanza Foundation, Padua); Gian Maria Piccinelli (Full Professor of Muslim Law, Second University, Naples).

Special thanks to Joan Rundo for the translation and revision of the text and for her helpful comments.

INTRODUCTION TO MUSLIM LAW

THE ORIGINS   1
THE PRESENT   7
CONCLUSION   9

THE ORIGINS

Life, including aspects of daily life, in Muslim societies has always been moulded and characterised by the precepts and values of Islamic Law, i.e. by the *Shari'a*. This can be defined as the "straight way – which Muslims must observe – revealed by God to regulate and evaluate human conduct (that is, the acts of the body, *a'mal al-badan*, which are carried out externally, and not the acts of the heart, *a'mal al-qalb*, which concern the interior) limiting, for the usefulness of man in his earthly and after-earth life, the original liberty of human actions". In other words, the *Shari'a* is the religious Law of divine origin revealed and structured in an almost omni-comprehensive system according to the definitions of jurists (*fuqaha, ulama* and *muftis*) from the first centuries onwards.[1] Whilst in Western culture the law regulates only some sectors of behaviour, the influence of the *Shari'a* is far more extensive in the sphere of the private, social, political and religious life of the believer. The result is the totalising character of Islam as a life system that interweaves religion and politics, the sacred and profane, the material world and the spiritual sphere in ways that are difficult to understand according to Western conceptual and juridical categories. For Muslim law, each human act belongs to one of the following five categories: compulsory (*fard, wagib*), recommended (*mandub, mustahabb*), free (*ja'iz, mubah*), reprehensible or unadvised (*makruh*) and forbidden (*haram, mahzur*).

The *Shari'a* has a considerable capacity of adaptation to situations as they vary; a peculiarity made historically possible above all by the absence of a supreme authority supervising orthodoxy. This flexibility was one of the elements which contributed to the expansion of Islam over the centuries to completely different peoples and geographical regions, without destroying the local cultures but only instilling the few fundamental principles of the faith, of worship and of Muslim law. The principal truths of the faith are listed in Koran 2.177 and 4.136: "O ye who believe! Believe in God and His Apostle, and the scripture which He hath sent to His Apostle and the scripture which He sent to those before (him).

---

[1] Castro F., Diritto musulmano e dei paesi musulmani, in *Enciclopedia Giuridica Treccani*, Roma, Istituto Poligrafico e Zecca di Stato, 1989, Vol. II, 1–17. Id., Diritto Musulmano, in *Digesto delle discipline privatistiche*, Vol. VI, IV Edizione, Torino, UTET, 1990, 1–66.

Dariusch Atighetchi, Islamic Bioethics: Problems and Perspectives.
© Springer Science+Business Media B.V. 2009

Any who denieth God, His angels, His Books, His Apostles, and the Day of Judgment, hath gone far, far astray."[2] In addition, there are the five pillars of Islam: the profession of faith, the five ritual daily prayers, legal alms, fasting in the month of Ramadan and the pilgrimage to Mecca.

At the basis of the *Shari'a* we find the four "roots of the law" (*usul al-fiqh*) or sources, from which the principles and rules of Muslim law have been drawn. These "roots" are the Koran, the *Sunna* (Tradition), *igma'* (unanimous consensus) and *qiyas* (reasoning by analogy); only the first three are of divine origin.

1. The Koran, the supreme source of the religion and law, is the direct and literal word of God revealed to Prophet Muhammad (d. 632) through Archangel Gabriel and not a simple text "inspired" by God as is believed for the Old and New Testaments. For Islam, Muhammad is the last and most important of a long series of Prophets bearing a divine message; in order of importance and not chronological order, Muhammad is followed by Jesus Christ (Ar. Isa) and Moses, all of whom are related with a Holy Book from which the definition of "religions of the Book" derives, which unites the three monotheistic faiths.[3] In this context the Koran is the last and perfect Revelation given by

---

[2] Koran 2.177: "but it is righteousness – to believe in God and the Last Day, and the Angels, and the Book, and the Messengers." As in the rest of the book, the translation of the Holy Text by a Muslim author has been preferred, namely Yusuf Ali, A. (Amana Corporation, Maryland, 1983).

[3] Apart from belonging in both cases to the group of Prophets and Envoys of God, there are ineliminable differences in the interpretation of the figure of Jesus in the two religions. For Christianity, Jesus is a God (hence the very term of "Christianity"), and the Son of God. For Islam, however, in the respect of monotheism, there exists exclusively one, unique and transcendent God. For Muslim apologetics, Christianity is not a pure monotheism. This criticism is directed above all at the concept of a triune God, i.e. faith in three divinities: Father, Son and Holy Spirit, which is associated with the worship of other divinities such as Virgin Mary. In two key passages, the Koran explicitly admonishes Christians against these false beliefs. Verse 4.171 says:

O People of the Book! Commit no excess in your religion: nor say of God aught but the truth. Christ Jesus the son of Mary was (no more than) an apostle of God, and His Word, which He bestowed on Mary, and a Spirit proceeding from Him: so believe in God and His apostles. Say not "Trinity": desist: it will be better for you: for God is One God: Glory be to Him: (far Exalted is He) above having a son.

In 5.116 this criticism also includes Virgin Mary:

And behold! God will say: "O Jesus the son of Mary! Didst thou say unto men, worship me and my mother as gods in derogation of God?" He will say: "Glory to Thee! Never could I say what I had no right (to say). Had I said such a thing, Thou wouldst indeed have known it. Thou knowest what is in my heart, though I know not what is in Thine".

Every justification of the faith and adoration of other beings outside the One God is refused. These, like other beliefs, are said to be the work of the alterations made by the Apostles, i.e. the first followers of Christ, to the original Christian message which was in line with the pure monotheism of Abraham. The Catholic Church itself, according to Muslim tradition, is said to have deliberately confirmed these mystifications.

The totalising character of Islam is personified by the figure of Muhammad who simultaneously embodied (contrary to Christ) the roles of political and military leader and social guide, as well as that of the Prophet (*nabi*) and Envoy of God (*rasul*).

God which "updates" and definitively "supersedes" the previous revelations altered by their respective followers, restoring the original and pure Abrahamic monotheism. In addition, as the direct word of God that has reached us unaltered compared to its first written version, dating back to the times of the third Caliph Uthman (644–656), the Koran is the perfect and inimitable monument of Arabic language and literature. For all these reasons, the Koran is a divine miracle, which explains the tremendous role it continues to play in the spiritual and social life of Muslims, perhaps to a greater extent than that of the previous Holy Books of Jews and Christians.

2. The *Sunna* (Tradition) is the set of rules based on the words, actions and tacit assents of the Prophet. Its importance is dictated by the Koran that requires the faithful to imitate the example of Muhammad. These testimonies are expressed in the *ahadith* (sing. *hadith*), i.e. the accounts or "sayings" of the Prophet, numbering some dozens of thousands, which at times are in contrast with one another and were compiled into various collections after his death,. Their greater or lesser authenticity is attested through two criteria: the authority of the testimony of the transmitters forming an uninterrupted chain which reaches the person who first heard the Prophet, and the examination of the content of the collection or Tradition; on these bases, the *ahadith* are distinguished into authentic or healthy (*sahih*), good or acceptable (*hasan*) and weak (*da'if*). As a consequence, these accounts possess a different normative value; therefore the *ahadith* present in the six most authoritative collections (Bukhari, Muslim, Abu Dawud, al-Tirmidhi, al-Nasa'i, Ibn Maja), but above all in those of Bukhari (d. 870) and Muslim (d. 875) (considered *sahih*), act as models to be imitated by the faithful.[4] The *Sunna* represents the perfect supplement to the Koran, completing it when the latter says nothing and interpreting it authentically when it is ambiguous or incomplete; for example, some Koranic passages introduce prayer without establishing the ritual, or prescribe ablutions without specifying the ways to perform them and only the *ahadith* have the authority to fill this shortcoming. At the same time, many juridical rules are also based on the *Sunna* and not the Koran, for example the rules on the division of booty, or those for the fast, the pilgrimage, and so on. This explains why the words of the Prophet make the second source of law of vital importance for the organisation of life in Muslim society and, indeed, from the juridical point of view, due to the variety of issues dealt with, it has become a source that is of greater importance than the Koran itself.

3. The Prophet said: "Allah has protected you from three things: . . ., that those who follow what is false should not prevail over those who follow the truth, and that you should not all agree in an error."[5] In fact, *igma'*, or consensus

---

[4] Santillana D., *Istituzioni di diritto musulmano malichita con riguardo anche al sistema sciafiita*, Roma, Istituto per l'Oriente, 1926–1938, Vol. I, 36–37.

[5] Abu Dawud, *Sunan, Trials and Fierce Battles* (*Kitab Al-Fitan Wa Al-Malahim*), Book 35, no. 4240, in www.usc.edu/dept/MSA/fundamentals/hadithsunnah/abudawud/satintro.html

of the community on issues of a ritual, legal and religious nature, when it is "continuous and unanimous" is valid as a source of Law that is the equivalent to the Koran or a *hadith* reflecting, in this case, a position inspired by God. According to the philosopher and jurist Averroes (d. 1198) two degrees of consensus should be distinguished: (a) regarding elementary duties, the *igma'* involves the consensus of all the faithful; (b) for the rules of worship and law, the legal schools generally consider the consensus of the doctors of Muslim Law sufficient as only they are capable of "correctly understanding the law".[6] In the opinion of some scholars, *igma'*, although defined as the third source in order of importance, is of decisive importance as it guarantees the authenticity and authority of the first two sources of law.[7] Indeed, it is thanks to the continuous and unanimous consensus of the faithful that the Koran is recognised as the authentic word of God and the *ahadith* as the word of the Prophet.

4. When the first three sources of divine inspiration fail to provide a clear rule of conduct in a given situation, it becomes lawful for the majority of the legal schools and within very varying limits, to have recourse to *qiyas*, or reasoning by analogy. This is an instrument of logic that allows deducing, with the aid of specific criteria, new rules of conduct to handle unforeseen situations, taking as a starting point the prescribed rules in similar cases identified in the Sacred Sources.[8] Muhammad himself and his Companions, in the case of necessity, used this method of logic and deduction, as attested by the words (*hadith*) addressed by the Prophet to Mu'adh Ibn Jabal, one of his envoys in the Yemen. The Prophet asked: "How will you judge when the occasion of deciding a case arises? He replied: I shall judge in accordance with Allah's Book. He asked: (What will you do) if you do not find any guidance in Allah's Book? He replied: (I shall act) in accordance with the Sunnah of the Apostle of Allah (peace be upon him). He asked: (What will you do) if you do not find any guidance in the Sunnah of the Apostle of Allah (peace be upon him) and in Allah's Book? He replied: I shall do my best to form an opinion and I shall spare no effort. The Apostle of Allah (peace be upon him) then patted him on the breast and said: Praise be to Allah Who has helped the messenger of the Apostle of Allah to find something which pleases the Apostle of Allah."[9] There remains the fact that the method by analogy is an essential instrument,

---

[6] Santillana, op. cit., 42–44.

[7] Schacht J., *Introduzione al diritto musulmano*, Torino, Edizioni della Fondazione Giovanni Agnelli, 1995, 123.

[8] In other terms, according to the expression of Juynboll Th.W., *Manuale di diritto musulmano secondo la dottrina della scuola sciafeita*, Milano, Vallardi, 1916, 32, *qiyas* consists of transferring one precept from the root (that is, from the case expressly formulated in the Sacred Sources) to a branch (that is, to a new case not mentioned in the text).

[9] Abu Dawud, *Sunan, The Office of the Judge*, Book 24, no. 3585, in www.usc.edu/dept/MSA/fundamentals/hadithsunnah/abudawud/satintro.html

due to its flexibility, for the adaptation of the classic juridical heritage to the challenges of modernity. At the same time, it is exposed to the criticism of traditionalists, according to whom there already exists in the Koran and in the *ahadith* everything that man needs without having to have recourse to fallacious human logic.

The first three "roots" or sources of the law and religion (the Koran, *Sunna* and *igma'*) represent the only unquestioned authorities in Islam, which does not have a supreme juridical-religious authority, nor an official teaching capable of guiding the faithful in all the circumstances on which the three sources are lacking in precise and exhaustive directives. The absence of this central authority is currently felt in particular regarding the need to rapidly provide authoritative answers valid for the whole of the Muslim world in the face of the ethical challenges presented by biomedicine.

When the Law is silent or it is impossible to decide its real meaning, despite recourse to interpretation by analogy, it is possible to turn to two subsidiary rules of law, namely the criterion of utility (*istislah*, the principle of utility = *maslaha*),[10] used above all by *Malikites and Shafi'ites* and the criterion of equity (*istihsan*). The *Hanafites* in particular turn to the latter when deduction by analogy means excessively rigid or potentially dangerous positions whilst the criteria of utility and equity allow identifying the best or preferable solution.

In fact, after the death of the Envoy of God (d. 632) it was evident that a mere literal knowledge of the Koran and of the "sayings" of the Prophet was not sufficient to orient, in practice, the conduct of the faithful due to the lack of indications on many problems, but also due to the emergence of contrasting interpretations of the existing contents and precepts. On the initiative of expert technicians (the jurists) capable of consulting the sources, the "science of canonical prescriptions" (*ilm al-fiqh*), or science of the law, began to develop, which has become the most important of the sciences of Islam, even greater than direct knowledge of the Koran and of the *Sunna*, as only their correct interpretation provides the faithful with the valid directives to follow.[11]

During the early centuries after Muhammad, i.e. until approximately the 10th–11th centuries AD, the great jurists and their students could use the analogical method (*qiyas*) with considerable freedom in their effort to "personally"

---

[10] Santillana, op. cit., 71. The important impact of this juridical criterion is underlined by al-Qarafi (d. 1285) when he points out that in Islam the precepts have multiplied with the passing of time:

  Indeed, we see the Companions of the Prophet introducing many provisions because they had recognized their usefulness; they decided to put the Koran into writing, whilst nothing authorized them to do so in the past laws, they minted coins, built prisons and so forth. For the same reason, i.e. usefulness, the law subjected testimonial evidence to strict conditions that were not at all to be found in Tradition (*hadith*).

[11] Juynboll, op. cit., 15.

interpret (*igtihad*),[12] explain and comment upon the "roots" of the law, consequently drawing on them for the principles and rules of the *Shari'a*. This period coincides with the "creative" phase of the Law, thanks to the work carried out by experts on the law, including the founders of the canonical juridical currents or schools of *Sunni* Islam which take their names: Abu Hanifa (d. 767) with the *Hanafite* school; Malik Ibn Anas (d. 795) initiator of the *Malikite* school; Muhammad al-Shafi'i (d. 820) who founded the *Shafi'ite* school and Ibn Hanbal (d. 855) with the *Hanbalite* school. These schools, which are still active, are considered the canonical schools as they share the juridical guidance of Muslim orthodoxy in *Sunni* Islam (which accounts for about 90% of Muslims), whilst the *Shi'ites* (the remaining 10%) have their own juridical schools. The presence of four canonical *Sunni* schools brings us to the conclusion that the different positions they sustain cannot concern essential elements such as dogmas or the principles of the faith. However, when schools agree on some specific point, it is stated that this is the effect of *igma'* or consensus and, therefore, "compulsory doctrine" for believers. When, on the other hand, there exist differences, the rules of each school have particular importance for the followers of each school.

Today, the *Hanafite* school is the most widely diffused in the Muslim world since it was imposed, as the official school, in the territories of the Ottoman Empire. The *Malikite* school is present today in the Maghreb, in sub-Saharan Africa and in Egypt. The *Shafi'ite* school is active in the Indian subcontinent, east Africa, Egypt and Yemen. The *Hanbalite* school is currently dominant in Saudi Arabia.

Since the 10th–11th centuries, a decisive date in the development of Muslim law, in *Sunni* Islam (not in *Shi'ite* Islam) recourse to the effort of personal interpretation (*igtihad*) of the sources has no longer been legitimate (except formally) for at least two fundamental reasons: (a) according to the scholars of the period, the main legal problems had already been solved by previous masters; (b) it was no longer possible to acknowledge prestige and authority to any jurist similar to that granted to those masters of the past.[13] With the "closure of the door of *igtihad*" (accepted conventionally by the *Sunni* schools) jurists are no longer recognised as having the authority necessary to "base themselves directly on the Koran or on Tradition, or to depart from *igma'*, consensus, nor to personally apply the methods of logical deduction. All this is given to them already in the books of their school with their countless glosses and commentaries, and there is nothing else to do except follow the path already laid out."[14] Therefore, the main function of the jurists, after about the 11th century, was to be limited to imitation (*taqlid*)

---

[12] Santillana, op. cit., 70 specifies the characteristics of *igtihad*:

> *Igtihad* is therefore not free will, a subjective and personal opinion, but cautious will, the juridical conscience of the interpreter, refined and trained by intense and profound meditation of the law as a whole, and therefore leads to finding that reason of usefulness that is the informing spirit of the whole juridical body, in order to apply it to the specific case put to the interpreter.

[13] Macdonald D.B., Idjtihad, in *Encyclopédie de l'Islàm*, Paris, 1971, Vol. III, 1052.
[14] Pareja F.M., *Islamologia*, Roma, 1951, 402.

of the way previously marked out or to a "technical" interpretation of the classic provisions aimed simply at adapting them to the new concrete situations (without the possibility of formulating original juridical solutions); to explanation of the rules contained in law; to the task of classifying and commenting upon the codes, principles and relative previously established interpretations.

All this was accompanied by the formulation of an impressive amount of rules, norms and interpretations of differing values, defined from the Sources by the jurists, the only ones historically qualified for this purpose, who made it problematic for the faithful to make any attempt at direct access to the Sources.

From the 11th century there thus began a long historical phase of more restrictive juridical reflection, lacking in the flexibility and creativity that had characterised it in the early centuries, a phase of decline which lasted until 1900 when, on contact with the West, the need for an awakening of the Muslim world in all fields, including the juridical one, became evident. The many theories put forward to promote this reawakening include two main trends: (1) that which encourages a direct return to the Sacred Sources, interpreting them "freely" (remaining faithful to the spirit of Islam) without remaining bound to the traditional mediation of jurists with their interpretations; (2) opposed to this, we find those who ask for a return to the literal application of the *Shari'a* as only it contains everything that is useful for guaranteeing a return to the dominance and successes of the golden age of Muslim civilisation.

THE PRESENT

What is to be done when the sources of Muslim law, with their canonical interpretations, drawn up before the 11th century, do not express themselves or do not express themselves clearly on new situations caused by the continuous historical and social changes over the centuries (in particular after the closure of the "door of *igtihad*")? On these (very frequent) occasions, the faithful call on the doctors of Muslim law (*muftis, fuqaha, ulama*) who, using the analogical method (*qiyas*) and personal interpretation (*igtihad*), issue *fatawa* (sing. *fatwa*), i.e. juridical opinions with which they explain to the faithful the prescriptions contained in the Sources, indicating how to behave in the case in point. Once again, we can recall that through the *fatwas*, the jurists should not offer new juridical solutions to problems (they may be accused of departing from the *Shari'a*) but present "technical" interpretations of already existing rules to model them on the new situations. In spite of this, depending on the authority and the preparation of the figure issuing them, the *fatwa* represents, at least morally, the rule of conduct to be followed for the faithful of the same juridical school. On the contrary, from the juridical point of view, the *fatwa* (as an opinion) is not binding for the faithful[15] (except possibly

---

[15] See the *fatwa* by *sheikh* Mahmud Shaltut with the significant title: *La fatwa n'oblige pas* [*The Fatwa does not oblige*], in Borrmans M., Fatwa-s algériennes, in Scarcia Amoretti B. and Rostagno L. (eds.), *Yad Nama*, Roma, Bardi, 1991, Vol. 1, 83–107, 83 note 2.

for the faithful of the same school) and may be contradicted by the opinions of other jurists, all the more so as a jurist can modify his own opinion in space and in time.[16]

Two "types" of *mufti* can be distinguished historically, the private ones (almost all of them) who attained this function thanks to the consideration won amongst the faithful due to their juridical-religious competence "acquired freely and almost without a licence"[17] and *muftis* appointed by the State (very few individuals), for example the so-called *muftis* of the Republic (or similar authorities). Through this appointment, governments intend to obtain specific indications on the religious validity of determined positions, by a single qualified interlocutor and who is in some way a representative of the community of the faithful. In Egypt, for example, the *fatwas* of three authorities are of particular value for the State: the Grand *mufti* of the Republic, Ali Gumaa; the *sheikh* of Al-Azhar, Muhammad S. Tantawi and the Committee of *Fatwas* of the University of Al-Azhar. However, the intrinsic value of a *fatwa* does not change if it is issued by a private *mufti* or by a state *mufti*, as it remains a juridical opinion that can be challenged by other jurists.

Even if the authority of the *mufti* is now predominantly moral (it is the State that makes the laws and it is not obliged to implement the advice of jurists or even of the official *muftis* of the State) their influence on society must not be underestimated. This influence is in fact due in the first place to the prestige of the jurists which is spontaneously acknowledged by believers, but also to the technical characteristics of the *fatwa*, an opinion with which a doctor of the Law publicly (orally or in writing) answers questions of different kinds asked by private citizens or by various institutional bodies. In the present day, this has become constant recourse to mass media, thanks to special columns in newspapers and magazines, as well as specific slots in radio and television programmes.

Today, to provide more adequate answers to the challenges of modernity and to problems that go beyond the traditional competences of the jurists, Muslims increasingly have recourse to the opinions provided by meetings, conferences, congresses and academies that bring together jurists and experts from various parts of the Muslim world.

1. The first purpose is that of collectively examining the issues to express decisions-resolutions and/or recommendations (the former with a prescriptive value) which, in any case, essentially preserve the value of opinions that can be modified and challenged by other juridical-religious subjects and bodies. Even if Muslim law has never known councils or assemblies of doctors to solve dubious juridical or theological questions, these meetings are legally legitimised (using reasoning by analogy) if it is possible to find similar precedents in the life of

---

[16] Abu-Sahlieh S.A., L'Institution du mufti et de sa Fatwa/Décision en Islam, *Praxis Juridique et Religion*, 1990, 7, 125–148.

[17] Castro, Diritto Musulmano, in *Digesto delle discipline privatistiche*, art. cit., 14.

the Prophet and his Companions. In the case in question, the meetings by the Companions of Muhammad to solve the doubts of the faithful are taken as a juridical precedent.[18] In divergences between doctors of the law, there does not exist a superior authority that can decide; the dispute is solved only by the passing of time or by the dominant opinion which thus comes to represent the consensus of the community as a whole (*igma'*).[19]

2. A second purpose of these meetings consists of the attempt to outline a clear and univocal position, the expression of the community of the faithful (the *umma*) and, in this case, with a strong normative value (referring to the *igma'*).

In fact, the problem of the effective representation of similar bodies, in the absence of a supreme authority and official teaching, remains unsolvable for at least two reasons: (1) Because the largest organisations in the arena[20] are sponsored and financed by groups of different Muslim countries, lending themselves to the accusation of being their political-religious instrument. (2) Because the number of Muslims in the world is constantly growing (at present they number almost one and a half billion) whilst the differences within it make it increasingly composite and fragmented to the point that it is very difficult to identify common concrete and operative programmes beyond the acceptance of the principles of the faith and defence of Islam.

CONCLUSION

At the conclusion of this juridical introduction, it would appear useful to summarise some important items of information which we will constantly meet in subsequent chapters and which significantly influence the course and quality of the debates on science and on the development of medical science in particular:

1. The absence of a supreme juridical-religious authority and of official teaching.
2. The substantial value of "opinion" of each pronouncement expressed both by individual doctors of the Law and by meetings, academies, etc. of jurists and experts on all the new problems requiring answers that are clear as well as coherent with the principles of Islam. Each of these positions can always be potentially challenged by other meetings or by other jurists. In other words, if the sacred texts do not express themselves exhaustively on a given problem, each competent Muslim, whilst expressing himself in the name of Islam, only exposes his own opinion and interpretation with which other Muslims may more or less agree. None of these opinions, although deeply rooted by deduction in

---

[18] Assemblée de l'Académie de Droit Islamique, in *Etudes Arabes*, Dossier, 1989, 1, 76, 113–121.

[19] Castro, Diritto Musulmano, in *Digesto delle discipline privatistiche*, art. cit., 14.

[20] Starting from the three main ones: the *Rabitat al'Alam al-Islami* (Muslim World League – MWL) based in Mecca; the *Munazzamat al-Mu'tamar al-Islami* (Organisation of the Islamic Conference – OIC) based in Jeddah and the *Mu'tamar al-'Alam al-Islami* (World Muslim Conference – WMC) based in Karachi.

the Sacred Sources, may be considered equivalent to the eternal and unchangeable positions of Muslim law.

3. The traditional role of the Muslim doctors of the law as the sole qualified interpreters of the Koran and of the *Sunna* is now going through a critical period. The causes for this include the increased literacy rate which leads many people to bypass the interpretative mediation of the Muslim jurists (classical and modern) to go directly to the sources, interpreting them personally. If we consider that this phenomenon involves the tens of thousands of *ahadith* with a different normative value (and at times contradicting one another), we can understand the reason for the proliferation of theories, opinions, and judgements that are more or less anchored to the Sacred Sources but often with a contradictory tone characterising the present-day Muslim world, in particular on the issues that we will be dealing with.

4. Today, the constitutions of Muslim states contain an article in which it is specified that the *Shari'a* forms one of the sources or the main source of the Laws of the state. Nevertheless, the majority of these countries are governed by codes, institutions and legal systems inspired by Western systems. The attempt to reconcile these influences with the oldest institutions of the *Shari'a*, has been translated, more often than not, into the isolation of the latter. At the same time, the jurists mainly represent only a moral and religious authority against the power of the State, the only one authorised to issue laws; in addition, the various State-appointed *muftis* are government employees and, as such, solicited to justify, from the juridical-religious point of view, the political and social programmes of the government. However, as far as bioethics is concerned, probably due to the delicacy of the subjects in question, state legislations generally give great consideration to the opinions of the most authoritative doctors of the Law.

All these preliminary remarks help explain why the different positions taken by the different countries on issues of bioethics can easily find precedents which are more or less authoritative in classic juridical reflection such as to legitimise them from the religious point of view. This is, for example, the case of population control policies and of legislations on abortion where the various positions of contemporary jurists are linked with the differentiated reflections of the past whilst, in their turn, the state laws can refer to both.

The link of continuity that can often be identified between the positions of the various states with respect to the juridical conceptions of the past represents the reason why we will have remarks on classic formulations alongside the definitions of contemporary doctors of the law and the national laws. At the same time, we can note those areas in which regulation by classic juridical doctrine is shortcoming or absent (e.g. brain death criteria) giving the opportunity to present the course taken by jurists with the aim of religiously legitimising these modern criteria.

The aim of the work is to offer a first attempt at a unitary reflection on three levels (juridical reflection, national laws and conduct of the faithful) on a representative part of the aspects involved in the debate on "bioethics" in the Muslim world, taking for granted the coverage of issues that evolve and change on a daily basis. Particular attention will be paid, out of all the Muslim countries, to Saudi Arabia and Egypt both due to the prestige of their respective juridical-religious bodies and the important representative role played by these two countries in the Muslim context.

FEATURES OF ISLAMIC BIOETHICS

THE VALUE OF THE DIFFERENT POSITIONS   13
THE PROBLEMS OF ETHICAL PLURALISM   14
ALGERIA, TUNISIA, PAKISTAN AND IRAN   16
THE DEPENDENCE ON MUSLIM LAW   18
THE POLITICAL DIMENSION OF ISLAMIC BIOETHICS   19
THE PRINCIPLES OF BIOETHICS   21
CULTURAL SENSIBILITIES AND MEDICAL ETHICS   23
BIOETHICS AND SOCIETY   26
BIOETHICS AND APOLOGETICS   28
BIOETHICS AND MUSLIM COUNTRIES   28
CONCLUSION   28

This chapter will indicate some of the problems and/or features emerging from a reflection on the topics of bioethics developed to date in the Muslim world. These reflections are those of a wide number of subjects, including doctors, jurists, religious authorities, sociologists and so forth, but the parties most involved are doctors and Muslim jurists. For this reason, the chapter will shift constantly between their different arguments.

The general reflections on bioethical issues in Muslim medical-juridical literature are based on the foundations of the Muslim faith and law. This explains why reference to the direct and literal Word of God (the Koran), divine law (*Shari'a*) and the most important interpretations that developed over the centuries still plays such an important role in Muslim bioethics.

THE VALUE OF THE DIFFERENT POSITIONS

At present, in order to formulate positions that are as representative as possible of the vast community of the faithful on new problems (i.e. those on which tradition has not clearly expressed itself or on which it has made no pronouncements), dependence is often on responses from pan-Muslim congresses and conferences (e.g. the Muslim World League, the Organisation of the Islamic Conference) as well as the various national *fatwas* committees etc. Their value, however, remains essentially that of a legal opinion (*fatwa*) which may be challenged by other juridical subjects.

Alongside these responses, documents issued by the biomedical bodies should also be taken into consideration; disagreement between medical positions and those of the religious authorities is not infrequent. It has to be remembered that the modernisation of health care in Arab-Muslim countries has mainly been

Dariusch Atighetchi, Islamic Bioethics: Problems and Perspectives.
© Springer Science+Business Media B.V. 2009

promoted by doctors and the political authorities, generally followed by the approval of the "official" (i.e. dependent on the State) religious authorities (after having been called on to express themselves on the subject).

In recent decades, the authority of the doctors of Muslim law, although prevalently limited to the moral field, has found new grounds of application on bioethical issues. In this regard, the "true religion" intends to play a decisive role on the basis of its own moral primacy in the face of a West perceived as an invading, but above all ethically disoriented, innovator.

At the same time, the new problems confronted by bioethics stimulate a modernisation of Muslim law although it is increasingly overlooked by the positive law in force in the individual countries. This action of bioethics, moreover, is parallel and contemporary with the influence exercised by Western juridical institutions and models concerning penal, civil law, etc. The attempt at modernisation instigated by bioethics represents an opportunity to update certain rules and values of the past but also the instrument to definitively bury the validity and topicality of other rules and classic values.

It should be remembered that in Muslim law any rule not expressed in the Koran or in the "sayings" of the Prophet (*ahadith*) or not legitimately inferred from these, is *bid'a*, i.e. "innovation", a word that very early on became synonymous with heresy and is sometimes used by conservative authors to challenge or refute any new idea or modification of the classic rules. However that may be, any new juridical element or new practice, in the absence of regulations which can be traced back to the Sacred Sources or to the juridical tradition, may be accepted by the various juridical-religious bodies, but with decisions the authority of which will always be weak and questionable.

THE PROBLEMS OF ETHICAL PLURALISM

When there are no clear indications by Tradition, a considerable amount of different and even contradictory responses may be seen; this may lead to the impression of a sort of "pluralism" intrinsic to the Muslim communities. In fact, Islam perceives itself as the perfect monotheism founded on the last and perfect Revelation (the Koran). At the same time, the *Shari'a* is the perfect and comprehensive divine Law capable of providing an appropriate answer to the problems of human society at any moment in history. One of the consequences is that any opinion regarding ethics should remain anchored to the Sacred Sources to be legitimised; all the variations must remain within Islam, otherwise they risk losing credit, even losing the possibility of being presented in public.

On this very subject, a *hadith* of the Prophet considers the multiplicity of opinions existing in the community as a blessing desired by God, whilst a second *hadith* specifies that the Muslim community would never reach agreement on an error. There are, however, differences that must not modify the basics of the faith and of worship. Classic cases are represented by the multiple positions taken by Muslim jurists regarding the lawfulness of contraception, abortion before the infusion of the soul or, going on to contemporary issues, xenograft and explantation from a corpse. The term "pluralism" therefore appears unsuitable to describe the current situation whilst more appropriate are expressions such as "differences and variety of positions within a common religious context" and other similar ones.

It must also be borne in mind that the political and ideological "pluralism" could easily be associated, in the public opinion, with an ethical relativism which is unacceptable according to the Muslim point of view. Islam, as the faith of the perfect Revelation (the Koran) and regulated according to the precepts of the ultimate divine Law (the *Shari'a*) does not require any ethical pluralism. The truth and the values are already present in the Sacred Sources and only have to be retrieved or the Sources have to be "interpreted" correctly. In addition, many authors underline how the *Shari'a* is not rigid at all, as diversified opinions emerge – allowing divine Law to adapt itself to highly varied situations – until they come into conflict with the spirit of the primary sources.

In Muslim countries (e.g. Egypt), in any discussion on bioethics, there is a tendency to look towards religion. As a result, many professionals seek out the opinion of theologians or personally interpret the religious doctrines to find answers to the new bioethical issues. However, as these can be based only on analogies and not on "explicit doctrines" (as the Sacred Sources do not mention modern technologies) the conclusions of these processes only represent personal interpretations which are potentially different from one another.[1]

In the Northern and Western world, i.e. in a pluralistic cultural and political context, the National Committees of Bioethics (and similar bodies) appear, in turn, characterised by a multiplicity of ideological positions (e.g. secular, atheist, religious) with "world views" that are at times antithetic; this may be translated into contrasts on many problems: when human life begins, experimentation on embryos, when human life ends, euthanasia, etc.

The National Committees of Bioethics and/or the National Committees of Medical Ethics (and similar bodies) existing in many Muslim countries (some countries have both, others have a plurality of Committees at national level whilst local committees on ethics are being developed) present widely varying specialisations and competences of the participants in the clinical, social and philosophical-religious context, according to different numerical balances. In the cases where Islam is the official dominant religion, it is taken for granted that every member is a Muslim, although with varied accents and positions on individual issues. Experts from other religious minorities may sometimes be present. However, the possibility of ethical pluralism where the religious element represents only one of the points of ethical reference amongst many others (secular, atheist, etc.) seems very far off at present.

---

[1] Hamamsy (EL) L., Le Langage de la bioéthique et les cultures, *Journal International de Bioéthique*, 2000, 11 (3–4–5), 159–164.

In Europe, there are very many opportunities for inter-religious meetings on the issues of bioethics. An example of "interreligious pluralism" is a collection of papers coordinated by the Council of Europe (published in 1990 and in a summarised version but updated in 1998) with the participation of Catholic, Protestant, Jewish, Muslim, Buddhist and atheist doctors, ethicists and jurists. In a strictly religious context, a similar study would encounter obstacles, in several Muslim countries, not only due to the presence of religions other than those "of the Book" (Islam, Christianity and Judaism) but, above all, due to the presence of atheists, whose ideas were presented on the same level as those of the "true religion".

ALGERIA, TUNISIA, PAKISTAN AND IRAN

Regarding the peculiarity of some of these bodies, in Algeria the *Conseil National de l'Éthique des Sciences de la Santé*[3] was established by Decree no. 96–122 of 6 April 1996 (on 6 July 1992 the Algerian Medical Professional Code was passed by Decree no 92–276) with 20 members, of whom at least 15 are doctors (including a military doctor); the others are representatives from the Ministry for Justice, the Higher Muslim Council, etc.. This predominance of doctors has been criticised as representing an obsolete conception of ethics, still understood as a natural extension of medicine and biology.[4] This opinion underlines how doctors and scientists do not have a specific competence in the moral field, which is why the development of ethics by technicians – accompanied by a marginalisation of religious and legal personalities– would damage a real inter-disciplinary reflection. Article 10 of the decree does not specify the tasks of the Committee (perhaps taking them for granted) unlike the rules establishing other Committees of Medical Ethics and of Bioethics (e.g. Tunisia, Lebanon). In addition, the Algerian Committee tries to fill a legislative and doctrinal gap, with the exception of responses from Muslim jurists.[5] Lastly, the decree does not lay down the importance of the Committee's pronouncements.

The National Committee of Medical Ethics (CNEM) of Tunisia is a multidisciplinary body with a consultative, informative and advisory role with the main aim of allowing Parliament and the government to pass laws or issue rules which do not hinder the progress of medical science. Established by Decree no. 94–1939

---

[2] AA.VV., *The Human Rights, Ethical and Moral Dimensions of Health Care-120 Practical Case Studies*, Strasbourg, Council of Europe Publishing, 1998.

[3] Ossoukine A., Un Conseil d'Éthique ou de Déontologie (BIS), *Revue Algérienne des Sciences Juridiques, Économiques et Politiques*, 1997, XXXV, 1, 263–255. Ouyahia A., Décret exécutif no 96–122 du 18 Dhou El Kaada 1416 correspondant au 6 avril 1996 portant composition, organisation et fonctionnement du Conseil National de l'Éthique des Sciences de la Santé, *Revue Algérienne des Sciences Juridiques, Économiques et Politiques*, 1997, XXXV, 1, 254–252.

[4] Ossoukine, art. cit., 259.

[5] Ibid., 255.

on 19 September 1994[6] it is made up (art. 3) of a Chairman; a member of the Constitutional Council; a member of the Higher Muslim Council; a member of the Committee for Human Rights and Fundamental Liberties; an advisor from the Supreme Court; one from the Administrative Court; a professor of philosophy; one of sociology and one of law; a representative of the Secretariat of State; the Presidents of the National Council of the Order of Doctors, of Dentists, Veterinary Surgeons and Pharmacists; the Deans of the Faculties of Medicine and Pharmacy; three figures belonging to the health sector and one figure with expertise in social problems. Tunisia has been one of the most active Muslim countries in setting up local "hospital" ethical committees in several university teaching hospitals in the respect of international indications (CIOMS, 1993) which invites doctors-researchers to request the opinion of an ethical committee for in-depth examination of certain biomedical-clinical problems or relating to research. The Tunisian CNEM has requested[7] that similar committees be set up at all hospitals and teaching hospitals. Their task does not consist of taking decisions in the place of health professionals but only has a consultative and educational function and, in particular: (a) to help hospital staff to reach decisions of an ethical nature concerning medical practice; (b) to supervise the application of the ethical rules of the protocols of research programmes (in the respect of informed consent of the patient, of an acceptable risk/benefit ratio and the pertinence of the scientific purpose of the research); (c) to take an active part in training health personnel in the ethical field. The opinions of the local ethical committees are not binding.

The Aga Khan University in Pakistan introduced biomedical ethics as a subject for its medical students in 1984. In 2001 the Pakistan Medical and Dental Council (PMDC) decided to include bioethics in the medical curricula although this is currently not the case in the majority of medical colleges. Permanent Hospital or Clinical Ethics Committee remain few and far between. However, there is a growing interest in the subject because the majority of international funding agencies and many peer-reviewed scientific journals "require ethical reviews of research protocols and manuscripts for acceptance".[8]

The seventh paragraph of the 2001–2002 Code of Ethics of the PMDC is devoted to bioethics, recommending its inclusion in the syllabi of doctors.[9]

In January 2004 the Pakistani government approved the formation of the National Bioethics Committee with 20–21 members chaired by the Director General of Health at the Ministry of Health.

---

[6] République Tunisienne, Décret no. 94–1939 du 19 septembre 1994, fixant les attributions, la composition et les modalités de fonctionnement du Comité National d'Éthique Médicale, *Journal Officiel de la République Tunisienne*, 27/9/1994, no. 76, 1590.

[7] Comité National d'Éthique Médicale, République Tunisienne, Avis no. 2, Avril 1997: *Rapport sur les Comités d'Éthique Locaux*, CNEM, Tunis, 18–23.

[8] Moazam F. and Jafarey A.M., Pakistan and Biomedical Ethics: Report from a Muslim Country, *Cambridge Quarterly for Healthcare Ethics*, 2005, 14, 249–255.

[9] www.pmdc.org.pk/ethics.htm, 6–7.

In Iran, the forthcoming National Committee on Bioethics should include the following representations as permanent members of the Committee[10]: the Ministry of Science, Research and Technology; the Ministry of Health and Medical Education; the Organisation for the Protection of the Biological Environment; the Ministry of Agricultural Jihad; the Legal Medicine Organization of Iran; the Hozeh Elmieh of Qom (Seminary); the Iranian Academy Centre for Education, Culture and Research (ACECR); the Medical Council of Iran (as an NGO). Other permanent members will be two specialists in philosophy and ethics, two lawyers, two biotechnologists and two biologists. Lastly, there will also be one specialist from each of the fields of immunology, genetics, pharmacology, biochemistry, psychology and epidemiology in Iran.

In the meantime, the current organisations on bioethics are the Ministry of Health and Medical Education, the Office of Study for Humanistic and Islamic Science on Medicine and Medical Ethics; the Iranian National Commission for UNESCO; local ethics committees are present in 85 research centres involved in molecular and cellular biology, biotechnology, etc.

The Charter of the Lebanese Committee of Bioethics of the Order of Doctors[11] establishes the role of the Committee in providing simple opinions on moral questions relating to clinical practice; the document clearly expresses always wishing to consider the "cultural and religious" diversities historically present in the country (art. 4). The Committee should consist of a majority of doctors (art. 7).

In Egypt, a Ruling of 25 November 1996 of the Ministry of Education set up the National Committee of Bioethics including members of the scientific community, figures from the academic world and representatives of civil society. At its first session (9 January 1997) Prof. J. Badran was elected Chairman. National days of ethics are held each year.

THE DEPENDENCE ON MUSLIM LAW

To date in literature on the Muslim side, few attempts appear to have been made at a critical interpretation of the reflection produced by Muslims on the issues dealt with by bioethics (principles and individual issues) in particular on their contrasting positions as well as their development on the diachronic and synchronic levels. Despite the presence of a wide variety of opinions, Muslim authors tend not to give excessive importance to these differences. In other words, perhaps fearing that diversification appears synonymous with ethical "relativism", the preference is to provide a rather "monolithic" image of Muslim bioethics,

---

[10] Zali M.R., Shahraz S., et al., Bioethics in Iran: Legislation as the main Problem, *Archives of Iranian Medicine*, 2002, 5 (3), 136–140. *Current Situation of Bioethics in Genetic Research in Iran*, in *The Experiences and Challenges of Science and Ethics: Proceedings of an American-Iranian Workshop*, 2003, The National Academy of Sciences, 77–85.

[11] Comité de Bioéthique de l'Ordre des Médecins du Liban, *Journal International de Bioéthique*, 1998, 9 (1–2), 139–140.

especially when addressing non-Muslims. Moreover, the believer wants clear and exhaustive answers from the "true religion" rather than arbitrary philosophical reflections or uncertain orientations, which would be plentiful in the West. It does not appear casual that this type of bioethics is inclined towards apology, i.e. the exaltation of the truth of the Koranic principles and of those of the Tradition, especially with reference to the chaotic situation in the West, which produces modernity but does not have the capacity to guide it ethically.

The apologetic approach has, to date, hindered bioethics from expressing a significant inclination towards denouncing the serious shortcomings on the social, economic, civil and political level that exist, to the detriment of the protection of the patient, in the fair distribution of health resources and of the principle of justice in Muslim countries. The relative lack of political freedom could contribute to creating bioethics that show little criticism of the official power.

In the absence of an equivalent in Arabic and in the languages of other Muslim countries, the term "bioethics" is still used in the English or French versions (bioethics, bioéthique) or with the phrase "medical ethics" (Ar. *akhlaq tibbiyyah*)[12] and similar.

The majority of scientific contributions on Muslim bioethics deal with individual issues (so-called "special bioethics"): contraception, abortion, transplants, etc.; on the other hand, the approaches regarding "fundamental bioethics", i.e. a reflection on the epistemology of bioethics, seem disregarded. As a consequence there are few studies of a certain depth on the methodology and on the "status" of Muslim bioethics, for this reason it is not yet characterised as an independent and separate discipline compared to every other discipline, in particular with respect to Muslim law of which, for the time being, Muslim bioethics appear to be a derivation. This fact is important as this reflection is still prevalently oriented towards identifying rules and answers to the problems rather than developing independent reflection with respect to the clinical decision-making phase. One of the results is the reduced speculative (i.e. philosophical) inclination of bioethics in the Muslim world.[13]

THE POLITICAL DIMENSION OF ISLAMIC BIOETHICS

By political dimension of Muslim bioethics, we refer to all those positions on the individual topics of bioethics (e.g. contraception) influenced by the desire to protect and/or expand the Muslim community as a whole (i.e. on the demographic, religious, political, moral, scientific, economic and military level) which are often in competition with other religions. The political perspective represents one of the many currents and/or keys of interpretation present in the rich and complex bioethical debate in Islam.

---

[12] De Caro A., La terminologia della bioetica islamica: calchi, neologismi e nuove accezioni, *Il Traduttore Nuovo*, 2001, LVI (1), 93–99.

[13] Aksoy S. and Elmali A., The Core Concepts of the Four Principles of Bioethics as found in Islamic Tradition, *Medicine and Law*, 2002, 21 (2), 211–224.

In Muslim law the legal status of an individual is based on three opposing positions: between Muslim and non-Muslims; between free individuals and slaves; between men and women. Regarding the first, there are two positions in the Koran: (1) the equality of all human beings before God; (2) the fundamental juridical distinction between Muslims (faithful in the "true religion"), *dhimmis* (or "protected", i.e. Christians and Jews) and polytheists. Muslim medical ethics refers principally to the first position; Muslim law develops and mainly elaborates the second position. The "political" approaches of Muslim bioethics refer in particular to this latter position. In fact, precisely because of the great importance of Muslim law in regulating the entire life of the faithful, the two perspectives will accompany us throughout the book, although the first position (i.e. the egalitarian one) appear predominant.

The many rules of worship and rules of Muslim law that require different behaviour with regard to infidels compared to Muslims do not concern us when these rules do not have a direct influence in the field of bioethics.

Amongst the examples of the political approach, in the following chapters, we will quote the positions against population control when these have the purpose of guaranteeing the expansion and strength of Islam; we will quote some religious authorities (but also doctors) in favour of penal mutilation as preference is given to damaging the physical integrity of a single corrupt individual for the higher aim of avoiding the moral and physical corruption of the whole community; we will quote some jurists (Muslim scholars) who are in favour of donating organs only between Muslims or by non-Muslims to Muslims in order to reinforce the health of the *umma* (the Muslim community); jurists who prefer blood transfusions between Muslims, etc.; jurists who are in favour of abortion for seriously handicapped foetuses not to weaken the health of the Muslim community; jurists in favour of the abortion of Muslim women raped by infidels for the purpose of not numerically reinforcing the latter; homologous in vitro fecundation acceptable only between Muslim partners or between a Muslim man and a monotheist woman for the purpose of increasing the number of the faithful; the tendency to reveal the identity of a person with AIDS in order to protect the healthy spouse, colleagues and the community as a whole; we will quote the jurists who are in favour of female genital mutilation (FGM) as, if carried out properly, it protects the moral health of the community without "damage" for the woman, etc.

One case that can be included in the political dimension is that where the protection of the individual and the protection of the community come into conflict. Every state has priorities which do not coincide with those of the individual. According to Muslim law, the protection of the community of the believers has priority over the communities of the unfaithful and the *dhimmis* (Christians and Jews); in addition, the protection of the community of the faithful (*umma*) has priority over that of its individual members, including in the management of health resources (especially when community budgets are limited), imposing choices which are often dramatic. On the contrary, the ethical approach could maintain that fundamental values (the protection of human life) have precedence over the evaluation of costs. For example, care for the weak, the elderly, the terminally ill and people without help could be values to which time and money are to be sacrificed, without limits. In this case, the protection of the individual takes on contrasting characteristics according to whether they are applied to medical ethics or to Muslim law.

An interesting case where the superiority of the community is not respected is that of transplants from a corpse. In this case, if during his lifetime the deceased had not expressed himself in favour of the post-mortem donation of organs, in Muslim countries it is not possible to automatically explant an organ in favour of the community on the basis of "presumed consent" or "silence-assent". Except in some legislations and in particular cases, the wishes of the family remain decisive.

THE PRINCIPLES OF BIOETHICS

Existing studies are frequently limited to tracing back to the Sacred Sources of Islam the principles of "Western" bioethics which are mainly taken from the "ethics of principles" or *principalism* of T.L. Beauchamp and J.F. Childress,[14] but without reference to the ethical theories justifying them and that is, to the "utilitarianism of the rule" and the "deontological theory". The principles in question are the principle of autonomy (with respect to the freedom of choice of a competent individual and protection of the incompetent person), beneficence (promoting the well-being of one's neighbour), non-maleficence (not doing evil) and justice (promoting a fair allocation and distribution of health costs and benefits which is connected with fair compensation for errors or shortcomings on individuals or groups). In fact, the identification of these four general principles in any monotheistic Sacred text is elementary, as God summons the believer to make a responsible choice according to his faith, to do good, to avoid evil and to apply justice. All these general principles can easily be found in the Koran and amongst the "sayings" of the Prophet.[15] The principle of autonomy can be found in the value of man as Vicar of God on earth and amongst the indicative passages we find 17.70: "We have honoured the sons of Adam; provided them with transport on land and sea; given them for sustenance things good and pure; and conferred on them special favours, above a great part of Our creation."[16] This is

---

[14] Beauchamp T.L. and Childress J.F., *Principles of Biomedical Ethics*, New York, Oxford University Press, 1994.

[15] See amongst others: Serour G.I., Islam and the Four Principles, in Gillon R. (ed.), *Principles of Health Care Ethics*, New York, John Wiley & Sons Ltd., 1993, 75–91. Aksoy and Elmali, op. cit., 211–224. Serour G.I., La Bioéthique dans la recherche biomédicale dans le monde musulman, *Revue Maghrebine d'Endocrinologie-Diabete et de Reproduction*, 1994, 1 (3), 143–152. Zaki Hasan K., Islam and the Four Principles: A Pakistani View, in Gillon R. (ed.), *Principles of Health Care Ethics*, op. cit., 93–103. Aksoy A. and Tenik A., The Four Principles of Bioethics as Found in 13th Century Muslim Scholar Mawlana's Teachings, *BMC Medical Ethics*, 2002, 3 (4).

[16] Amongst the other passages in which a reference to the principle of autonomy connected with the value of man can be found, we can quote 10.99. "If it had been thy Lord's Will, they would all have believed, all who are on earth! Wilt thou then compel mankind, against their will, to believe!"; 95.4. "We have indeed created man in the best of moulds"; 82.6–8; 14.32–34, etc.

an autonomy that is far from individualistic extremism in order to guarantee the superiority of the *umma*.[17]

On the promotion of good, Koran 3.104 suffices in stating: "Let there arise out of you a band of people inviting to all that is good, enjoining what is right, and forbidding what is wrong."[18] Not doing evil is also found in 16.90: "God commands justice, the doing of good, and liberality to kith and kin, and He forbids all shameful deeds, and injustice and rebellion."

In bioethics of Muslim origin, the four principles mentioned have yet to find an equilibrium with the general principles of Muslim law (e.g. necessity, public benefit and justice). This confrontation could challenge the importance of the four principles mentioned and the respective priorities.

By way of comparison, the different importance and significance that the principle of autonomy assumes in North American and European society compared to any Muslim society, especially when anchored to tradition, cannot be denied.[19] In the former case, the priority given to the principle of autonomy (starting from the hierarchy of the principles of bioethics) is none other than the projection into the bioethical context of the importance of the liberal-individualistic matrix characterising the juridical-political culture of the Western world, despite the strong differentiations within it. Vice versa, in the Muslim context, the primacy appears reserved for the principle of public benefit (*maslaha*) and the principle of justice as the collective interest takes precedence over that of the individual. For example, the Code of Ethics of the Pakistan Medical and Dental Council (2001, art. 7) states: "If secular Western bioethics can be described as rights-based, with a strong emphasis on individual rights, Islamic bioethics is based on duties and obligations (e.g. to preserve life, seek treatment). . . ."[20] Furthermore, the family often remains the decisive subject for the patient's decisions in relations with the doctor; this can be a limit to the patient's autonomy. The principle of justice and of public benefit are also variously applied to the allocation of resources for care for the chronically and terminally ill etc., but systematic speculation on these consequences is in its early days.

However, if universal principles of bioethics exist, diversities arise in their interpretation and application, i.e. in the attempt to adapt them to local cultures

---

[17] Aksoy and Elmali, *The Core Concepts of the Four Principles of Bioethics as Found in Islamic Tradition*, art. cit., 216–217. Saleh S., *Réponse de l'Islam aux défis de notre temps*, Beirut, 1979, 96–97, in Roussillon A., Science Moderne, Islam et Stratégies de Legitimation, *Peuples Méditerranéens*, 1982, no 21, 105–127.
  there will be no place in this for purely individual or purely material interest or for the strictly local interest of a generation or of a race at a specific period of time [il n'y aura pas de place dans celle-ci pour l'intérêt purement individuel, purement matériel ni pour l'intérêt strictement local d'une génération ou d'une race à une époque déterminée.]

[18] See also 99.7–8: "Then shall anyone who has done an atom's weight of good, see it! And anyone who has done an atom's weight of evil, shall see it".

[19] See the section "Autonomy and Consent of the Patient" in Chapter 3.

[20] www.pmdc.org.pk/ethics.htm, 7.

and customs. This problem is similar to that which arose with the issue of human rights. Therefore, confrontation with the values, mentality, local customs and religions is decisive to establish the best criteria with which to treat a patient. Greater attention than that generally paid to the principles of bioethics should be devoted to these fundamental aspects in caring for a patient.

Compared to the characteristics still in the making of Muslim bioethical reflection, Catholic "personalism", independently of the different currents, appears to be a model that is perhaps excessively rigid, indeed even extremist in parts as shown by the Church's position on therapeutic abortion, the refusal of the techniques of homologous assisted procreation, etc. compared with the flexibility and possibility of adaptation offered by the Muslim juridical doctrine, above all thanks to the principles of necessity and public benefit. Everything that has not been regulated by classic law is inevitably subject to oscillations in a fairly broad area, with the incentive of the absence of a supreme authority and of a magisterium as in the Catholic Church (teachings).

It is worthwhile underlining how Muslim authors, in dealing with the principles of bioethics, mainly refer to the now classic ones produced by Western authors of a secular orientation in English, whilst curiously, the principles developed on bioethics by Catholic bioethicists or of other religions, which could have a greater resemblance with Islam, are disregarded. This probably occurs as Muslim experts only intend to use these texts exclusively as a starting point or a first formal approach, then to immediately break away from them to make room for the juridical doctrine and the Muslim medical ethical tradition.

CULTURAL SENSIBILITIES AND MEDICAL ETHICS

The phenomenon of "globalisation" finds its most obvious expression in the health field. This takes place at two levels: the instrumental one and the conceptual-normative one.

At an instrumental level, it is opportune to recall that modern clinical, diagnostic, etc. instruments are not "neutral" technological products but can be assimilated with "cultural artefacts", that is, they are crystallisations of conceptions, mentalities, implicit values, characters, ideals of the culture that produces them. As such, they end up by conveying lifestyle models, ideals, values, etc. of a specific type of culture or types of culture. In fact, today we witness a generalised homogenisation of clinical-health instruments at international level which has no precedent in the history of humanity.

At the same time, we can see that progressively, at international level, the regulations, codes and health rules are increasingly similar and this similarity is again without precedent in human history.

Despite the structural technological dependence on the West (including in the health context) and the influence of the values and mentalities conveyed through these technologies (with the consequent tendency to the homologation of health principles and regulations), the sensibility reserved for many ethical problems may still be maintained autonomously to a certain extent as it is influenced by elements such as: mentality, culture, local tradition and religion. Even the application of the four classic principles of bioethics may be perceived differently and according to their cultural and operative context.

Two examples are shown below, the first concerning the Egyptian context and the second the situation in Saudi Arabia.

At least three factors have an effect on biomedical research in Egypt[21]:

1. The prestige of scientific research tends to be placed before the well-being of the subjects who are experimented on whilst the bodies or committees for the assessment of the protocols of research remain few in number; the ethical criteria to be followed may be overlooked, again due to a lack of internal cultural elaboration of the values and criteria to be followed in the research.[22]

2. The majority of subjects taking part in experimentation are poor, illiterate and come from state hospitals; at the same time the high social status of doctors may facilitate less attention towards the rights of these patients.[23]

3. There exists considerable reluctance on the part of doctors to criticise the work of their colleagues. In addition, doctors are assigned by the government to a university or institution where they generally stay for a long time which leads to avoiding disagreement with colleagues. It is possible to disagree privately but not publicly in front of individuals of a lower status. All this has encouraged the lack of regulations against unethical research.[24]

The fact that the figure of the doctor is socially very highly respected encourages a paternalistic approach.[25] Patients with little education expect the doctors to take decisions for the good of the patients and fear, if they behave otherwise, offending the doctors. These situations are translated into a reduced autonomy of the patients.[26]

Due to the low literacy rates, Egyptian women appear to become easily involved in research programmes. All over the world, for classes with little education, consent forms and written information are difficult to understand. In Egypt, the signature of a person has particular cultural significance[27]; in addition, the request to sign a consent form may imply lack of trust of the person who has already given consent verbally. This causes further difficulties in obtaining signed consent. For all these reasons, there is a tendency to request simple verbal consent.

---

[21] Lane S.D., Research Bioethics in Egypt, in Gillon R. (ed.), *Principles of Health Care Ethics*, New York, John Wiley & Sons Ltd., 1993, 885–894.

[22] Ibid., 888–889.

[23] Ibid., 889–890.

[24] Ibid., 890.

[25] Daoud R., Doctors in Egypt deal with Patients in their own Way, *British Medical Journal*, 8 January 2000, 320, 118 (Letter).

[26] Rashad A.M., MacVane Phipps F., et al., Obtaining Informed Consent in an Egyptian Research Study, *Nursing Ethics*, 2004, 11 (4), 394–399.

[27] This generally refers to problems of managing property or marriage, financial or legal problems, etc. See Rashad A.M., MacVane Phipps F., et al., 396.

To overcome such difficulties, the Ethical Guidelines in Human Research in the Islamic World[28] were drawn up by the University of Al-Azhar (Cairo, Egypt) in 1991, which took as the starting point international documents on the protection of the individual and anchoring them to the values of Islamic medical ethics. An Ethics Committee for Research in Human Reproduction was also established; this was presented as the first committee of its kind in the Islamic world.

A Saudi survey was carried out on the attention given to some problems of medical ethics to the detriment of others in the Kingdom's hospitals by two groups of doctors, namely Physician Executives[29] and clinicians; the study shows a sensibility towards the individual problems which in parts is very characteristic[30] and different in the two groups of reference. Four hundred and fifty-seven doctors from several hospitals in different parts of the Kingdom took part in the survey which was carried out between February and July 1997. Of these, 140 (30%) were Physician Executives (PEs) whilst 317 (70%) were clinicians; all the PEs and the majority of the clinicians were Saudi nationals. Regarding the concession of favours (private rooms or early appointments) to particular patients at their hospital, over 99% of the PEs do not consider this an ethical problem, whilst 83% of the clinicians saw it as such. The disclosure of a patient's personal clinical information was not deemed an ethical problem by 92% of the PEs. whilst almost 80% of the clinicians consider it such. "Discontinuing medical treatment for terminally ill patients" was not considered an ethical problem for about 81% of the PEs against 77% of clinicians who deemed it a problem.[31] Regarding the fact that "some clinicians receive gifts from patients or their families for providing favourable treatments" none (i.e. 140) of the PEs considered this an ethical problem in hospital nor did the majority (57%) of the clinicians. "Issuing false medical reports to individuals due to social pressures and favouritism" does not come under the heading of ethical problems for all the PEs and the majority (58%) of the clinicians. "Discriminating between patients due to social status, nationality or gender" is not perceived as an ethical problem by 80% of the PEs whilst 66% of the clinicians consider it an ethical problem. The indifference of doctors to the cultural and social rules of patients does not appear an ethical problem for 64% of the PEs whilst 57% of the clinicians deem it an ethical issue. Lastly, the fact that doctors show little attention to the clinical needs of their patients is considered an ethical problem by the majority both of PEs (51%) and by clinicians (69%).

---

[28] Serour G.I. and Omran A.R. (eds.), *Ethical Guidelines for Human Reproduction Research in the Muslim World*, Cairo, IICPSR Al-Azhar University, 1992, 19–40.

[29] Chief executive officer, chief of medical staff, head of clinical department, etc.

[30] Bin Saeed K.S., How Physician Executives and Clinicians Perceive Ethical Issues in Saudi Arabian Hospitals, *Journal of Medical Ethics*, 1999, 25, 51–56.

[31] Ibid., 52–53.

Which characteristics are recognised as ethical problems in Saudi Arabian hospitals? The presence of a multinational staff of doctors in the hospital[32] led exactly half the PEs to consider that this raised problems of an ethical nature, an interpretation denied by the remaining 50%; 78% of the clinicians were in favour of the first hypothesis. The remark that the "hospital management does not facilitate the conduct of seminars and workshops concerning ethical issues" was considered a possible cause of problems of an ethical nature by the majority of the PEs and of the clinicians (55 and 68.5%).[33]

One of the reasons for the different interpretation of the ethical value of the various situations by the PEs and by the clinicians (all doctors) can perhaps be attributed to the fact that the PEs assess the ethical problems in a broader administrative context, having to reconcile administrative, financial and bio-medical responsibilities.[34] For example, regarding the discontinuation of care for terminally ill patients, 77.6% of the clinicians deem it an ethical problem, perhaps following the principle of benefit according to which it is the doctor's duty to preserve life; on the contrary, 81.5% of the PEs do not deem it an ethical problem, having to bear in mind other obligations such as the possible scarcity of resources, therefore it is preferable to waive useless expenditure to devote it to patients with a greater probability of survival. In this case the doctor opts for the protection of society at the expense of the protection of an individual.[35] In other terms the PEs seem oriented towards some principles of utilitarian or teleological ethics.

BIOETHICS AND SOCIETY

The introduction and the significance of bioethics in the Muslim cultural context must not overlook any considerations of a social nature. In the land of election of bioethics (the USA) and in other technologically advanced countries, the so-called extreme cases of biomedical practice (coming under "borderline" bioethics) have often rapidly become frequent clinical events, such as to attract the interest and positioning of public opinion, the press, patients' associations, medical associations, political movements, trade unions, etc. The debate that has followed has induced a progressive reconsideration and updating at theoretical, professional and legislative level of the more general doctor–patient relationship.

---

[32] Saudi Arabia, the cradle of Islam and land of Prophet Muhammad, currently has many non-Muslim doctors caring for Muslim patients and the presence of a multinational medical staff is felt as the cause of ethical problems especially by the clinicians (78%) and by half of the PEs; to this must be added the presence of many foreign workers with a high number of non-Muslim patients. The numerous foreign doctors bring with them different cultures and values which may upset Saudi doctors and patients.

[33] The differences between the two groups on other attributes are less clear cut. See Bin Saeed, art. cit., 53–54.

[34] Bin Saeed, art. cit., 54.

[35] Ibid., 54.

These changes have been imposed especially thanks to the sensitivity and participation (to a greater or lesser extent depending on the countries and the issues) of public opinion in the issues under discussion.

In Muslim countries, the situation, although in a highly varied social and political context, appears slightly different. In the first place, interest in debates remains generally elitist; participation by civil society, even if extensive, at least from what can be seen from certain debates in the daily press (e.g. transplants and abortion), remains unorganised, therefore incapable of having a direct effect on the positions taken by governments and on national legislations.[36]

At the same time, the ethical-juridical problems opened up by new technologies may represent an unrepeatable opportunity for Muslim experts to return to the international debate at the highest level on technological modernisation.[37] The aim of these contributions is to reassert the modernity of Islam as capable of moralising on contemporary medicine and, indeed, solving its continually new contradictions. In addition, from this first step it is easy to maintain that Islam is the modern religion par excellence (see Chapter 12 "The Koran and Modern Science" and the section "Moderate Concordism" therein), a statement which is, at times, accompanied by a radical criticism of Western science where the absence of religious references has given rise to the modern scientific hegemony of the West and this hegemony is perceived as indifferent to religious values. Only Islam is believed to be able to fully restore the harmony between religion and science and between the sacred and the profane.

The lack of "free" debate (according to the Western pluralistic view) does not compromise the possibility of passing updated legislation. However, these legislations, more than the translation into law of solicitation from civil society, through the mediation of the political class, often seem to represent a set of decisions taken by the political and scientific leaders approved by the "official" religious authorities. One of the most obvious effects is that of the separation which can be seen between the local mentality and culture and the rules and regulations in force, which are increasingly modern and abreast with the orientations developed by international bodies. From this point of view, the state policies essentially take on a function of anticipation and guidance compared to a social reality which is often more backward and passive.

Lastly, medical practice requires a constant re-balancing between maximising the diagnostic-therapeutic performance of the health instruments available compared to their costs and the limits of expenditure imposed. In some Muslim oil-producing countries, such as Saudi Arabia or Kuwait, health care is to a great extent free of charge and the health budgets are very large, therefore particular ethical-financial problems may arise linked to a facile waste of analyses and tests

---

[36] Consider the case of referendums on abortion promoted by organisations of citizens, political parties, etc. and held in many European countries.

[37] Moulin A.M., Bio-éthique en terres d'Islam, in AA.VV. Bioéthique et cultures, Paris-Lyon, Vrin, 1991, 85–96.

that are not "strictly" indispensable for a diagnosis and therapy of the illness but easy to use due to a reduced awareness of the problem of health costs (a problem which on the contrary is dramatic in the poor Muslim countries), and health investments motivated mainly by requirements of image and prestige.[38]

BIOETHICS AND APOLOGETICS

Muslim bioethics is constantly oriented towards the exaltation of the principles and precepts of the Sacred Sources (the Koran and the *Sunna*). This is explained by the fact that Islam is not a monotheism in the same way as the others but, as it is based on the last Revelation and on the "sayings" of the last Envoy of God (Muhammad), Islam is perceived as the pure and perfect monotheism that supersedes and definitively updates the other monotheistic Revelations (whilst respecting them). In bioethics, this ethical and juridical primacy finds an opportunity of redemption in the face of Western biotechnological dominance. From this point of view, bioethics also become an opportunity for the rediscovery of a strong religious identity, especially in countries of immigration.

If this approach is morally legitimate, the scientific consequences cannot be ignored. Problems could be tackled in an abstract perspective whilst the contrasts between the different opinions tend to be underestimated.

BIOETHICS AND MUSLIM COUNTRIES

A further tendency is associated with apologetics: the widespread refusal of a confrontation with the positions of Muslim countries on bioethical issues (e.g. abortion, contraception) by laws, regulations, etc., which frequently differ from one to another. This means disregarding the fact that national laws, at least on bioethical issues, often reflect – at a more articulated and specialized level – the indications of religious authorities (individual or collective); in their turn these may differ from the positions of religious authorities in other countries. The discomfort in the face of this variety of positions on bioethical issues can be perceived as synonymous with "ethical relativism", a position that is alleged to belittle the value of Islam.

Lastly, the little interest in the state laws decreases the importance of the enormous work of rethinking classical law by contemporary experts in the different Muslim countries with the purpose of adapting Islam to modernity.

CONCLUSION

Muslim bioethics is rich in thought, reflections and diversified positions which all have the intention of being legitimised by the Sacred Texts. This is the result

---

[38] Umeh J.C., The Prescription for Health Cost Control: Increased Dose of Cost Consciousness in Saudi Public Sector Physicians, *Saudi Medical Journal*, 1996, 17 (3), 272–280.

of some features peculiar to Islam. On the one hand, it is a religion based on the direct, literal and perfect word of God (the Koran) and on its eternal and all-inclusive law (the *Shari'a*). On the other hand, there is no supreme juridical-religious authority or magistery, an element which inevitably leads to the multiplication of interpretations and positions. This plurality is also increased by the expansion of Islam to different peoples (there are approximately 1.5 billion Muslims in the world), cultures, geographical areas etc. that differ greatly from one another and this has been possible precisely thanks to the great capacity of adaptation to very different cultures and mentalities but without contaminating the principles of the faith. The instable balance that derives from the first centripetal (the constant reference to the Sacred Sources) and centrifugal force (adaptation to different contexts and innovations which is also due to the absence of a central authority and magistery) explains the oscillation between unity and variety that concerns all the manifestations of Islam as a whole.

On the majority of bioethical issues there exists a vast range of oscillations of positions, a singular wealth that has to be highlighted. Sometimes, instead of speaking of "Muslim bioethics" it would be more correct to speak of the "different types of Muslim bioethics" but all of which based on the divine Sources.

SOME ASPECTS OF MEDICAL ETHICS

INTRODUCTION   31
PRINCIPLES AND CHARACTERISTICS   33
THE DOCTOR–PATIENT RELATIONSHIP   37
MEN AND WOMEN   39
AUTONOMY AND CONSENT OF THE PATIENT   47
SAUDI ARABIA, LIBYA, TUNISIA, ALGERIA   51
THE PROBLEM OF PENAL MUTILATION   54
DOCTORS AND PENAL MUTILATION   57
CONCLUSION   63

INTRODUCTION

In consideration of the vastness of the field under examination, from both the diachronic and synchronic points of view, we prefer to restrict ourselves to a general overview of some issues in particular, but of considerable importance in the international debate on bioethics. At times, the Muslim approach to these issues appears strongly apologetic. We will dwell, in particular, on the doctor–patient relationship, on the presence of women in health structures, information and autonomy of the patient and penal mutilation.

Islam is presented as a salvific religion in the profound sense in that adhesion to the faith and application of the precepts of the *Shari'a* (the perfect divine Law) during existence ought to represent a viatic of happiness both during life and after death, except in the case of divine will decreeing otherwise. In this salvific work, medical practice has historically taken on a leading role, becoming the most important of the sciences in the Muslim world, second only to the study of the Law of God. Ibn Qayyim (d. 1350) and Ibn Qudama (d. 1223) report a *hadith* in which the faithful are invited to assist the ill day and night; those who look after an ill person are blessed by 70,000 angels who constantly pray for them.[1]

---

[1] Another "saying" collected by Muslim says: "Verily, Allah, the Exalted and Glorious, would say on the Day of Resurrection: O son of Adam, I was sick but you did not visit Me. He would say: O my Lord; how could I visit Thee whereas Thou art the Lord of the worlds? Thereupon He would say: Didn't you know that such and such servant of Mine was sick but you did not visit him and were you not aware of this that if you had visited him, you would have found Me by him? O son of Adam, I asked food from you but you did not feed Me. He would say: My Lord, how could I feed Thee whereas Thou art the Lord of the worlds? He said: Didn't you know that such and such servant of Mine asked food from you but you did not feed him, and were you not aware that if you had fed him you would have found him by My side? (The Lord would again say:) O son of Adam, I asked drink from you but you did not provide Me. He would say: My Lord, how could I provide Thee whereas Thou art the Lord of the worlds? Thereupon He would say:

Dariusch Atighetchi, Islamic Bioethics: Problems and Perspectives.
© Springer Science+Business Media B.V. 2009

The role of the doctor, in addition, ideally recalls the divine role as stated in one of the first Arabic texts on medical ethics by Ali al-Ruhawi, a Christian doctor of the 9th century AD, in which medicine is defined as an art given by God that imitates His curative role.[2] The Islamic Code of Medical Ethics (Kuwait, 1981) expresses itself coherently with this approach, stating that the physician "is a soldier for Life"; he is an instrument that God uses to relieve human suffering and, lastly, medicine is sacred as it brings one closer to the faith by means of the contemplation of what has been created.

In a *fatwa* of 1952 (but also in many medical-juridical documents) the former Grand *mufti* of Egypt Hasanayn Muhammad Makhluf[3] recalled that medicine in the *Shari'a* enjoys the status of *fard kifaya*, i.e. a duty that has to be performed by a sufficient number of Muslims. More precisely, in *Malikite* and *Shafi'ite* Muslim law, the *fard kifaya* is identified by a collective precept to which not everyone is obliged and, once performed by some believers, the others are exempted from it.[4]

Islam encourages its followers to take care of themselves and not to lose hope, while confirming that everything is predestined. Medicine is part of divine determinism. A *hadith* reported by Ibn Khuzayma (*Shafi'ite* scholar, d. 924) says that the use of drugs and medicines does not modify predestination as they are part of it. Seeking a cure does not contradict faith or the acceptance of divine will. Ibn Maja reports a *hadith* in which Muhammad is asked: "Do you think that talismans which we resort to and medicines which we use reduce something of God's power?", to which the Prophet replies, "They are part of God's power."[5] Our problems are predestined by God who asks us, however, to make an effort in order to overcome them. According to Ibn Qayyim (d. 1350) recourse to treatment and medicine does not contradict submission to the divine will just as we protect ourselves from hunger, thirst, the heat and the cold with the remedies that God Himself has provided (food, water, clothes, etc.).[6] The theologian al-Dhahabi (d. 1346) expressed his thoughts in a similar way in his *The Medicine of the Prophet*: medicines are a natural extension of food;

Such and such of servant of Mine asked you for a drink but you did not provide him, and had you provided him drink you would have found him near Me." See Muslim, *Sahih, The Book of Virtue, Good Manners and Joining of the Ties of Relationship*, Book 32, no 6232, in www.usc.edu/dept/MSA/fundamentals/ hadithsunnah/muslim/smtintro.html

[2] Nanji A.A., Medical Ethics and the Islamic Tradition, *Journal of Medicine and Philosophy*, 1988, 13 (3), 257–275.

[3] Rispler-Chaim V., *Islamic Medical Ethics in the 20th Century*, Leiden, E.J. Brill, 1993, 75.

[4] These obligations include (1) giving oneself a guide; (2) studying Muslim law; (3) taking part in the *jihad* (holy war); (4) fostering abandoned children. Santillana, *Istituzioni di diritto musulmano malichita con riguardo anche al sistema sciafiita*, Roma, Istituto per l'Oriente, 1926–1938, I, 17–74–88–306–307.

[5] Rispler-Chaim, op. cit., 110.

[6] Jeilani (Al) M., Pain: Points of View of Islamic Theology, *Acta Neurochirurgica*, Suppl. 1987, 38, 132–135.

the quest for them is based on instinct in a similar way to the hungry man who looks for food.[7] The religious justification for recourse to medical treatment is present in the well known *hadith:* "There is no disease that Allah has created, except that He also has created its treatment" (Bukhari).[8] Ibn Hanbal (founder of the *Hanbalite* legal school) considered medical care as permitted but not compulsory because it is better to trust in God rather than in the efficacy of drugs. On the contrary, according to the *Hanbalite* jurist Ibn Taymiyya (d. 1328) medical treatment is compulsory. However, for the majority of jurists, recourse to treatment seems indispensable in order to obtain recovery which is granted only by God, according to Koran 2680: "and when I am ill, it is He who cures me". A *hadith* collected by Muslim confirms: "There is a remedy for every malady, and when the remedy is applied to the disease it is cured with the permission of Allah".[9]

Islam does not have any sects which, on principle, oppose recourse to medicine. However, there existed individual *Sufi* ascetics with this opinion but the majority of them were favourable to the use of treatment. Opposition can be found in the illness of a woman in the 8th century, the *Sufi* Rabi'a al-'Adawiya who even refused to pray to God for her recovery as it was God Himself who wanted her illness.[10] *Shi'ism* too, whilst offering a positive evaluation of suffering, recognises that medical care is of great value.

In conclusion, the majority of the faithful appreciate modern medicine whilst accepting the faith in divine predestination. In order to avoid this inducing some patients not to seek treatment, the patient who leaves hospital should be given as much information as possible on the treatment to be followed involving, if possible, other members of the family.[11]

PRINCIPLES AND CHARACTERISTICS

The fundamental principles of Muslim medical ethics derive from the concentric action of several currents – Hippocratic ethics, classical medical ethics, the Holy Sources and juridical authorities:
1. Koran 5.32: "If anyone saved a life, it would be as if he saved the life of the whole people". This is the basic principle of medical practice in Islam according to which the doctor must not only heal but also prevent, in the respect of the words of Koran 2.195: "and make not your own hands contribute to (your) destruction".

---

[7] Rahman F., Islam and Health/Medicine: A Historical Perspective, in Sullivan L.E. (ed.), *Healing and Restoring*, New York, Macmillan, 1989, 149–172.
[8] Bukhari, *Sahih*, Medicine, Vol. 7, Book 71, No. 582, in www.usc.edu/dept/MSA/index.html
[9] Muslim, *Sahih, The Book on Salutations and Greetings (Kitab As-Salam)*, Book 26, No. 5466, in www.usc.edu/dept/MSA/index.html
[10] Rahman, Islam and Health/Medicine, art. cit., 156.
[11] Lawrence P. and Rozmus C., Culturally Sensitive Care of the Muslim Patient, *Journal of Transcultural Nursing*, 2001, 12 (3), 228–233.

2. An authentic "saying" of Prophet Muhammad (quoted above) states that "There is no disease that Allah has created, except that He also has created its treatment" (Bukhari), which is interpreted as an explicit invitation to doctors to discover cures and defeat pathologies.

3. The juridical principle of necessity says: "pressing needs even allow what is prohibited".[12] It is based – according to analogy and by extension – on various verses in the Koran, e.g. 6.145: "Say: 'I find not in the Message received by me by inspiration any (meat) forbidden to be eaten by one who wishes to eat it, unless it be dead meat, or blood poured forth, or the flesh of swine, – for it is an abomination – or, what is impious, (meat) on which a name has been invoked, other than God's.' But (even so), if a person is forced by necessity, without wilful disobedience, nor transgressing due limits, thy Lord is oft-forgiving, most merciful." Whenever the characteristic of necessity is found, one may, by analogy, decide in a similar way. As a consequence, the forbidden may become licit on the condition that a necessity has to be overcome; however, the results of the application of this criterion must be carefully evaluated, seeking an equilibrium with respect to the following *hadith:* "God has created for each disease a treatment, but do not use forbidden methods."[13] When the need "persists", it is possible to reach the state of "extreme necessity", that is, of mortal risk.[14]

4. Once the necessity is recognised as the exception to the rule of the right to avoid serious harm, another principle is derived, that of the "lesser evil". Amongst the classic applications in Muslim law we find that it is lawful to abjure Islam in appearance (Razi; Averroes) or drink intoxicating liquids (Bukhari) to save one's life or the life of a Muslim.[15] Similarly, the doctor may act on the physical integrity of a patient only for a higher therapeutic purpose, with the choice of the lesser of two evils becoming valid (e.g. the amputation of a limb with gangrene to save the individual). Avoiding the evil has priority over the acquisition of good. This is one of the principles traditionally used to justify therapeutic abortion.

---

[12] Santillana, op. cit., I, 38 and 68.

[13] For example, a famished Muslim can eat prohibited foods (e.g. the flesh of a dead animal or pork) if there is nothing else available, but only in a quantity sufficient to save his life and not out of mere pleasure. Recourse to insulin of animal origin (especially from pigs) was, until recently, the best treatment for many diabetics and, as such, inevitably accepted by Muslim doctors and patients. Animal insulin has been replaced by insulin produced from recombining DNA.

[14] In Muslim law, the state of necessity (*darura*) justifies those who use violence to obtain food they need urgently (Abu Yusuf), such as for example the requisition – in the public interest – of foodstuffs found from hoarders or murder for legitimate defence, Santillana, op. cit., 68.

[15] It is lawful to recognise a Muslim principle without the personal requisites required by the *Shari'a* or which is even unworthy, in order to avoid the greater evil of civil war. The suspension of the Holy War is lawful if Muslims are not capable of winning; in this case a truce is better even with unfavourable conditions in order to limit the damage (Khalil, Zurqani).

5. The principle of "public good" or "usefulness" (*maslaha*). In Islam, general usefulness (Zurqani) must always prevail over private usefulness. According to Ibn Nugaym: "Utility is the rule of the jurisconsult." The law can be perfected because needs and usefulness change.[16] Ghazali (d. 1111) identified five priorities in the protection of the collective interest: the protection of religion, of the soul, of the intellect, of children and of property. *Maslaha* is one of the principles that is useful in order to legitimise simultaneously and paradoxically the transplant of organs, penal mutilation (see below) and the death penalty. At the same time, the principle of "public good" remains closely connected with the altruism that has its roots in Koran 5.2: "Help ye one another in righteousness and piety."

6. The sacred character of the human being. God Himself says: "We have honoured the sons of Adam" (Koran 17.70). Life is a divine gift to be protected and on which man does not have total freedom. God also says: "if any one slew a person –unless it be for murder [a clear reference to the 'lex talionis'] or for spreading mischief in the land- it would be as if he slew the whole people" (Koran 5.32).

In the Muslim vision, man is not the bearer of inalienable and irreducible rights, ontologically connected with his nature (as this would represent a limit to absolute divine omnipotence). Man remains a simple creature (to be accurate a "servant of God"), even if the most perfect of them, whilst only the Creator possesses real rights. Everything belongs to God, including the body that man has received as a gift in trusteeship, as he is the substitute of God (Adam was the Caliph, i.e. the Vicar of God on Earth). Man cannot do as he likes with his body: on the contrary, he has the duty of respecting it and protecting it according to the dictates of the *Shari'a*, the best and most complete Law ever given to man in order to promote his well-being, harmony and happiness in application of Koran 2.185: "God intends every facility for you; He does not want to put you to difficulties."

In the light of absolute divine omnipotence, God is the source of every illness and recovery whilst the doctor remains essential in order to distribute remedies

---

[16] In this regard, the *Malikite* jurist al-Qarafi (d. 1285) said: "In the time of Adam, the condition of men was weak and miserable; therefore God was indulgent in many things. When society became richer and more numerous, the precepts were multiplied. This progress is visible in the very history of Islam. . . . Indeed, we see the Companions of the Prophet introduce many provisions because they had recognised their usefulness; they decided to put the Koran in writing, whilst nothing in the previous laws authorised them to do so, they coined money, built prisons and so on. For the same reason, namely usefulness, the law has submitted testimonial evidence to very strict conditions which are not to be found at all in the tradition (*hadith*), it broadly interpreted certain contracts, such as loans, the plantation contract, the commendams and other agreements. . . . All these changes and innovations, of which there are many in our law, were imposed by the change in general conditions and different times." Santillana, op. cit., 71.

and treatment. The figure of the doctor in the history of Muslim civilisation has been influenced by two complementary elements:

(a) a structural element, namely the totalising character of Islam (Islam is Religion and State) regulating all human acts.

(b) the figure of the *hakim* or great sage (e.g. Avicenna, Averroes) whose skills could comprise the whole of human knowledge (medicine, ethics, natural sciences, religions, law, etc.) reflecting a unitary vision of knowledge, the cosmos and man, taking the absolute uniqueness of God as the point of origin.

In the wake of the influence of these two elements, the sphere of action of the Muslim doctor traditionally tends to exceed the strictly clinical context to widen out into the social and religious sphere where doctors and patients act. Before being a good doctor, he should be a good Muslim, aware of and inviting respect, wherever possible, of the rules and precepts of the *Shari'a*, as well as of the five pillars of the religion (in particular fasting, praying and the pilgrimage). In this sense, he incarnates the role of mediator between religion and health, advising the patient on what it is best to do in order to respect the laws and rites of Islam.[17]

Amongst the consequences of the role of mediator between health and religion, the doctor is summoned to fight against the behaviour that undermines individual and collective health: alcohol, environmental pollution, smoking, etc. The multiple damage provoked by sexual freedom should be exposed by the doctors as stated apologetically by the *Islamic Code of Medical Ethics* (Kuwait 1981): the prophylaxis against venereal disease deriving from moral permissiveness "lies in revival of the human values of chastity, purity, self-restraint". The text maintains that in some "developed" countries (i.e. Western) the spread of gonorrhoea and syphilis has almost reached epidemic dimensions leading the health authorities to request the declaration of a state of national emergency. Nevertheless, doctors continue to state that it is not a scandal, that it is normal. The Western mass media do not dare to speak of chastity as one of the solutions to eliminate this disease: "In contrast with anti-pollution, anti-smoking, anti-saccharine, anti-fat and several other anti's, sexual license has been signalled out as the area where a doctor should not moralize but just treat".[18]

Therefore, the Muslim doctor could easily influence the context of the "personal" values and conduct of an individual (e.g. sexuality, alcoholism, smoking) in what, at least partially, is defined in the West as the sphere of personal autonomy. This perhaps appears to be taken for granted for a doctor who aims to cure the spirit and health in the widest sense as well as the body; a doctor who – in the footsteps of tradition – should also be an ethicist, a philosopher, a theologian etc. This "classic" figure of the doctor approaches, from several points of view, that of the Western bioethicist. This leads to one consequence: the figure of the contemporary Muslim doctor "suffers" from the increasingly

---

[17] *Islamic Code of Medical Ethics*, Kuwait, 1981, chapter 9 in www.islamset.com/ethics/code/
[18] Ibid.

accentuated specialisation in modern clinical practice. This specialisation appears to be to the detriment of the projection towards the transcendent that characterised the "classic" doctor.

The ideal legitimisation of medical treatment consists of identifying a similar or analogous practice in the Holy Sources or in some authoritative "classic" juridical text. As a consequence, doctors of the Law sift through the Sources in order to identify some allusion or affinity that legitimises a specific practice. In the absence of any reference, recourse is made above all to the juridical principles of necessity and that of public good (quoted above). According to these principles, Muslim medical ethics and the attitude of jurists traditionally appear characterised by substantial realism in agreement with the provisions of the *Shari'a*. For example, the Muslim Law presents numerous exemptions reserved for the sick, pregnant women, the elderly, the handicapped, travellers, soldiers: categories that cannot adequately respect ritual obligations – in particular prayer, fasting (e.g. Koran 2.184–185[19]) and the pilgrimage – especially when these obligations put their health at risk.[20]

THE DOCTOR–PATIENT RELATIONSHIP

With respect to the Hippocratic tradition where the interest of the doctor was limited to the relationship with the patient, in the world of Islam the fifth principle of medical ethics (*maslaha*), i.e. the principle of public good and of help and care for the needy and ill, has taken on preponderant importance, deeply affecting values and medical practice. With this contribution, medicine has become a social and moral practice and this activity has been conveyed above all by the hospitals or Bimarestan (from the 8th century onwards).

---

[19] With reference to Ramadan, Koran 2.184–185 says: "(Fasting) for a fixed number of days; but if any of you is ill, or on a journey, the prescribed number (should be made up) from days later. For those who can do it (with hardship), is a ransom, the feeding of one that is indigent. But he that will give more, of his own free will, it is better for him. And it is better for you that ye fast, if ye only knew. Ramadan is the (month) in which was sent down the Qur'an, as a guide to mankind, also clear (Signs) for guidance and judgment (between right and wrong). So every one of you who is present (at his home) during that month should spend it in fasting, but if any one is ill, or on a journey, the prescribed period (should be made up) by days later. God intends every facility for you."

[20] Many books on law include a chapter describing the concessions made to the sick for prayer (called *salat al-marid*, i.e. the prayer of the ill person), for example the possibility of shortening some prayers or combining two, or modifying the movements, adapting them to the physical condition of the individual. The *Shari'a* prescribes monetary compensation or compensation of another type to remedy the obligations not performed due to force majeure. For example, if fasting is harmful, it may be avoided by the elderly, pregnant women or nursing mothers, who can make up for this by giving alms in the form of food for each day missed, etc. Generally the concessions are made in offers of money or meals for the poor. See Atighetchi D., Ramadan e paziente musulmano: problemi etico-clinici, in Compagnoni F. and D'Agostino F. (eds), *Il confronto interculturale: dibattiti bioetici e pratiche giuridiche*, Milano, San Paolo, 2003, 391–412.

The "sayings" of the Prophet emphasise the spiritual merits connected with visiting the sick. Similar recommendations seem to imply that the visit contributes to improving the condition of the patient. The Traditions advise short visits and words of encouragement. Muhammad himself usually visited an ill person after three days as it was thought that only after this period the problem could be serious. Another *hadith* recalls that the patient is in a state of purity, a reason why his prayers are more easily answered by God; this means that visitors should ask the ill person to pray for them.[21]

In the times of the Prophet there were no hospitals in Arabia. These were established on a wide scale in the Abbasid period (i.e. from about AD 750). The first Muslim hospital director was Bakr al-Razi (d. 925).[22]

The Muslim hospitals existing in the period from 9th to 12th century were the largest and best organised in the world.[23] A description of the aims of the activity in the most important Egyptian hospital of those centuries, the Mansuri hospital (completed in 1284), is presented in one of its founding documents.[24] It was a place of treatment for the ill, independently of their gender, economic conditions, region of residence, race and religion. Major or minor disorders were treated there, including both physical or mental problems, as the protection of mental health is one of the aims of the *Shari'a*. The greatest attention was to be devoted to those who had suffered from the loss of their minds and honour, and individuals were to be admitted as well as entire groups until they were perfectly healed. All the costs were borne by the Hospital. The insistence on mental illnesses is to be underlined with the connection to physical disease, showing the importance of psychosomatic medicine in classic Islam. No patient could be rejected on the basis of race, religious faith, gender, etc.; there was a mosque and also a chapel for Christians.

This "egalitarian" trend for patients of different faiths can be accompanied by a different attitude well rooted in Muslim law, according to which the *dhimmis* (i.e. the faithful of other monotheistic religions), whilst protected, remain in a state of juridical inferiority compared to Muslims. This can be found in a historical episode. The head of the doctors of Baghdad, Sinan Ibn Thabit (d. 942), questioned the Grand Vizir or Plenipotentiary Minister or lieutenant of the Prince (Wazir at-Tawfid – Ali Ibn Isa), on the conduct to be followed by doctors concerning the majority of the inhabitants of Sawad (southern Iraq) who were Jews, recalling that in Baghdad Muslim and non-Muslim patients were treated in the same way. The Vizir replied that animals, Christians and Jews have to be treated but following a specific order: first Muslims are treated, then non-Muslims and lastly animals.[25]

---

[21] Rahman, Islam and Health/Medicine, art. cit., 157.
[22] According to tradition, Razi put pieces of meat in different parts of the city and built the hospital on the spot where, a few days later, the meat was less tainted.
[23] Hamarneh S., Developments of Hospitals in Islam, *Journal of the History of Medicine and Allied Sciences*, 1962, XVII (3), 366–384.
[24] Rahman F., *Health and Medicine in the Islamic Tradition*, New York, Crossroad, 1989, 70.
[25] Rahman, Islam and Health/Medicine, art. cit., 158–159.

Following the strictly juridical order, the Muslim doctor could first protect the faithful of the "true religion". For example, according to al-Qaradawi, "if there is a Muslim and a non-Muslim and both are in need of organ or blood donation, the Muslim must be given priority for Allah Almighty says in the Ko-ran (9.71): 'The Believers, men and women, are protectors one of another'."[26] Obviously, this juridical approach may go against the medical ethical approach, for example art. 3 (Chapter 2) of the last Islamic Code for Medical and Health Ethics of the IOMS (2004) which maintains that the physician should treat all his patients "without any discrimination based on . . . their religion . . . or their gender, nationality, or color".[27] The following art. 4 seems more problematic as it says: "A physician should . . . show respect for their [of the patients] beliefs, religions, and traditions when engaging in the process of examination, diagno-sis and treatment. He should avoid any violations of Islamic law, such as being alone with a member of the opposite sex or looking at the private parts ('awra) of a patient except in as much as the process of examination, diagnosis and treat-ment requires; in the presence of a third party; and after obtaining permission from the patient."

These aspects highlight a contrast between Islamic medical ethics and Muslim law, which in turn is rooted in the Koran where both interpretations are present at the same time: (a) the equality of the faithful and/or men before God; (b) the structural juridical distinction between Muslims (faithful to the "true religion"), *dhimmis* and infidels.

MEN AND WOMEN

When the doctor examines a patient, independently of gender, he should avoid looking at their private parts. The *Shari'a*, fulfilling the indications of Koran 24.30–31,[28] has listed the parts of the body to be protected (Ar. *'awra*) to protect the chastity of both sexes (in particular women) from the looks of others: in men, the anatomical parts from the navel to the knees are to be covered, while in the women all parts except the face and hands.

If the doctor and the patient are alone and of different sexes, the problem of *khalwa* arises, a situation where a man and a woman are in a closed room where

---

[26] Qaradawi (Al) Y., *Donating Organs to non-Muslims*, in www.islamonline.net, Fatwa Bank.

[27] See www.islamset.com/ioms/Code2004/index.html

[28] Koran 24.30–31: "30. Say to the believing men that they should lower their gaze and guard their modesty: that will make for greater purity for them: and God is well acquainted with all that they do. 31. And say to the believing women that they should lower their gaze and guard their modesty; that they should not display their beauty and ornaments except what (must ordinarily) appear thereof; that they should draw their veils over their bosoms and not display their beauty except to their husbands, their fathers, their husbands' fathers, their sons, their husbands' sons, their brothers or their brothers' sons, or their sisters' sons, or their women, or the slaves whom their right hands possess, or male servants free of physical needs, or small children who have no sense of the shame of sex; and that they should not strike their feet in order to draw attention to their hidden ornaments."

they risk having sexual intercourse. These individuals, if they are not married or not related by blood to a degree that forbids marriage (with reference to the woman, her father, grandfather, brother, uncle, nephew),[29] may be suspected of having had unlawful contacts or relations.

In this regard the Fiqh (Muslim Law) Academy of Mecca, the body of the Muslim World League, pronounced itself at the 8th session (19–28 January 1985). The document[30] stated that the Muslim woman is prohibited from undressing completely in front of a man with whom it is unlawful for her to have sexual intercourse; this is allowed only for a legitimate purpose recognized by the *Shari'a*. These include the case of a woman who requires medical treatment or care. Even in similar circumstances, however, the doctor should also be a Muslim woman or, alternatively, a non-Muslim female doctor; in the absence of both, the doctor must be a male Muslim or, in the absence of this, a non-Muslim. It is however prohibited for the male doctor and the female patient to be alone as the husband or, alternatively, any other woman, must be present.[31] On the basis of the usual "realism", in the case of grave necessity, these restrictions are widened (e.g. in births, orthopaedic fractures, etc.) so that the doctor must treat the woman whose life is at risk, on pain of infringement of the fundamental principle of necessity to save human lives (*al-darurat tubih al-mahzurat*). On some of these aspects, the leader of the Islamic Revolution in Iran, the *ayatollah* Khomeini (1902–1990) stated that if a woman is called to carry out a gastric lavage on another woman or on a man, or to clean their genitals, she must cover her hands so that they are not in direct contact with their genitals; the same precautions must be taken by a man regarding another man or a woman who is not his wife. If a man has to, in order to perform medical treatment, look at a woman other than his own wife and touch her body, he is authorised to do so, but if he can provide this treatment only by looking at her, then he must not touch her and if he can treat her only by touching her then he must not look at her. If a man and a woman are obliged, for treatment, to look at someone's genital organs, they have to do so indirectly in a mirror, except in cases of force majeure.[32]

Due to the emphasis on the separation of the sexes, doctors and nurses should refrain from touching a patient of the other sex except when under treatment. Even a touch on the arm or a pat on the shoulder to console or tranquillise a patient may, on the other hand, cause discomfort whilst there are no problems with the same sex.[33]

[29] Qaradawi (Al) Y., *The Lawful and the Prohibited in Islam*, A.T.P., 1994, 150–151. In addition, Prophet Muhammad also warned against the husband's relatives (e.g. brother or cousin) who can have easy access to the house of the man's wife.

[30] Borrmans M., Fécondation artificielle et éthique musulmane, *Lateranum*, 1987, LIII, 1, 88–103.

[31] Ibid.

[32] *Ayatollah* Khomeiny, *Principes politiques, philosophiques, sociaux et religieux*, Paris, Éditions Libres-Hallier, 1979, 132.

[33] Lawrence and Rozmus, art. cit., 230.

The *khalwa* of a male psychiatrist with a female patient appears more complex, as therapy may require the two to remain alone, in a confidential relationship and for a lengthy period. In this case, the solution imposes that the psychiatrist be a devout Muslim.

The roles within the Muslim family remain traditionally diversified. For the Koran, men and woman are equal but at the same time different as their natures are different. The prime task of the man is to sustain the family economically; the primary role of the woman is, on the other hand, to raise the children. These convictions affect aid for patients. The man (generally elderly) has to take the decisions whilst the woman bears above all the burden of care.[34]

Similar considerations inevitably recall the issue of health organisation with that of mixed male and female personnel. *Sheikh* 'Atiyya Saqr in a *fatwa* (in *Minbar Al-Islam*, 1988) repeated the priority of the principle of necessity as a criterion to legitimise therapeutic interventions between individuals of different sexes on the example of women such as Om Emara Nassiba, defined by tradition as the "first nurse in Islam"[35] who, in the times of Prophet Muhammad, cared for the ill and wounded. Some Muslims of a "radical" tendency maintain that it is lawful for men and women to work side by side on condition that women wear Islamic dress and work exclusively in female wards. To the objection that in the times of the Prophet, women looked after the men in battle, the Egyptian *sheikh* 'Abd Al-Rahim replied that this is true but the type of dress worn is not known.[36] Some modernists criticise these obstacles as the result of sexual discrimination. However, a document by the International Islamic Center for Population Studies and Research of the University of Al-Azhar (Cairo) reconfirmed the indispensable conditions for a woman to enter the world of work in respect of Muslim Law[37] and these conditions could include nursing and health work: (1) before undertaking any work, the woman must have the permission of her husband or legal guardian; (2) juridical Tradition prohibits a woman from using perfume when she leaves the home; (3) a woman must wear decent dress and behave in a decent fashion; (4) she must cover her body with the exception of her face and hands in respect of the Word of God (Koran 24.30–31); (5) in the event that she is eligible for marriage she must not perform work that requires her to work on her own with a stranger as this isolation may lead to undesirable consequences[38]; (6) her work must not induce her to mingle with men in undesirable ways; (7) the type of work must correspond to the female physical and psychological constitution.

---

[34] Moazam F., Families, Patients and Physicians in Medical Decisionmaking: A Pakistani Perspective, *Hasting Center Report*, 2000, 30 (6), 28–37. Lawrence and Rozmus, art. cit., 230.

[35] International Islamic Centre for Population Studies and Research (IICPSR), *Islamic Manual of Family Planning*, Cairo, 1998, 61–62. Rispler-Chaim, op. cit., 112.

[36] Rispler-Chaim, op. cit., 64.

[37] IICPSR, op. cit., 64–66.

[38] As told by Ibn Abbas the Prophet is believed to have said: "A man and a woman should not stay together alone except with a 'mehrem' (a relative whom she cannot marry). Similarly, a woman should not travel except with a 'mehrem'." in IICPSR, op. cit., 66.

Apologetics are one of the characteristics of Islamic medical ethics (but not only of medical ethics) therefore Muslim scholars often highlight the many reports in the press in the West regarding sexual relations that take place between doctors and patients. These relations are believed to be caused by the very widespread laxity on the problem in the West, whilst the rules of Islam allow safeguarding the peace and purity of society[39] in this sector as well. For Muslim juridical literature, adultery is a grave sin and the doctor who establishes an unlawful relationship with a patient does not enjoy any exceptions with respect to the harshness of the penalty foreseen by the *Shari'a*, in the case that such penalties are envisaged by the legal system in force in the country concerned.

It goes without saying that the many Muslim countries, whilst having a common religious faith, differ in their health organisation, including a different regulation of the presence of female staff and in the social habits regulating relations between patients and medical staff of different sexes. The rules and habits in force in Tunisia and in Saudi Arabia, for example, are somewhat different.

In Tunisia, there is a "feminisation" of all the medical and para-medical staff (in as early as 1990, 52% of the doctors in Tunisian public structures were women)[40] with a growing willingness by female patients to accept being examined by male doctors.

In Saudi Arabia, on the contrary, the traditional customs and the rigorous application of the *Shari'a* give rise to a model which is far from that of the West but also from that common in other Muslim countries.[41] These social rules imply inevitable consequences in hospital practice and, if the circumstances allow it, various means of separation between the sexes could be used, from the waiting rooms to the interpreters.[42] There is a small number of hospitals with wards common to both sexes, but the rooms remain single-sex.[43] Both the female patients and their husbands generally insist on the woman being looked after by a female doctor and not only for gynaecological examinations. This trend can lead to curious situations. For example, in the event that the female doctor is foreign whilst only a male translator is available, it may happen that the doctor speaks to the translator who (as a male) must remain behind the curtain, the translator speaks to the patient's husband who in turns asks his wife the question.[44] If there is more than one patient per room, the curtains are drawn to guarantee sufficient

---

[39] Abbas M., Medical Ethics and Islam, *The Muslim World League Journal*, 1995, 22 (9), 40–43.

[40] Labidi L. and Nacef T., *Deuil impossible*, Tunis, Edition Sahar, 1993, 5–6.

[41] In the Kingdom of Saudi Arabia it is unconceivable for a woman (whether rich or poor) to live independently without a male guardian. She cannot even drive alone or with men who are not her relatives. The realm of women is the home and whenever they appear in public they must be almost totally veiled.

[42] Pirotta M., Cultural Aspects of Medical Practice in Saudi Arabia, *The Medical Journal of Australia*, 1994, Vol. 161, 153 and 156–159.

[43] Shahri (Al) M.Z., Culturally Sensitive Caring for Saudi Patients, *Journal of Transcultural Nursing*, 2002, 13 (2), 133–138.

[44] Pirotta, art. cit., 156.

privacy during visits by relatives and friends of the opposite sex.[45] The Saudi Arabian woman's sense of decency may be so strong as to lead her to refuse being examined, even if her condition is serious, in casualty wards if there is only a male doctor, preferring to return when a woman doctor is on duty. Even in front of a women doctor, showing her abdomen may be an arduous task if male relatives are present.[46]

The frequent mediation by a male relative in communication between the female patient and the male doctor today appears in the West as an affront to the equality and dignity of the woman.[47] However, Muslim families that agree with this habit do not seem to judge male mediation as an abuse of the woman's rights. The Saudi Arabian woman appreciates the social security provided by the family as the male has to legally and culturally meet the woman's needs (food, clothes, home, health needs, etc.). However, this social relationship "helps to abolish the tendency for independency and autonomy among females".[48] The consequences include the tendency by women to delegate to male relatives the signature of a form of consent despite their right to take independent decisions.

Verbal mediation or the simple presence of the male relative (starting from the husband) in the relationship with the male doctor are requests that are not infrequent amongst Muslim immigrants in Europe.

Instructions for Kuwaiti doctors are close to the precautions listed so far[49]: during a vaginal examination, the patient should remain covered as far as possible by a sheet; if the patient is being examined by a male doctor, a nurse or other woman must be present. When the patient is being examined by two male doctors, the one not examining must remain at the level of the patient's head. During labour[50] the pregnant woman may instinctively grip the doctor's hand or arm to obtain psychological comfort. However, the doctor must not encourage such behaviour but be replaced by a nurse. Relations with female patients must be characterised by kindness but taking care not to encourage relations, especially outside the hospital.

Regarding the implant of contraceptive devices (e.g. the intrauterine device) the former Grand *mufti* of Egypt, *sheikh* Jad al-Haq (*fatwa* of 1980), required the intervention of a woman doctor; there can be recourse to a male doctor when this is impossible.[51]

As can easily be imagined, Muslim women, especially in traditional countries, are reluctant to be examined by men even in the presence of another

---

[45] Shahri, art. cit., 136.

[46] Pirotta, art. cit., 156.

[47] Shahri, art. cit., 136.

[48] Ibid., 137.

[49] Abu Zikri A., Ethics in Obstetric Practice, *Journal of the Kuwait Medical Association*, 1976, 10, 191–196.

[50] Ibid., 193.

[51] IICPSR, op. cit., 18 note 8.

woman. A survey at the gynaecological clinic of two major hospitals in Riyadh (Saudi Arabia) in 1992 on some 500 female patients (two-third of whom were Saudi Arabian) confirmed this trend.[52] 66% of the Saudi Arabian patients refused an internal examination by a gynaecologist (those most strongly against the vaginal examination were students) against 32% of non-Saudi Arabians (who may nevertheless be Muslims); 48% of the Saudi Arabians refused an external examination by a male doctor, against 22% of non-Saudi Arabians. Regarding the reasons given to justify the preference for women doctors, religion was the main reason for 69% of the Saudi Arabians against 50% of the non-Saudi Arabians; the conviction that a woman knows the problems of female patients better was accepted respectively by 39% and 48% of the women; embarrassment in front of a male doctor concerned 22% of the Saudi Arabians[53] and 18% of the non-Saudi Arabians; lastly, opposition by their husband involved 15% of the Saudi Arabian women against 5% of the non-Saudi Arabians. To the question of whether the patients would have revealed all their problems to a male doctor, only 27.7% of the Saudi Arabians were willing to do so, whilst 2% would have said something, 32% nothing, 37.5% were uncertain (on the other hand, the non-Saudi Arabians answered respectively: willing to disclose everything 51%, something 2%, nothing 14%, uncertain 32%).[54]

The importance of the religious factor is obvious in a very recent survey on the behaviour of 508 patients (95.6% of whom were Muslims) of the obstetric and gynaecologic departments of Al-Ain Hospital, United Arab Emirates. In this case, 86.4% preferred women doctors, 12% had no particular preference, whilst 1.6% preferred male gynaecologists. The main reasons for the clear preference for female gynaecologists were privacy during examinations (89.1%), religious beliefs (74.3%), counselling (68.8%) and cultural traditions (45.3%).[55]

---

[52] Faris (Al) E., Hamdan (Al) N., et al., The Impact of the Doctor's Gender on the Doctor–Patient Relationship in a Saudi Obstetric and Gynaecology Clinic, *Saudi Medical Journal*, 1994, 15 (6), 450–455.

[53] Again limited to Saudi Arabia, a survey of 372 long-term hospital patients (average length of stay eighteen months) in six hospitals, in the period January–June 1994, indicated a female percentage of hospital admissions of about 38% (therefore the majority of the hospital patients were male). This low female percentage can be interpreted as a consequence of the local culture which prefers to look after women for as long as possible in their own homes. See Shammari (Al) S., Jarallah J.S., et al., Long-Term Care Experience in Saudi Arabia, *Social Science and Medicine*, 1997, 44 (5), 693–697.

[54] According to a survey carried out in Bradford (GB) at the end of the 1980s, 62% of Pakistani women were against being examined by male general practitioners against 21% of Indian women. The great resemblance between Saudi women and Pakistani women can be attributed to the religious factor. See Faris (Al) and Hamdan (Al), art. cit., 454.

[55] Rizk D.E.E., Zubeir (El) M.A., et al., Determinants of Women's Choice of their Obstetrician and Gynecologist Provider in the UAE, *Acta Obstetricia Gynecologica Scandinavica*, 2005, 84, 48–53.

Regarding the situation in Iran, in 1998 the Parliament approved a law laying down the separation of the sexes in health care, including pharmacies.[56] This progressive separation implies a curious advantage for the women, that is, an incentive to take up a career in medicine. Since the 1980s, when the *ayatollah* Khomeini recalled that gynaecological examinations by men infringed religious laws, no new male gynaecologist (apart from the ones already practising) has been accepted into the speciality whilst the number of male gynaecologists is decreasing.[57]

The greatest effect of this sexual revolution can be seen at the educational level. In order to guarantee a sufficient number of female doctors for the more than 35 million women in the country, thousands of women doctors will have to be trained in the next few years. Whilst in 1992, only 12.5% of medical students were women, at in 2001 one-third of the students of medical schools are women.[58] Taking all the fields of health as a whole, female students account for 54%. In order to train a sufficient number of female specialists, the periods of postgraduate specialisation of doctors in obstetrics and gynaecology in the hospitals are reserved for women, as are half the positions in internal medicine, general surgery and cardiology.[59] In other specialist areas, at least 25% of the places are guaranteed for women.

Female students have protested against segregation from men, complaining of a lower level of medical education compared to that offered to male students; the female students complain that the best professors and instruments are provided only for the training of male doctors. A medical school in the holy city of Qom has been opened for women only whilst in Teheran the universities impose separate lectures for men and women or they must sit in opposite areas of the lecture theatre.[60]

As male–female "contact" in health structures cannot be eliminated, these contacts can create social disapproval for medical personnel and for female health care workers (nurses).[61]

A recent study on breast examination shows a tendency that differs from that promoted by the Iranian political authorities. It concerned a sample of 410 women with a high level of education, employed by seven health centres in

---

[56] At first, in October 1998, the Guardian Council opposed the bill passed by Parliament as the creation of separate hospitals for men and women was too expensive for the State. REUTERS, 14 October 1998.

[57] Azarmina P., In Iran, Gender Segregation becoming a Fact of Medical Life, *Canadian Medical Association Journal*, 2002, 166 (5), 645.

[58] Malekzadeh R., Mokri A., et al., Medical Science and Research in Iran, *Archives of Iranian Medicine*, 2001, 4 (1), 27–39. Azarmina, art. cit.

[59] Azarmina, art. cit., 645.

[60] Abdo G., *Iran's Female Students protest at Segregation: Medical School Sit-ins reflect Growing Demands for Sexual Equality*, www.zan.org/news, 28 January 2000.

[61] Nikbakht N.A., Emami A., et al., Nursing Experience in Iran, *International Journal of Nursing Practice*, 2003, 9 (2), 78–85.

Teheran (Iran).[62] The majority of the sample, 58% (239), prefer that the breast examination is performed by a female doctor; 28% prefer a man, whilst 14% are uncertain. However, the relative majority, i.e. 47% (192), consider that a breast examination by a male doctor does not contradict the principles of Islam; on the contrary, 33% deem that this examination is contrary to these principles and 20% are uncertain. The authors of the study maintain that – despite certain precautions – the sample may be representative of the prevailing opinion amongst Iranian women.

A survey of 524 Turkish doctors in the Ankara area in 1995–1996 shows an apparently more secular approach to the patient of the opposite sex. 98.3% of the sample (515 doctors) deem that a doctor must not discriminate sexually.[63] Amongst the main reasons, for 277 doctors, the gender is important only for illnesses linked to the patient's sex; for 242 the medical activity must be separate from the personal beliefs of the doctor; for 236 the gender of the patient does not require any special attention in the modern medical approach.

Another approach in the male–female relationship would appear to be followed in Malaysia, a country with a Muslim majority (62% of the population). The Code of Medical Ethics of 2002, in discussing cases that involve intimate examinations, amongst the other conditions, always and peremptorily require the presence of a chaperon as a neutral witness, without specifying a gender.[64] Similar indications were already present in the rules of "Good Medical Practice" adopted by the Malaysian Medical Council in 1999 on any examination of a patient[65]: "A doctor must always examine a patient, whether female or male, or a child, with a chaperon being physically present in the consultation room, with visual and aural contact throughout the proceedings." The purpose of the presence of a chaperon is to protect doctors from accusations of misconduct or physical abuse without particular references to religious traditions.

Article 11.1.9 of the Code of Ethics (2001–2002) of the Pakistan Medical and Dental Council appears just as general with regard to the gender of the witness in the case of intimate examinations: "For any intimate examination the patient irrespective of age, is entitled to ask for a attendant to be present. Such requests should be acceded to whenever possible."[66]

However, the vague formulation on the gender of the witness in the last two Codes (Malaysian and Pakistani) is solved with the respect of local customs.

---

[62] Montazeri A., Haji-Mahmoodi M., et al., Breast Self-Examination: do Religious Beliefs Matter? A Descriptive Study, *Journal of Public Health Medicine*, 2003, 25 (2), 154–155.

[63] Pelin S.S., and Arda B., Physicians' Attitudes Towards Medical Ethics Issues in Turkey, *International Journal of Medical Ethics*, 2000, 11 (2), 57–67. An in-depth study on doctor-patient communication in gynaecology in a Turkish hospital shows a more paternalistic approach by the male doctor compared to the woman doctor, see Uskul A.K. and Ahmad F., Physician–Patient Interaction: a Gynecology Clinic in Turkey, *Social Science & Medicine*, 2003, 57, 205–215.

[64] Persatuan Perubatan Malaysia, *Code of Medical Ethics*, February 2002, Section II (14).

[65] Majlis Perubatan Malaysia, *Good Medical Practice*, November 1999, 9 and 11.

[66] See the Code in www.pmdc.org.pk/ethics.htm

AUTONOMY AND CONSENT OF THE PATIENT

The brilliant examples of Muslim medical tradition on the protection of the autonomy and respect of the consent of the patient[67] today appear far from common practice.

In Muslim contexts, the stimulus and impact of contemporary Western models are above all the factors that lead to a reinterpretation and a retrieval of specific passages of Muslim medical ethics and the Holy Texts to be used for a greater protection of the rights and autonomy of the patient. In order to contextualise this trend, at least five obstacles which have their origin in interacting factors in the cultural background and in the social articulation of Muslim communities have to be assessed.

1. Traditionally, the sphere of influence of the doctor should not be limited to the therapeutic phase.[68] He should not only heal but also prevent illness. As a good Muslim, he must first of all fight against that behaviour which may damage Muslim society as a whole and individuals. This may have an influence on the general behaviour of the individual. Projected into the contemporary situation, this implies a potential increase in the situations of conflict between doctor and patient, beyond the strict sphere of diagnostics-therapy, which could be translated into a limitation of the autonomy of the patient.

2. A *hadith* (collected by Bukhari) stated: "If one organ complains, all others share its complaint, suffering sleeplessness and fever."[69] It can be deduced that the disease of a part, the more serious it is, the more it involves the whole person at various levels (physical, psychological, etc.). This very modern outlook can however lead to the recognition of a state of psycho-physical inferiority of the person in such a way as to diminish the value of their requirements whenever they go against habit (e.g. a request for euthanasia). Two clinical examples[70]:

(a) a 45-year-old man, with amyotrophic lateral sclerosis and with serious disorders of phonation and deglutition, asked for active euthanasia. For Muslims, the psychological state of the person requesting euthanasia cannot be considered due to the psycho-physical inferiority of the patient;

(b) a 42-year-old patient with children, following a viral infection, had cardio-respiratory insufficiency in a terminal phase which required a heart and lungs

---

[67] For example: Ajlouni K.M., Values, Qualifications, Ethics and Legal Standards in Arabic (Islamic) Medicine, *Saudi Medical Journal*, 2003, 24 (8), 820–826. Aksoy S. and Elmali A., The Core Concepts of the Four Principles of Bioethics as found in Islamic Tradition, *Medicine and Law*, 2002, 21 (2), 211–224.

[68] See above the section on "Principles and Characteristics".

[69] Sherbini (Al) I., Life and Death between Physicians and Fiqh Scholars, in Mazkur (Al) K., Saif (Al) A., et al. (eds), *Human Life. Its Inception and End as viewed by Islam*, Kuwait, IOMS, 1989, 321–333. Another translation says: "when one part of it [the body] is affected with pain the whole of it responds in terms of wakefulness and fever", in Ebrahim A.F.M., The Living Will (wasiyat al-hayy): A Study of its Legality in the Light of Islamic Jurisprudence, *Medicine and Law*, 2000, 19, 147–160.

[70] Massué J.-P. and Gerin G. (eds), *Diritti umani e bioetica*, Roma, Sapere 2000, 163 and 282.

transplant. Thanks to his consent, he was put on the list of applicants for organs. On the day of the explantation, whilst still conscious, he refused the operation. His wife and children asked the doctors to operate regardless. Here too the answer was that due to his state of psycho-physical inferiority, the patient's refusal could not be taken into consideration, therefore, if the doctors deemed the operation useful for saving his life, they had to operate, as well as on the basis of the previous consent by the relatives and the patient himself. In these cases, the real judge remains the doctor (Chapter 10 "The End of Life").

3. The irreplaceable role taken in the past by "medical paternalism" in the doctor–patient relationship. By inertia, paternalism remains alive even when the current professional codes follow a more balanced perspective in the doctor–patient relationship. It is true that, according to the Prophet Muhammad, the patient must represent the guide for the doctor and not the contrary (a very modern statement), however, collective mentality and medical practice of the past have left the doctor with very great powers in decision-making to which the low cultural level of the patients contributed, especially when the doctor has great religious or moral prestige.[71] In previous centuries, the poor intellectual level of the patient was accompanied by the absence of awareness of his rights. In addition, and this applies in particular today, claiming rights and the patient's autonomy and sanctioning them through specific regulations, is not sufficient as long as there are few bodies to which the patient can turn in order to take legal action against the doctor, or as long as the action of these bodies appears ineffective. The current Professional Codes of Muslim countries tend to highlight with multiple variants the duty of the health personnel to inform – as far as possible – the patient of his conditions (with all due caution, if necessary) but also to avoid abrupt notification or completely avoiding it when considered harmful. These Codes express the desire to have the patient take part in the therapies that will follow.[72] This, in concrete reality, in particular in disadvantaged social contexts, is often translated into recourse to simple verbal consent to the tests and therapies suggested by the doctor[73]; however,

---

[71] Moazam F., art. cit., 31.

[72] In a study on Arab immigrants in San Francisco, it has been reported that Middle Eastern patients were confused and uncertain when given a choice of therapies offered by American doctors. This behaviour was interpreted as a signal of the lack of preparation of the doctor, see Lane S.D., Research Bioethics in Egypt, in Gillon R. (ed.), *Principles of Health Care Ethics*, New York, John Wiley & Sons Ltd., 1993, 885–894.

[73] This was, for example, the solution adopted in involving the volunteers of categories "at risk" in taking part in certain tests in Pakistan: Bali S., Nab N., et al., HIV Antibody Seroprevalence and Associated Risk Factors in Sex Workers, Drug Users and Prisoners in Sindh, Pakistan, *Journal of AIDS and Human Retrovirology*, 1998, 18, 73–79. A similar solution was adopted for the 23 patients who underwent a kidney transplant (Riyadh, Saudi Arabia) before Ramadan and took part in research to ascertain the extent of the modifications in biochemical parameters on plasma and urine as a result of fasting, see Abdalla A.H., Shaheen F.A., et al., Effect of Ramadan Fasting on Moslem Kidney Transplant Recipients, *American Journal of Nephrology*, 1998, 18, 101–104.

it is easy to imagine how the poor comprehension of the doctor's words and the scarce autonomy in the patient's choices invalidate this consent. In the last place, in paternalistic societies the doctor–patient relationship, more than a relationship of collaboration, is a relationship in which the doctor tends to guide the patient and his relatives.

4. The fourth issue is linked with the previous one. We may ask whether the information and autonomy of the patient, the respect for his decisions and desires in the relationship with the doctor, are a phenomenon in themselves (i.e. unrelated to social conditioning) or, on the other hand, whether they are the projection into the health context of the more general autonomy, freedom and possibility of information of the "average" citizen of a specific country.[74] A geopolitical observation comes into play in favour of this theory: in the contemporary period, the countries where informed consent has begun to be systematically discussed and is best asserted (obviously in relative terms) are some democratic nations in the West, starting with Germany, Switzerland, Austria, France and the USA. In other words, the recognition and respect of specific rights of the patient have been proven to be the result of wider social dialectics within liberal democratic systems in which the protection of the individual's rights has taken on a dimension which has no precedent in the entire history of humanity. The result is that a real development of the autonomy of the patient in specific political and social contexts can come up against supplementary obstacles. Furthermore, in different cultural and social contexts, the principle of autonomy tends to take on different meanings; inevitably the meaning of informed consent changes (e.g. Malaysia).[75] These differentiated cultural approaches are very important to effectively handle the local patient. Lastly, it is pointless concealing that the distribution of affluence assumes a major role in this subject, but it is reductive to indicate only in these shortcomings certain imbalances in the doctor–patient relationship which, in actual fact, are also rooted in cultural and social values.[76]

5. In many Muslim countries (Egypt, Pakistan, Turkey, etc.) the desires of patients are often subordinate to those of the family and/or of the social group to which the patient belongs.[77] In developed and secular Western societies, the

---

[74] In this regard, see the section "Cultural Sensibilities and Medical Ethics" in Chapter 2.

[75] Blum J.D., Talib N., et al., Rights of Patients: Comparative Perspectives from Five Countries, *Medicine and Law*, 2003, 22 (3), 451–471.

[76] In Pakistan, for example, there is a shortage of instruments to monitor and record the quality of care. The state authorities are inefficient and pressure groups do not exist to intervene and challenge the types of treatment given to the patient. Legal proceedings against doctors are almost non-existent; similarly ethical problems on treatment and experimentation are not raised. The majority of hospitals do not have a Committee for the Protection of the patient whilst demand for informed consent on the treatment to be given and research to be experimented appears infrequent. Malik I.A. and Qureshi A.F., Communication with Cancer Patients, *Annals of the New York Academy of Sciences*, 1997, 809, 300–308.

[77] Lane, art. cit., 891.

autonomy of the patient represents the fundamental element of medical ethics; the competent patient is considered a rational individual, capable of self-determination. On the contrary, in the majority of the world's population, the family (or other intermediate social categories such as clans, social groups, etc.) still often has a decisive role in the decision-making process on the health of one of its members, especially for the purpose of offering psychological protection. In traditional contexts, the individual member is bound by a strong internal hierarchy and strong social bonds. The elder members have a central role in decisions, whilst women are responsible for caring for and looking after the patient. In Muslim contexts, reverence for the figure of the doctor, deemed an instrument of divine compassion, can also be added.

In other words, personal identity can take secondary position with respect to the collective family identity; the expression "You are your family, and your family is you" may be valid here.[78] Alternatively, the decision can easily be delegated to the most authoritative figure in the context in which it is taken, e.g. father, doctor.[79] When the decision-making power of the family is strong, the interests and desires of the "weak" members may be overlooked. For example, when the inequality of power between men and women is strong, the subordination of the woman to her husband's wishes or to those of another male, is very likely. In very poor and/or disadvantaged contexts, the survival of a single family member may be sacrificed due to economic demands, excessively expensive treatment etc. for the survival of the group.

Medical specialisation and depersonalising technology tend to weaken the close relationships with the family members and the patient, replacing these relationships with a contractual model similar to that dominant in the West.[80] However, this type of contractual relationship requires an informed and educated population accompanied by effective controls by professional, legal, governmental bodies etc., which is a situation that is virtually absent in many developing countries.

In the last place, the patient, when seriously ill or without hope, rarely takes autonomous decisions regarding a possible therapy to be followed (see Chapter 10 "The End of Life"); in addition, it is generally the family that decides who to consult. The family has a role of mediation between the patient and the doctor but with a tendentially subordinate position to the latter.

In the light of the five points listed here, it often still appears problematic for doctors to accept open dialogue with the patients and their concrete participatory and decision-making role. However, precisely because of the historic authority of the figure of the doctor has in Muslim areas, doctors must make a greater effort to involve the weakest elements in the decision-making process, even at the cost of rebelling against dominant cultural norms.

---

[78] Moazam, art. cit., 30.

[79] Rashad A.M., MacVane Phipps F., et al., Obtaining Informed Consent in an Egyptian Research Study, *Nursing Ethics*, 2004, 11 (4), 394–399.

[80] Moazam, art. cit., 36.

SAUDI ARABIA, LIBYA, TUNISIA, ALGERIA

The contradictory nature of the situation in Saudi Arabia between the trends oriented to increasingly protect the autonomy of the patient (starting with the woman) and current medical practice anchored in traditional values and paternalism is emblematic. The Royal Decree 22941 of 19 November 1392 (1972) was issued following the refusal of a doctor to operate on two patients in a state of unconsciousness, following road accidents, as there were no relatives to express their consent for the operation. The decree established for all emergencies the lawfulness of the operation on a male or female patient even without their signed consent (if they were incapable of signing) and in the absence of the substitutive consent of the relatives.[81] Article 21 of the Ministerial Resolution 288/17/L of 23 January 1990 makes a similar statement, promulgating the rules implementing the Regulations on the Practice of Medicine and Dentistry of the Royal Decree M/3 of 2 October 1988. Resolution 288 takes up Royal Circular 4/2428/M of 29 April 1984 in its turn based on Resolution 119 of the Committee of Senior Ulama of 26 May 1404 (27 February 1984). The latter states that before health treatment or an operation, it is essential to obtain the consent of the patients, if they are capable of signing it, whether male or female. In the event that the patient's capacity of judgement is altered, consent must be given by the patient's legal representative. In addition, the doctor must give the patient or the guardian all the information on the treatment or operation that he intends performing.[82]

In actual fact, there still appeared to exist in 1985 the widespread practice amongst doctors in the Kingdom not to recognise any right to consent for treatment or invasive tests that were essential for the health of female patients[83]; in these cases, the doctors contacted the guardian of the patient (generally the husband or a brother) so that he could sign the form of informed consent. A similar approach has, inevitably, given rise to serious situations such as that of a pregnant woman with a foetus in a transversal position on which the doctors deemed it indispensable, because of the prolonged duration of labour as well, to perform a Caesarean section. They asked the husband to sign the form of consent, but he refused a Caesarean section (as it had not been necessary in the previous deliveries), despite the explanations of the doctors and the director of the hospital. Whilst waiting for the intervention of the higher authorities to convince the husband to give his consent, the uterus ruptured with the subsequent death of the mother and of the foetus. The woman had never been asked for her opinion on the matter.[84]

---

[81] Abu Aisha H., Women in Saudi Arabia: Do they not have the Right to give their own Consent for Medical Procedures? *Saudi Medical Journal*, 1985, 6 (1), 74–77.

[82] In *International Digest of Health Legislation (IDHL)*, 1992, 43 (1), 27–28.

[83] Abu Aisha, art. cit., 75.

[84] Ibid., 75.

The director of the hospital subsequently requested an explanation from the Legal Medical Board of Riyadh, obtaining the following answer: a mentally healthy mature woman has the right to accept or refuse the medical treatment offered to her and does not require the approval of her husband, father or brother before giving her assent. Despite strong social resistance, the current laws follow this last arrangement without making any difference between men and women as shown by art. 60 of the rules of the hospital administration of the Saudi Arabian Ministry of Health, according to which written consent (to anaesthesia or surgery) must be signed by the patient if he/she is mature and mentally healthy, in the presence of two legally acceptable witnesses who then sign the form in their turn. In the case of minors, unconscious patients or mentally retarded patients, the form must be signed by the guardian or close relatives in the presence of two witnesses. If these cannot be found, the governor or the police may replace the guardian in the respect of the rule of the *Shari'a*: "The governor is the guardian for he who has no guardian."[85]

Resolution 67/5/7 on medical treatments by the Academy of Muslim Law, Jeddah, 9–14 May 1992, specified that if the legal guardian does not want to authorise the treatment that is indispensable for the patient under his guardianship to recover, consent must be obtained from another guardian and, in the last place, by the ruler. The Resolution also stated that medical research on a patient of diminished capacity is not allowed even when there is the consent of his legal guardian.

At present, although the regulations require the previous consent of mature and mentally healthy women even for "invasive" X-rays, there still seem to exist many Saudi Arabian doctors who ignore or deny this right.[86] As mentioned earlier, Saudi Arabian society does not seem to judge male mediation as an infringement of women's rights precisely due to its specific and characteristic family model.

The lack of communication between the doctor and the patient or relatives is becoming one of the main reasons that can lead to legal dispute. The guidelines in medical practice (Resolution 288 of 1990) recognise three types of responsibility: (a) civil responsibility, (b) penal responsibility and (c) disciplinary responsibility that can be recognised by one of six legal-medical Committees[87] of the Kingdom following legal action by patients. Appeal can be made against the decision to the Board of Grievances within 60 days of the verdict.

(a) Civil responsibility for negligence can be identified,[88] amongst other cases, for lack of knowledge of technical questions generally known to doctors of the same speciality; for errors in treatment; use of instruments without adequate

---

[85] Ibid., 76.

[86] Abd El Bagi M.E., Thagafi (Al) M.E., et al., Saudi Women's Consent for Invasive Radiological Procedures, *Annals of Saudi Medicine*, 1997, 17 (2), 261–262 (Letter).

[87] These bodies are made up of judges from the Ministry of Justice and doctors from the Ministry for Health and the Faculties of Medicine. Hajjaj (Al) M.S., Medical Practice in Saudi Arabia: the Medico-Legal Aspects, *Saudi Medical Journal*, 1996, 17 (1), 1–4.

[88] Hajjaj, art. cit., 2.

knowledge or without taking adequate precautions in order to avoid damage when they are used. In these cases, the doctor has to pay monetary compensation, the amount of which is established by the Legal-Medical Committee on the basis of the *Shari'a*.

(b) Penal responsibility concerns infringements of the rules of the Ministry for Health even if the patient has not been harmed and is also valid for the hospitals which do not belong to the Ministry for Health. Infringements include recourse to diagnostic and therapeutic methods which are not scientifically recognised; failure to report to the police or to the health authorities examinations on patients suspected of having been affected by contagious pathologies; refusal to treat a patient without a valid motive; an abortion without the certification of a medical committee on its legitimacy to save the life of the mother. Penalties can be up to a fine of 50,000 Saudi riyals and/or six months' imprisonment. The penalties for any injury to the patient are added according to what is established by civil responsibility.

(c) The recognised disciplinary responsibility entails some types of penalties: a warning, a fine of up to 10,000 Saudi riyals or "revoking of licence to practise and removal of name from the list of licensees".[89]

Law no. 17 of 3 November 1986 of the Republic of Libya concerning the responsibility of the doctor under art. 6 states, amongst other things, that the doctor cannot: (a) damage the physical integrity of the citizen, except where the law establishes otherwise; (b) treat a patient against the latter's will, except where his condition prevents him expressing his will or threatens public health or is contagious, or when a medical committee decides that refusal of treatment may cause further complications, making treatment difficult or impossible.[90] Point (a) could be used to practise penal mutilation according to the rule that we will quote in the following paragraph.

Regarding the consent of the patient, the Professional Code of the Republic of Tunisia[91] of 17 May 1993 under art. 35 recalls that in emergencies the doctor must carry out the necessary treatment on a minor or incompetent individual when it is impossible to obtain legal consent in time. On therapeutic experimentation, art. 103 maintains the freedom of the doctor to use a new therapeutic approach if he deems there exist valid hopes to "save life, restore health or reduce the suffering of the sick person". For this purpose, "he must, as far as possible and bearing in mind the psychology of the patient, obtain his free and informed consent and, in the event of legal incapacity, the consent of the legal representative replaces that of the patient". As far as non-therapeutic experimentation is concerned, any experimentation on man requires the free and informed consent of the person directly concerned (art. 107) who must be in a psycho-physical and legal condition such as to allow him to choose freely (art. 108) and express his consent in writing (art. 109).

---

[89] Ibid., 2–3.
[90] *International Digest of Health Legislation*, 1992, 43 (1), 94.
[91] *Journal Officiel de la République Tunisienne*, 28 May-1 June 1993, 40, 766.

Limited to experimentation, law no. 90–17 of 31 July 1990 of the Republic of Algeria[92] under art. 168/2 considers compulsory in experimentation on human subjects "the respect of the moral and scientific principles which regulate medical practice"; experimentation must obey "the free and informed consent of the individual or, failing that, the consent of his legal representative. This consent is always required".

THE PROBLEM OF PENAL MUTILATION

As stated in Chapter 2 on the characteristics of Islamic bioethics, at international level there is a strong tendency towards homogenisation of medical practice which can be found both in clinical-therapeutic techniques and in the increasingly numerous affinities (and "literary" identities) present in the Codes of medical ethics, regulations and state laws produced in different countries and cultures. In many cultures, however, these codes and regulations may remain connected with precepts, moral, religious and philosophical values of the original culture. The result is that what is ethical in one society may not be so in another. This contrast between a complete and articulated system of classic rules and values (the *Shari'a*) and modern-Western ones finds an example in the penal law of the *Shari'a*.

Muslim Law in the Penal Context

A brief synthesis of the provisions of the *Shari'a* in the penal context appears indispensable.

In Muslim law, the distinction between offences is based on punishments: (a) lex talionis; (b) punishments established by the Koran (*hudud* punishments); (c) punishments depending on the discretionary power of the judge (*ta'zir*).[93]

Lex talionis (*qisas*) or retaliation punishes acts that harm life and the physical integrity of a person, giving the latter or his legal representative the right to apply them, to the same extent, on the body or life of the offender, under the control of a judge (e.g. Koran 17.33; 2.178–179; 4.92).[94] The injuried party or, in

---

[92] *International Digest of Health Legislation*, 1991, 42 (2), 230.

[93] See Castro F., Diritto musulmano e dei paesi musulmani, in *Enciclopedia Giuridica Treccani*, Roma, Istituto Poligrafico e Zecca di Stato, 1989, Vol. 11, 1–17.

[94] The Koran 17.33 says: "Nor take life – which God has made sacred – except for just cause. And if anyone is slain wrongfully, We have given his heir authority (to demand Qisas or to forgive): but let him not exceed bounds in the matter of taking life; for he is helped (by the Law)". Koran 2.178–179: "Ye who believe! The law of equality is prescribed to you in cases of murder: the free for the free, the slave for the slave, the woman for the woman. But if any remission is made by the brother of the slain, then grant any reasonable demand, and compensate him with handsome gratitude. This is a concession and a Mercy from your Lord. After this whoever exceeds the limits shall be in grave penalty. In the Law of Equality there is (saving of) Life to you, O ye men of understanding; that ye may restrain yourselves."
Koran 5.38–39: "As to the thief male or female, cut off his or her hands: a punishment by way of example, from God, for their crime: and God is Exalted in Power. But if the thief repent after the crime, and amend his conduct, God turneth to him in forgiveness; for God is Oft-forgiving, Most Merciful."

the event of death, the closest relative, can, at his own discretion, replace the lex talionis by a monetary punishment –the *diyya* or blood price – which is generally paid by the guilty party, if alive, or by his clan (*aqila*, i.e. by the relatives on the paternal side who are jointly responsible for an offence committed by the offender); together with the *diyya* there may be expiation (*kaffara*) consisting of the freedom of a Muslim slave in the absence of which the offender must fast for two months.

Recourse is made to the punishments indicated by the Koran (*hadd* or *hudud*: death, amputation, lashing, banishment) to prosecute theft, unlawful sexual relations, the false accusation of unlawful sexual relations, drinking wine, highway robbery, rebellion and apostasy.

The crimes punishable by discretionary penalties (*ta'zir*) – the great majority – are those which are not included in the *hudud* category or offences previously mentioned but where one of the conditions is lacking to be prosecuted with the *hudud* punishments. The Koran recommends avoiding, whenever possible, the harshest punishments (mutilation and death) in favour of settlement and expiation.

As far as the *hudud* punishments are concerned, highway robbery and armed insurrection[95] are punished according to Koran 5.33–34 and the subsequent elaborations by the jurists: "In truth the reward of those who fight God and His Messenger and corrupt the earth is that they will be massacred or crucified or their hands and feet amputated on opposite sides or banished from the earth; this will be for their ignominy in this world and in the future world they will have immense torment, (34) except those who repent before you master them. But know that God is compassionate and indulgent".

Unlawful sexual relations impose, according to the Koran, the punishment of flagellation (24.2 and 4–5) whilst death by lapidation is based on a "saying" of the Prophet. Muslim legal literature foresees recourse to lapidation on certain conditions: complete or incomplete sexual relations, absence of constriction, possession of mental capacity, four male witnesses; the legal schools do not agree on whether the guilty parties must both be married. The unproven accusation of adultery is punishable by eighty strokes for the accuser.

The death penalty for apostates is not based on the Koran but on a "saying" of Muhammad. Muslim law foresees death for the male, recognising however, that he has the possibility to return to Islam within three days. The woman apostate must be lashed every three days until she once again embraces Islam.

The Koran does not indicate punishments for taking inebriating substances, unlike the *Sunna* which lays down eighty strokes of the lash if the intentional nature of the act is proven.

---

[95] See Abu-Salieh S.A.A., *Les Musulmans face aux droits de l'homme*, Bochum, 1994. We will refer to this text on several occasions in the next few pages. Castro, op. cit. Schacht J., *Introduzione al diritto musulmano*, Torino, Edizioni della Fondazione Giovanni Agnelli, 1995. Qaradawi (Al) Y., *The Lawful and the Prohibited in Islam*, Plainfield, 1994.

Muslim Law and the Universal Declaration of Human Rights

The fact that the *Shari'a* inflicts the death penalty, the lex talionis and flagellation for certain offences introduces us to the contrast between this penal system
with respect to the Universal Declaration of Human Rights and the countless
international documents against the death penalty, torture and all types of violence (e.g. mutilation and flagellation). Islam is a salvific religion and the respect
of the *Shari'a* represents, according to the Tradition, an indispensable premise
for salvation after death (save the will of God) but also to produce a fair and
orderly society on earth. Divine compassion and clemency do not contradict
the possible recourse to harsh penalties considered at times indispensable for
the purpose of preventing, dissuading and punishing crime, protecting the community as a whole at the price of the sacrifice of individuals or their physical
integrity. The equilibrium that is required between the salvific and the punitive
inclination is exemplified in Koran 5.32 when it says:

"if anyone slew a person – unless it be for murder or for spreading mischief in
the land – it would be as if he slew the whole people: and if anyone saved a life, it
would be as if he saved the life of the whole people". This arrangement, aimed at
safeguarding life and the integrity of the individual as long as it does not violate
the precepts of the divine Law, is present in the documents on the protection of
man in Islam of a traditionalist orientation. For example, art. 2 of the Declaration of Human Rights in Islam approved in Cairo by the Organisation of the
Islamic Conference on 5 August 1990 declared[96]:

(A) Life is a God-given gift and the right to life is guaranteed to every human being. It is the duty
of individuals, societies and states to protect this right from any violation, and it is prohibited to
take away life except for a *Shari'a*-prescribed reason. (C) The preservation of human life throughout
the term of time willed by God is a duty prescribed by *Shari'a*. (D) Safety from bodily harm is a
guaranteed right. It is the duty of the state to safeguard it, and it is prohibited to breach it without
a *Shari'a*-prescribed reason.

Similarly, the Declaration proclaimed on 19 September 1981 at UNESCO by
the Islamic Council of Europe stated under art. 1 (in the Arabic version) after
having quoted Koran 5.32: "This sacred character of human life may be disregarded only in the name of the authority of Muslim Law and according to the
measures it lays down".

The documents which do not come into the conservative-traditionalist context
respond to different viewpoints although without refusing the fundamentals of the
faith. This is the case of the "Arab Charter of Human Rights" drawn up in 1994 by
the "Committee for Human Rights of the League of Arab States",[97] a document
of liberal and reformist inspiration as shown by art. 19: "The people are the source
of all power" rather than God as in the texts of an Islamic tendency. In addition,
the text recognises the validity of the principles of the "Universal Declaration

---

[96] *International Human Rights*, Wien, 1993, 325.

[97] Pacini A. (ed.), *L'Islam e il dibattito sui diritti dell'uomo*, Torino, Edizioni della Fondazione
Giovanni Agnelli, 1998, 229–263.

of Human Rights" of the Charter of the United Nations but simultaneously the principles of the "Declaration of Cairo" of 1990 are also reasserted.

DOCTORS AND PENAL MUTILATION

It is indispensable to specify that the defence of the integrity of an individual and his life are not absolute values (and it is the same in all religions) as the death penalty and physical penalties are ordered by the Koran (2.178; 5.38–39; 5.32; 5.45) and by the *Shari'a* for certain offences (The Cairo Declaration on Human Rights in Islam, see above) whilst penal mutilation – in the countries where, as well as being theoretically envisaged, it is also actually applied (e.g. Saudi Arabia, Iran, Sudan, Somalia) – can be at times carried out by doctors in hospitals. In principle, this participation is legitimised by the absolute superiority of the Divine Word and Law compared to any human code including codes of medical ethics. In the spirit of Muslim law, a guilty individual is mutilated or sentenced to death to protect the physical and moral integrity of the whole Muslim community (*umma*) therefore for a superior salvific purpose. In other words, this action takes place on two levels: in the physical protection of the community in this world and, at the same time, in the conservation of the faithful from sin in view of future life (Religion and State intersect here too). The community appears as a living organism (Prophet Muhammad said: "The faithful in their mutual love and compassion are like the body: if one member complains of an ailment all other members will rally in response" or "The faithful to one another are like the blocks in a whole building: they fortify one another")[98] from which a diseased part has to be eliminated or put in such a condition as not to cause harm according to the explicit precepts of Koran 2.178–179 and 5.38–39. It is important to point out how the "sayings" of the Prophet (*ahadith*) shown above, thanks to the analogy between the human body and the Muslim community, can contribute to supporting simultaneously, for example, both the practice of the transplant of organs and penal mutilation, without any contradictions.

Those in favour of recourse to Muslim penal law – those governments which, to a certain extent, apply it and various radical Islamic currents – tend to object to the universality of the principles contained in international documents on the protection of "inalienable" human rights as Western philosophical formulations. These documents and authors (e.g. A. Maududi) consider that for Muslims the rights of individuals are already all contained in the Law of God and only its observance allows a real respect of human dignity and effective social justice. From this point of view, human rights end up by contemplating recourse to corporal punishment (assimilated by international bodies as torture) including capital punishment, even although on their current application there exist discordant opinions amongst the Muslim "scholars" themselves. The Egyptian

---

[98] *Islamic Code of Medical Ethics*, op. cit., Chapter 10.

sheikh Yusuf al-Qaradawi, one of the most authoritative contemporary *Sunnite* Muslim jurisconsults, is in favour of the application of both measures but recalls that these penal laws have significance only in a society that is already profoundly Islamicised. Regarding the death penalty, al-Qaradawi follows traditional doctrine according to which it must be claimed for unjust murder, for unlawful sexual relations (fornication and adultery) and for apostasy[99]. The Lebanese *sheikh* al-Alayli considers correct recourse to the harshest provisions only in extreme cases in line with the Koran itself (in which, for example, the punishment of lapidation does not exist) which calls for forgiveness whilst the *Sunna* underlines the need to find evidence before punishing.[100] *Ayatollah* Khomeini uses a very common comparison amongst Muslim jurists: society is like a human body; at times to correct society it is indispensable to punish the person even with death if the criminal is incorrigible and before he corrupts society, in the same way that a tumour is removed before it invades the whole of the human body. Penal mutilation is a "divine punishment", the benefits of which are certain.[101] At the same time, the analogy between society and human body underlines the priority of the protection of the Muslim community (the *umma*) compared to the protection of the rights of the individual.

In 1992, the Academy of Muslim Law of Jeddah issued Resolution 58/9/6 (14–20 March) which prohibited the surgical reattachment of a limb that had been amputated due to application of retaliation (*qisas*) or the *hadd* penalty (e.g. to punish theft) as the punishment's effect of deterrence is annulled. However, there exist two exceptions allowing it in the case of a limb amputated out of retaliation (*qisas*): (1) the aggrieved person amputated by the culprit grants permission to the latter to reattach the limb after the punishment of amputation has been carried out on the culprit; (2) the aggrieved person has been able to restore his own limb. The Resolution also deemed it lawful to reattach a limb amputated by mistake during the execution of the previous penalties.[102]

With regard to penal measures, the *Muslim Code of Medical Ethics* (Kuwait, 1981) stated that life is sacred at every phase and must not be taken voluntarily except in the cases provided for by Muslim Law. In any case, it should be recalled that at present the majority of the member states of the Organisation of the Islamic Conference (OIC) prohibits recourse to physical punishment within the legal context.

The main problem concerns the lawfulness for a doctor to actively take part in, or to witness, penal mutilation or flagellation applied by the State in which he works. If, on the one hand, the doctor after total or partial anaesthesia, damages the integrity of the criminal by mutilating him, at the same time however, he limits the damage, preventing the consequences of a clumsy mutilation which may lead to serious complications or the death of the condemned man. In addition, from

---

[99] Qaradawi, op. cit., 326–327.
[100] Abu-Salieh, op. cit., 68–69.
[101] Khomeini, *A Clarification of Questions*, London, Westview Press, 1984, 428.
[102] www.islamibankbd.com/page/oicres.htm#58/916

the historical point of view and in Muslim law, it was not a doctor who performed the mutilation but, in observance of the rules of the lex talionis, a close relative of the victim of the criminal or the "executioner". The activity of a "mutilating doctor" therefore appears a contemporary and unparalleled element in Muslim law and one that may be diffused in those countries where the *Shari'a*-prescribed laws are actively applied following the principle of the lesser evil and necessity.

The controversy in the West on the active or passive participation of doctors in the application of *Shari'a*-prescribed punishments (flagellation and mutilation) in Pakistan after 1979 makes explicit the points of contrast on the criteria of the World Medical Association. The accusations of Amnesty International reported not only the possible activity of doctors but their presence at flagellations; this would be against both the Universal Declaration of Human Rights (1948) and the Declaration of Tokyo (1975) of the World Medical Association which forbids the participation of doctors in similar practices[103] as their presence would not reduce the cruelty of the punishment but on the contrary increase it. In particular, the Declaration of Tokyo stated that doctors must not participate with any action in a penal mutilation; they must not provide the place, instruments or other material to perform it nor must they be present at the practice. The responses from the Muslim side played on cultural relativism and the respect for the different cultures. Thus N.H. Naqvi[104] recalled that in Pakistan and Saudi Arabia abortion was murder whilst in Denmark and the USA it was lawful for social reasons if performed by doctors. However, he added, Saudi Arabian and Pakistani doctors do not seem to have the right to appeal to the Declaration of Tokyo of 1975 against their Western abortionist colleagues. In addition, the author has recourse to cultural relativism when he recalls that male circumcision, practiced by both Muslims and Jews, may be performed by doctors with anaesthesia but for critics it remains torture or violence. Lastly, capital punishment is present in many countries in the world (including the USA); therefore, each people is entitled to its own laws. According to K.Z. Hashmi[105] no society can prosper if it uses the "Declaration of Human Rights" for the purpose of protecting criminals whilst Saudi Arabia applies the *Shari'a* enjoying the lowest rate of criminality in the world (1979 data).[106]

Limitedly to the death penalty, the discourse also applies to penal mutilation, the concepts expressed by the representative of the Arab Republic of Syria, M. Wahba,

---

[103] Lykkeberg J., Doctors and the Islamic Penal Code, *The Lancet*, 24 February 1979, 440 (Letter).

[104] Naqvi N.H., Doctors and the Islamic Penal Code, *The Lancet*, 17 March 1979, 614 (Letter).

[105] Hashmi K.Z., Doctors and the Islamic Penal Code, *The Lancet*, 17 March 1979, 614 (Letter).

[106] Amnesty International has recorded 82 individuals mutilated between 1981–1995 including 4 crossed amputations (hand and foot); in actual fact, there would appear to be far more cases than this. The amputations are mainly limited to the right hand for theft, the same hand plus the left foot for highway robbery. Flagellation is widely practised, even on minors, almost always for sexual offences but also, at the discretion of the judge, as a substitute measure or in addition to another punishment.

to the Third Commission of the General Assembly of the United Nations, appear significant (1999). He maintained that the real question was not so much that of the application or not of the death penalty, but rather that of the sovereignty of each country in choosing its own political, judicial, social and cultural system. Asking countries not to apply the death penalty means precisely asking for a change in historical, cultural, religious and political connotations of a country. The discussion of the death penalty converges on the concern to protect the human dignity of the person condemned to death, whilst it completely denies the human dignity and the rights of the victim who, all the more so, deserves obtaining justice.[107] The Pakistani Representative, M.S. Bhatti, presented the following concepts on the same occasion: he stated that the sovereign right of other countries to implement legislative choices and the fight against crime they deem most suitable for their domestic situation should be respected. Values on other cultures and societes must not be imposed, otherwise this would give the impression that someone wants to control the others, and this causes resentment. The ideas of the Egyptian and Sudanese representatives were very similar.

At times the countries that perform penal mutilation use medical staff. The Libyan law of 11 October 1972 restores the *Shari'a*-prescribed punishments for theft and highway robbery according to the *Malikite* school.[108] Article 21 specifies the context in which amputation will be carried out:

1. The execution must be preceded by a medical examination of the condemned person in order to avoid the punishment in the presence of illness, pregnancy or similar, in which case the doctor must establish the date of postponement of the execution;
2. The amputation may be performed in a hospital or in the infirmary of the prison by a specialised doctor, also with recourse to anaesthesia;
3. The amputation of the hand is done from the wrist and that of the foot from the joint of the tibia;
4. After mutilation, the criminal remains under clinical observation in order to prevent any complications.

After the unification of Yemen, Law no.12 of the New Penal Code of 1994 lays down the penalty of flagellation and amputation. In line with the *Shari'a*-prescribed tradition, flagellation punishes calumny, the drinking of alcohol and sexual offences. If the offender is a bachelor, arts. 263–264 prescribe 100 lashes for fornication. In the case of adultery, there is lapidation. Regarding the consumption of alcohol and calumny, arts. 283 and 289 prescribe 80 lashes. Amputation is laid down for theft and highway robbery; art. 298 lays down the amputation of the right hand in the case of theft. If the offender is repeated the left foot up to the ankle is added; if the theft is committed by several people art. 298 prescribes the amputation for all of them, independently of the roles played. Highway robbery is punished by the amputation of the right hand and the left foot. See Amnesty International Report–MDE, March 1997, *Yemen Ratification without Implementation: The State of Human Rights in Yemen* and Amnesty International-REPORT–MDE, November 1997, *Saudi Arabia Behind Closed Doors: Unfair Trials in Saudi Arabia*.

[107] Nessuno Tocchi Caino, *La pena di morte nel mondo*, Venezia, Marsilio, 2000, 247 and 232.
[108] Minganti P., Ricezione di pene "hadd" nella legislazione della Repubblica di Libia, *Oriente Moderno*, 1974, LIV (4), 265–274, in particular 271.

It would appear that these *Shari'a*-prescribed punishments have not been applied for many years, both because they require the confirmation of the sentence by the Supreme Court (art. 19 of the law of 11 October 1972) and because the punishments should be decided by judges trained according to Western legal criteria and therefore not enthusiastic about ordering penal mutilations.[109]

The situation appears to have changed in recent years. The International Secretariat of the World Organisation Against Torture (OMCT) reported that Libya had begun to apply the penalty of amputation. On 13 October 2001, the Criminal Court of Misurata sentenced Ali Mansour Mohammed al-Guinaidy to the amputation of his right hand. The amputation was carried out on 23 June 2002. In 2002, the official Libyan press agency, JANA, reported the arrest of four Libyans accused of armed robbery on 30 April 2002. The four were sentenced to ten years' imprisonment and amputation of the right hand and left foot. On 25 June 2002, the Supreme Court confirmed the sentence and on 3 July the amputations were performed.[110]

Regarding the Islamic Republic of Iran, Amnesty International has stated that all the possible *Shari'a*-prescribed corporal punishments have been applied with the possible exception of crucifixion.[111] There are numerous eye witness accounts of penal mutilation, lapidation and flagellation, the last of which is often performed before capital punishment. As the *Shari'a* imposes strong limitations on the application of the *hudud* punishments requiring, amongst other things, eye witness accounts and confessions that have not been extorted with force (in which case they would not be valid), the doubt on the respect of the *Shari'a*-prescribed laws for the high number of sentences executed remains compulsory.[112] In Iran too, recourse to the health structures seems frequent in order to perform

---

[109] Peters R., The Islamization of Criminal Law: A Comparative Analysis, *Die Welt des Islams*, 1994, 34, 246–274.

[110] Amnesty International, *Libya: Time to make Human Rights a Reality*, 2004, April, 36–37.

[111] Article 207 of "The Hudood and Qisas act and relevant rules" (25 August 1982) says: "Crucifixion of a *mufsid* or *muharib* [corruptor or rebel] shall take place with the following conditions: (A) He should be tied in a way that it may not cause his death. (B) He should not be left on the cross for more than three days, but if during the three days he dies, his dead body may be brought down after the death. (C) If after three days he is still alive, he shall not be killed" in Government of Iran, Hudud and Qisas Act of Iran, *Islamic Studies*, 1985, 24, 521–556. The same Penal Code under art. 117 and 119 regulates the conditions for lapidation: the man must be buried in a pit up to his waist and the woman up to her bust. The stones must not be too big in order to avoid one or two being sufficient to kill, nor too small.

[112] The number of capital punishments carried out in recent years according to the data of Amnesty International is as follows: 558 (1979); 955 (1980); 2690 (1981); 698 (1982); 473 (1983); 735 (1984); 470 (1985); 115 (1986); 158 (1987); 3200 (1988); 1500 (1989); 750 (1990); 775 (1991). In 1996 110 death sentences were ascertained as having been performed (at least two by lapidation) compared to 50 in 1995. In 1997 at least 143 executions were reported to the international bodies but the real number would appear to be much higher. The number of deaths by lapidation is as follows: 2 (1979); 0 (1980); 1 (1981); 0 (1982–1983–1984); 2 (1985); 8 (1986); 4 (1987); 0 (1988); 43 (1989); 10 (1990); 1 (1991). In 1995 one person died after lapidation whilst two sentences were not carried out due to international protest. In 1997 there were at least seven deaths by lapidation.

penal mutilation. In particular in order to facilitate the amputation of fingers, a special device was made in 1986 thanks to the consultancy and collaboration of the Faculties of Medicine of the Universities of Teheran and Beheshti.[113] In January 1998 a man was sentenced by an Iranian court to be blinded as he had previously blinded another individual with acid. The punishment was to have been performed after medical examination but no doctor was available for the operation. Amnesty International has been unable as yet to ascertain whether the operation was performed. In August 2000 a court in Teheran condemned G. Aryabakhshahyesh to the surgical removal of an eye in application of the lex talionis laid down by the Penal Code due to a quarrel in 1997 in which the offender had struck the victim, depriving him of sight in one eye.

In Sudan corporal punishments require the presence of a doctor. The Penal Code of Pakistan (Text of *Qisas & Diyat* Ordinance) under art. 337-P (1) specifies that the *qisas* punishment must be performed in public by an authorised medical officer who has, preliminarily, the task of examining the condemned individual and providing the appropriate treatment so that performance of the lex talionis does not cause death and the injury caused by the mutilation does not exceed that provoked by the offender on his victim. In the event that the condemned person is a pregnant woman, the court may, with the consent of the medical officer, postpone the performance of the *qisas* up to 2 years after the birth of the child (337-P-3). Between 25 and 27 January 2001, five men accused of armed robbery underwent "cross amputation" (right hand and left foot) with the application of the 1991 Criminal Act. A doctor supervised the amputations.

The situation in Iraq[114] (during Saddam Hussain's regime) was somewhat anomalous, where Decree no. 59 approved by the Council of the Command of the Revolution in June 1994 prescribed for someone guilty of theft the amputation of the right hand followed by the left foot if he were to steal a second time. If the thief is armed or if someone dies as a consequence of his act, the offender will be sentenced to death. A subsequent decree of the Council on 18 August 1994 provides that a tattoo be applied to the forehead of every individual whose hand has been amputated. Recourse to *Shari'a*-prescribed punishments appeared instrumental in the desire to limit the wave of offences caused by the continuation of the economic sanctions of the United Nations against the Iraq of Saddam Hussain; to this end there is also the measure of tattoos which has no precedent in Muslim law. Doctors can be summoned to perform the mutilations; anyone who refuses risks being arrested or disappearing. The doctor finds himself between two intolerable choices: complicity with the government or heroism.[115]

In twelve Muslim states of the north of Nigeria, the *Shari'a* has been re-applied (but with different conditions). The Nigerian Medical Association has defined

---

[113] Peters, art. cit., 262. Amnesty International, *Rapporto 1994*, Torino, Sonda, 266–270.

[114] Court C., Doctors in Iraq face Amputation Dilemma, *British Medical Journal*, 24 September 1994, 309, 760.

[115] Ibid.

mutilation of the limbs as an "unethical and inhuman" practice and has threatened to expel doctors who perform it from the Association. Despite this condemnation, there have been some reports of amputations performed by doctors in hospitals in the past few years.[116]

The case of doctors employed by international health organisations and requested by the government of the country where they operate to carry out penal mutilation is different. This was the case in Afghanistan in 1997 with a doctor from the Red Cross who was working in a hospital of the Ministry for Health from where he was collected by the authorities and taken to a square where he amputated the hand of a condemned individual under application of the local law. The same year, in another Afghan hospital, a local team amputated the right hand and left foot of two brothers, in execution of the *Shari'a*-prescribed punishments laid down for robbery, on premises shared with the organisation Médecins sans Frontières. Similar episodes have created doubts on the extraneousness of these bodies to the performance of punishments and controversy. In reply, the International Committee of the Red Cross has expressed guidelines for its doctors working in areas where penal mutilation is applied[117]: (a) the Committee opposes *Shari'a*-prescribed amputations but may not do so publicly (the international health organisations do not want their humanitarian activity to be judged in any way as promoting Western or anti-Islamic values); (b) it does not authorise its doctors to take part in amputations; (c) it refuses to provide the premises, knowledge and instruments for these operations; (d) it undertakes to protect doctors who refuse involvement. In addition, the Committee reconfirms the duty of doctors to help the mutilated without this being equivalent to any involvement or tolerance regarding the practice. Lastly, the local authorities have accepted an agreement according to which the assistance of hospitals or the medical staff of foreign health organisations will no longer be requested for their involvement in the practice of mutilation.[118] In early 2000 the active participation of local doctors in the amputation of the right hand and left foot of six individuals and of the left foot of a seventh person in the stadium of Kabul was reported.

CONCLUSION

In the rich and democratic part of the West, attention for individual autonomy in all sectors of social and personal life has taken on an unprecedented role in human history and also in the context of health.

---

[116] Human Rights Watch, *Political Shari'a? Human Rights and Islamic Law in Northern Nigeria*, 2004, 16 (9 A), 1–113 (PDF).

[117] Perrin P., Supporting Shari'a or providing treatment: the International Committee of the Red Cross, *British Medical Journal*, 14 July 1999, 319, 445–446.

[118] Nolan H., Learning to express dissent: Médecins sans Frontières, *British Medical Journal*, 14 July 1999, 319, 446–447. Perrin, art. cit., 445.

Many cultures do not agree with this individualistic approach to social and health issues; relations between the individual and society follow different models according to local social rules and values. In addition, other social "subjects" (the family in the first place) continue to have often a decisive role in managing the illness and care for the patient or woman, especially when in a serious condition.

This is why the concepts of patient autonomy, informed consent and many other concepts should not be defined in the abstract as is often the case in codes of ethics but should be adapted to the different local, cultural, religious and national realities.

CONTRACEPTION AND POPULATION CONTROL

SOME CLASSIC FORMULATIONS   65
THE CONTEMPORARY DEBATE:
THE PRO-CONTRACEPTION JURISTS   71
STERILISATION   78
THE OPPONENTS OF CONTRACEPTION   79
THE SOCIO-POLITICAL CONTEXT   82
THE CASE OF IRAN   86
CONCLUSION   89

SOME CLASSIC FORMULATIONS

Since the times of Prophet Muhammad, coitus interruptus (Ar. *'azl*), practised by the ancient inhabitants of the Arabian Peninsula, has been the object of contrasting opinions amongst Muslim scholars.

The Koran does not deal, even indirectly, with contraceptive techniques, even if jurists in favour or against the practices will attempt to extrapolate various passages to support their personal interpretations. More explicit, but partially discordant, are the "sayings" (*ahadith*) of Prophet Muhammad, despite the *hadith* "Marry and multiply". The most authoritative *ahadith* on contraception tolerate or do not prohibit the practice of coitus interruptus. The absence of unanimity has however given rise to the differing juridical opinions formulated over the centuries.

A companion of Muhammad reported: "We used to practise coitus interruptus during the time of the Prophet while the Quran was being revealed. The Prophet, adds Muslim, came to know of this but did not forbid us (doing it)" and, in another version, "If this were something to be prohibited, the Koran would have prohibited us (doing it)"[1] was added.

Some *ahadith* take on particular importance. These include one with the Prophet's answer to a man concerned that his slave might become pregnant: "Practise coitus interruptus with her if you wish. What is pre-ordained for her will certainly befall her."[2] The second *hadith* reported by Abu Sa'id: "A man came to the Prophet to ask about the practice of *al-azl* with his mate. He added

---

[1] Glossary of Hadiths, in Nazer I., Karmi H.S., et al. (eds.), *Islam and Family Planning: A Faithful Translation of the Arabic Edition of the Proceedings of the International Islamic Conference held in Rabat (Morocco), December 1971*, Beirut, International Planned Parenthood Federation, 1974, Vol. II, 557.

[2] Ibid., 551.

'I do not like her to get pregnant and I am a man who wants what other men want. But the Jews claim that *al-azl* is minor infanticide'." The Prophet denied such a contention by the Jews and referred to destiny.[3] The two passages anticipate one of the reasons for the tolerance for *'azl* over the centuries, namely, the conviction that this technique is not an obstacle to divine omnipotence, in line with the words of the Prophet reported by Anas "If the semen out of which the child is formed were to be spilled on to a rock, God would produce a child, or a child be produced out of it."[4]

The Prophet's Arab contemporaries were aware of the unreliability of coitus interruptus to avoid pregnancy. However, faith in absolute divine omnipotence has transformed *'azl* from an obstacle to His will – as maintained by Jews and Christians – to an instrument subjected to it (Principle of Predestination): if God wants a child to be born, it is; otherwise it is not born. The Arab theologian and mystic Ibn al-Arabi (d. 1240) stated that if God wanted to create a living soul, He could make sperm which is deposited "outside" its natural place reach it in any case.[5] This interpretation is coherent with the words of the Koran 42.49–50: "He [God] creates what He wills (and plans). He bestows (children) male or female according to His Will (and Plan), or He bestows both males and females, and He leaves barren whom He will."

In general, it can be stated that each new technique in the context of procreation, i.e. both the techniques aimed at limiting procreation and those introduced to circumvent the problem of sterility (assisted medical procreation), tend to be tolerated from the juridical point of view, due to these techniques being believed as not representing an obstacle but, on the contrary, an instrument with respect to absolute divine omnipotence.

Obviously, there are some "sayings" of the Prophet opposing population control. There are two that are the most frequently quoted. The first is also used today by religious figures in favour of increasing the Muslim population and states that the Prophet will be proud, on the day of the Resurrection of the dead, of the number reached by the community of believers compared to that of other communities and to this end, he exhorted the faithful to multiply.[6] The second is a *hadith* considered less authoritative, narrated by Judama Bint Wahb al-Asadiyya according to whom coitus interruptus corresponded to secret infanticide or minor infanticide.[7]

The Koran itself moreover, on the one hand assimilates riches and children to "allurements of the life of this world" (18.46; 63.9; 64.15) and, at the same time, descendants are considered a divine benediction. To encourage the latter

---

[3] Omran A.R., *Family Planning in the Legacy of Islam*, London and New York, Routledge, 1994, 132.

[4] Glossary of Hadiths, op. cit., Vol. II, 544.

[5] Borrmans M., Islam et contraception, *Se Comprendre*, 1979, 79/9, 1–9, 7 note 6.

[6] The *hadith* says: "Marry among yourselves and multiply, for I shall make a display of you before other nations on the Day of Judgement", in Glossary of Hadiths, op. cit., 547.

[7] Omran, op. cit., 130–134.

pro-populationist approach of the Koran which continues the Bible's approach – *sura* (chapter) 2.223 is often picked out: "Your wives are as a tilth unto you; so approach your tilth when or how ye will."

The pro-populationist interpretation, which became established later in tradition, was favoured by the relative absence of problems connected with the demographic growth in the times of the Prophet when, on the contrary, the need was felt to increase the number of fighters for a forcefully expanding Islam.[8]

For the sake of accuracy, Muslim pro-populationism should be defined as "moderate" in virtue of the observation that neither the Koran nor Muhammad prohibited contraceptive practices, which was then translated, on the level of daily life, into tolerance for this custom.

When the Prophet suggested procreating, he often subordinated this action to the parents' ability to provide an adequate upbringing and sustenance so that their children could be of benefit to the family and society.[9] If children have the possibility of becoming virtuous, civil and healthy, the nation can count on them. If they grow up in precarious conditions, they become corrupt and a source of harm for the nation. The Prophet said: "The most gruelling trial is to have plenty of children with no adequate means."[10]

A classic synthesis of the reflections by jurists in the first centuries of Islam on the problem of '*azl* is contained in some pages of a work by Ghazali (d. 1111), the great jurist of the *Shafi'ite* school, as well as one of the figures whose thought has had the greatest influence on Islamic "orthodoxy", entitled: "Rivivification of the religious sciences". These pages are still a mandatory reference for anybody who is interested in contraception in these areas.

In examining the different interpretations of the doctors of the Law who were his contemporaries, Ghazali lists four positions[11]: (a) those who deem that '*azl* is always lawful; (b) those who always prohibit it; (c) it is allowed only after the prior consent of the woman; (d) it is allowed exclusively with a slave. For this author, coitus interruptus is a permitted but reproachable practice, i.e. according to the classification of juridical doctrine, a *makruh* action, similar to the reprehensible, but not prohibited, act of a Muslim who sits in a mosque without

---

[8] There is no lack of accounts to the contrary such as a *hadith* that reports the words of the conqueror of Egypt and companion of Muhammad, Amr Ibn al-As (d. 663) who, in a sermon in a mosque of Fustat (at the time the main centre of Egypt) invited the faithful not to bring too many children into the world.

[9] Madkur M.S., Sterilization and Abortion from the Point of View of Islam, in Nazer I., Karmi H.S., et al. (eds.), op. cit., Vol. II, 263–285.

[10] *Glossary of hadiths*, IPPF, 548. Abu Hanifa said to Abu Yusuf: "Do not marry until after you know that you can meet all your wife's needs. Refrain from having anything to do with women before you have attained knowledge, lest you should waste your time and have too many children, since too much money and too many children disturb the peace of mind", Madkur, art. cit., 284.

[11] Ghazali, *Le Livre des Bons Usages en Matière de Mariage*, (Extrait de *l'Ihiya Ouloum ed-Din, or Rivivification of the religious sciences*), Paris and Oxford, Maisonneuve-Thornton, 1953, 88–89.

praying or meditating, or of a believer who lives in Mecca but does not make the pilgrimage to the holy places every year.

Why is it reprehensible? Because for the jurists, procreation was and remains a meritorious action, abstention from which however, does not imply a sin; Ghazali himself specified "It presents a negligence of that which it would have been more fitting to do." Reasoning by analogy (*qiyas*), Ghazali stated that withdrawal from sexual intercourse, before the two seeds have mixed,[12] is legitimate and analogous with the act of withdrawing an offer – in the middle of a commercial negotiation – before the counterpart has accepted it.

In weighing up the motives that justify recourse to coitus interruptus, Ghazali holds three as valid:

1. *'Azl* is authorised in order to keep a slave in her condition. This was necessary as, according to the *Shari'a*, the slave who became pregnant by her master could not be sold and, on the death of her "lord", is freed.
2. To safeguard the beauty and health of the woman from the damage caused by pregnancy and giving birth.
3. To avoid a numerous progeny if this may lead to material poverty such as to oblige the parents to undertake illicit activities in order to maintain their children.

On the economic motive, the Koran 6.151 orders: "Kill not your children on a plea of want; We provide sustenance for you and for them"; 17.31 says: "Kill not your children for fear of want: We shall provide sustenance for them as well as for you. Verily the killing of them is a great sin." These passages have at times been interpreted as evidence of the opposition of the Holy Text to contraception, whilst, in actual fact, they are compatible with *'azl* as they refer to the prohibition of causing the death of a child after birth, i.e. actual infanticide.

The three points of Ghazali (especially 2 and 3) have provided the guiding criteria for scholars who until today have supported the use of contraceptive techniques.

Ghazali was also interested in the diffusion of *'azl* amongst his fellow citizens, making the observation that the majority were justified in their recourse to the practice,[13] precisely on the basis of economic motivations (Ghazali's third point). Nor did the jurist overlook the presence of a minority of the faithful, belonging to mystic groups, for whom contraception, justified by Ghazali's third reason, is illegitimate since God, as He created the soul, will not leave the new being devoid of sustenance. "There is no moving creature on earth but its sustenance dependeth on God."(Koran 11.6 and the aforementioned 6.151 and 17.31).

---

[12] According to the conceptions of the period, the woman also emitted semen which coincided with menstrual blood.

[13] Rahman F., Islam and Medicine: A General Overview, *Perspectives in Biology and Medicine*, 1984, 27 (4) 585–597. Id. *Health and Medicine in the Islamic Tradition: Change and Identity*, New York, Crossroad, 1987, 115.

Following on this, Ghazali turned his attention in particular to the reasons that would legitimise recourse to contraception. After ascertaining the legitimacy of the practice in the light of the case studies set forth, his opinion on the method or instrument used passes into the background. The greater attention devoted to the analysis of the reasons justifying recourse to contraception to the detriment of attention paid to the instruments has remained the prevailing trend to the present day.[14]

Muslim juridical doctrine allows sexual activity only with wives (up to four in number) and slaves-concubines. As today slavery (regulated by the Koran) is a practice that has declined, sexuality is allowed only between married partners.

Marriage mainly satisfies two needs[15]:

1. It satisfies the natural desire of the partners allowing them physical and spiritual union. The words of God are particularly delicate in 30.21: "And among His Signs is this, that He created for you mates from among yourselves, that ye may dwell in tranquillity with them, and He has put love and mercy between your (hearts)."

2. The second function lies in procreating according to the Koran 16.72: "God has made for you mates (and Companions) of your own nature, and made for you, out of them, sons and daughters and grandchildren, and provided for you sustenance of the best." The Koran 18.46 specifies, however, that riches and children are the "allurements of the life of this world". Therefore, children represent a divine benediction and the blessed fruit of marriage but this does not induce the Koran to explicitly forbid contraception.

As 'azl is a male act which limits the two rights mentioned, the majority of jurists have circumvented the obstacle with the precautionary consent of the wife to the husband, following the hadith that says: "Coitus interruptus is not to be practised with a free woman without her consent."[16] The strongest objections to this approach seem to be put forward by Zahirite jurists but there were divergences in each juridical school.

In the Hanafite[17] school, the dominant position allowed coitus interruptus with the consent of the wife; with the passing of time, some jurists waived this

---

[14] Amongst the exceptions is the Hanbalite theologian Ibn Taymiyya (d. 1328) in his Al-Fatawa al-Kubra in which he commented on the lawfulness for a woman to insert a pessary (diaphragm) before sexual intercourse so that the sperm would not reach her uterus and on the lawfulness of removing this "medicine" before the ritual prayer and fast after the compulsory ablution (ghusl). The text indicates the lawfulness of the ritual prayer and of the fast with the "medicine" inserted but with the clarification that, although without condemning the contraceptive practice, it would be more prudent not to have recourse to this, perhaps due to some risk of harming the woman. See Ebrahim A.F.M., Biomedical Issues-Islamic Perspective, Mobeni, IMASA, 1988, 56–57.

[15] In Ghazali's opinion, there are at least four advantages in marriage: (1) it satisfies carnal desire; (2) it increases the number of believers; (3) the son prays for his late parents; (4) if the son dies before his parents, he will intercede with God on their behalf. Ghazali, op. cit., 16–23.

[16] Glossary of Hadiths, op. cit., 538.

[17] For the schools, we refer to Omran, op. cit., 153–166.

condition in the case of "difficult times" and in the fear of seeing delinquent children being raised. Amongst the advocates of the classic position were Abu Hanifa, Abu Yusuf, Muhammad al-Shaybani (8th–9th centuries AD), al-Tahawi (d. 933), al-Kasani (d. 1198) and al-Marghinani (d. 1197). Ibn al-Humam (d. 1457) preferred to waive female consent in the case of "difficult times"; Ibn Nujaim (d. 1562) and Ibn Abidin (d. 1836) also expressed similar views, with the latter adding the condition of travellers to be included amongst the situations valid for avoiding consent.

The *Malikite* school agrees with the classic position for which the woman's consent is essential but some jurists deem it lawful to pay monetary compensation to the woman in exchange for her consent. Amongst the former, Khalil b. Ishaq (d. 1374 or 1365) in his *Al-Mukhtasar* (Abbreviated Treaty) specified[18]: "The husband of a slave wife may, with his wife's and her master's consent, practise *al-azl* with her. The same with a free wife, provided she consents." In the minority group, al-Dardir (d. 1786) stated "The free wife [may] consent to her husband [to practise coitus interruptus] free or for some small or large compensation."[19]

The prevailing position amongst the *Shafi'ites* considered coitus interruptus lawful without the wife's consent; several jurists maintained the classic position whilst others disapproved of coitus interruptus. Al-Shirazi (d. 1083) allowed coitus interruptus after the wife's consent and, in its absence, listed the most followed opinions in his juridical school: those who allowed it as the woman has the right to intercourse but not to ejaculation; a second opinion disapproved of it as it prevents procreation. Nawawi (d. 1277), in his commentary on *Sahih Muslim*, deemed *'azl* reproachable (*makruh*) independently of the wife's consent as it represents a limit to procreation, however he did not define it a prohibited practice. According to al-Hafez al-Iraqi (15th century) and his father Abdel Rahman Ibn al Hussein al-Iraqi, *'azl* is allowed both with the consent of the wife and without. Ibn Hajar al-Asqalani (d. 1449) agreed with the idea according to which the woman is entitled to sexual intercourse but not to ejaculation therefore *'azl* is lawful without her consent. Al-Qastallani (d. 1517) refers to *'azl* as a means that is not prohibited in order to limit the number of children for economic reasons; however, the wife's consent is preferable. Al-Sharani (d. 1565) agreed with the liberal attitude towards the subject.

The majority of the *Hanbalite* doctors of the law deemed *'azl* lawful with the wife's consent but this could be waived in certain situations, e.g. when in enemy territory. Ahmad Ibn Hanbal (d. 855) allowed *'azl* with the consent of the wife. In his *Al-Mughni*, Ibn Qudama (d. 1223) considered *'azl* reprehensible as it reduces the number of children and female sexual pleasure. If justifications exist, e.g. when in enemy territory, the husband can overlook his wife's consent; in this case, coitus interruptus is reprehensible but not prohibited. Ibn al-Najjar (d. 1564) deemed coitus interruptus without the wife's consent prohibited except in enemy territory

---

[18] Omran, op. cit., 158.
[19] Ibid.

where it becomes, on the contrary, recommended. Al-Bahuti (d. 1641) defined '*azl* without the wife's consent prohibited whilst it became compulsory in enemy territory to avoid a possible child ending up as a slave.

For the majority of *Duodecimal Imamite Shi'ites*, '*azl* is licit with the consent of the wife. However, this could be obtained on one occasion when the marriage contract was drawn up, remaining valid even when the woman changed her mind. For some jurists, the husband who ignored his wife's consent was to pay her monetary compensation, called *diyya*, fixed at 10 dinars for the loss of the *nutfa* (drop of sperm). Al-Ajali (d. 1202) deemed '*azl* reprehensible because it would represent negligence with regard to more meritorious conduct (procreating) therefore the man must pay a *diyya* of 10 dinars for the loss of the *nutfa*. Ja'far Ibn al-Hasan al-Hilli (d. 1277) asserted that the payment of the *diyya* by the husband for the loss of the *nutfa* was compulsory if recourse to '*azl* had not been established in the marriage contract or if the woman had not given her consent. On the other hand, in intercourse with a slave, the master does not pay anything, although '*azl* remains reprehensible.[20]

For *Zahirite* jurists, '*azl* is prohibited because it is considered the equivalent of concealed infanticide, in reference to the *hadith* narrated by Judama. This is the thesis of Ibn Hazm (d. 1063), the best known jurist of this school as well as the most authoritative opponent of contraception amongst classic jurists.

In conclusion, it is worth recalling that the most important pre-modern medical texts on birth control techniques were Arabic. Avicenna, in his Canon Medicinae of the 9th century AD, already claimed knowledge of a good number of contraceptives, including a spermicide plant, another that acted as a pessary, contraceptive talismans, etc., and he also made the distinction between contraceptives and abortifacients.

THE CONTEMPORARY DEBATE: THE PRO-CONTRACEPTION JURISTS

Today, the *Shafi'ite* school generally allows contraception for the husband without the wife's consent as the woman can put forward rights only on an orgasm but not on children. In the other *Sunni* juridical canonical schools, (*Hanbalite, Malikite, Hanafite*), many "scholars" accept contraception between spouses exclusively with the consent of both of them.

A very tolerant position was that expressed by the *mufti* of Egypt, the *Hanafite sheikh* Abdul-Majid Salim, in his *fatwa*[21] of 25 January 1937: after obtaining the agreement between the spouses, the husband can ejaculate outside the vagina or the woman can block the entrance to her vagina. In certain cases, the husband can have recourse to coitus interruptus without the permission of his wife if he

---

[20] Querry A. (Par), *Recueil de Lois concernant les Musulmans Schyites*, Paris, Imprimerie Nationale, MDCCCLXXII, Vol. II, 666.

[21] Schieffelin O. (ed.), *Muslim Attitudes toward Family Planning*, New York, The Population Council, 1974, 11–12.

fears, e.g. that an abnormal foetus will be born or if, having to set off on a journey, he fears for the fate of the foetus; similarly, the woman can close the opening
of her uterus without her husband's consent if she has reasons to do so.

Since the end of the Second World War, the impressive population growth in
many Muslim countries has rekindled the interest of governments in limiting
their populations.

The Egyptian *sheikh* A. al-Sharabassi, a professor at Al-Azhar, in a book published in 1964–1965 on family planning, deemed contraception lawful in order
to avoid the spread of infectious diseases to the offspring; if the woman's health
is damaged by her pregnancy; to space out pregnancies; to preserve the woman's
beauty and to avoid economic difficulties.[22]

Thanks to reasoning by analogy and due to the permissiveness shown by the
Sunna, the Muslim world currently appears the advocate of recourse to various contraceptive practices which are justified according to some reasonable
motives. At the same time, as Islam is a "Religion-State" system, the numerical expansion, the force and the defence of the community (number is power)
remain fundamental values shared by both those in favour of contraception
and its opponents. The political dimension is often decisive on this issue in
establishing the "bioethical" position.

In 1964, the Grand *mufti* of Jordan, al-Qalqili, considered marriage a strong
requirement in Islam, especially to increase offspring, as their multiplicity implies
"power, influence and invulnerability".[23] Procreation is a divine blessing encouraged by the Prophet whilst marriage remains subordinate to the possession of
means allowing offspring to be fed and brought up. In any case, in his opinion,
tradition accepts the methods of voluntary control of procreation ('*azl*), agreed
by all four canonical *Sunni* juridical schools.

In a *fatwa* of 22 August 1964 the then Rector of Al-Azhar, *sheikh* Hasan Mamun subordinated the attitude of Islam on population control to an evaluation
of the historic-economic context: at the start of the Arab-Muslim expansion it
was essential to increase the number of Muslims, as they had to confront dominating and threatening enemy forces whilst today the situation is completely
different. Overpopulation represents a threat for humanity and the *Shari'a* – according to Mamun- does not hinder contraception on condition that there is a
real need, it is the result of a free choice and the instruments are legitimate.

The commonest attitude amongst the believers on the growth of the population was made explicit in point 3 of the document, on this subject, produced by
the Islamic Research Academy of Al-Azhar in 1965: Islam desires the multiplication of its children to reinforce the Islamic nation from the social, economic and
military point of view, in order to increase its prestige and make it practically
invincible.[24] However, in the case of necessity, parents may freely limit births.

---

[22] Omran, op. cit., 209–210.
[23] Schieffelin, op. cit., 3.
[24] Resolutions and Recommendations of the High Islamic Research Academy of Al-Azhar, Cairo,
The Second Conference, in Nazer I., Karmi H.S., et al., op. cit., Vol. II, 532; Omran, op. cit., 215.

It is important to specify that the advocates of population control do not question the need to increase the power and numerical and military strength of the *umma* but they want to reconcile it with the need to limit, in unfavourable socio-economic periods, uncontrolled population growth which would impoverish the community as well as inhibit its economic development. The Moroccan *sheikh* M. Mekki Naciri speaks from this perspective when he states that families can use contraceptives if the woman's health is precarious, the husband's earnings are insufficient or there is a slump in the national economy. In similar cases, couples must stop births until the period of crisis is over.[25]

The Saudi *sheikh*, Sayed Sabiq, expressed himself in a fairly similar way: "Islam desires the increase of progeny, for this is a sign of strength and power for nations and peoples."[26] In his opinion, Islam does not prohibit birth control in any circumstances using contraceptive drugs or other means. Birth control is allowed when a father is poor, has many children and cannot bring them up in a decorous manner and also when the wife is weak or pregnant or when the husband or the wife suffer from contagious diseases. Contraception is also lawful to protect a wife's beauty.

Islam refuses a "weak multitude" and one of the ways of avoiding it consists of applying a programme of population control respecting a number of points[27]:
1. Temporary contraception to be used during breastfeeding.[28]
2. "Total" contraception to be applied if the spouses or one of the parents suffers from hereditary incurable pathologies; if they refuse, the State should intervene to protect the nation and citizens from potential harm.
3. Population control is recommended for parents who are not in a position to raise and look after their children or to assume their responsibilities in observance of the juridical principle: harm must be avoided as far as possible.

---

[25] Mekki Naciri M., Aperçu sur la Planification Familiale dans la Législation Islamique (Séminaire National sur les problèmes de Planification Familiale, 1967), *Oriente Moderno*, 1979, 399–410.

[26] Arab Republic of Egypt, Ministry of Waqfs, *Islam's Attitude Towards Family Planning*, Cairo, 1994, 64.

[27] Ibid., op. cit., 19–26.

[28] On the subject Koran 2.233 and 46.15 invite mothers to continue breastfeeding for at least two years which, added to the months of pregnancy, would allow the woman to space out pregnancies every three years. Today, it is a known fact that breastfeeding causes the release of prolactin, a hormone that naturally blocks ovulation. Lastly, Ibn al-Sakan reported a *hadith* according to which Muhammad is alleged to have invited his followers not to kill children through intercourse with a woman whilst she was breastfeeding because the child would be born with a weak constitution. Koran 2.233 says: "The mothers shall give suck to their offspring for two whole years, if the father desires to complete the term. . . . If they both decide on weaning, by mutual consent, and after due consultation, there is no blame on them. If ye decide on a foster-mother for your offspring, there is no blame on you, provided ye pay (the mother) what ye offered, on equitable terms"; Koran 46.15: "We have enjoined on man kindness to his parents: in pain did his mother bear him, and in pain did she give him birth. The carrying of the (child) to his weaning is (a period of) thirty months".

In a *fatwa* which appeared in the Algerian newspaper *Al-Sha'b* (The People) on 23 April 1968 issued by the Higher Islamic Council of Algeria,[29] birth control is religiously licit on condition that it is practised with the assent of the subjects directly concerned and it is justified by the need to avoid emergencies concerning the mother (her life is endangered) and present and future children (scarcity of maternal milk and economic difficulties for sustenance, all motives which refer to Ghazali). The *fatwa* does not make any pronouncement on the preference of specific contraceptive methods compared to others as these are choices regarding the freedom of the couple.

The final document of the Congress of Rabat (1971) on "Islam and family planning"[30] considered family planning lawful for the spouses after a freely made agreement, using safe and legitimate means in order to delay or to decide upon a pregnancy according to what is most appropriate for their health, socio-economic circumstances and in the context of their responsibility to their children and to themselves. The Congress, whilst wondering how to reconcile a more numerous *umma* (community of the faithful) with the need for population control, deemed it currently a priority, for Islam, to dedicate greater energy to the challenge of improving the quality of life of its members rather than limit the discourse to the numerical increase of the population.

Resolution no. 39 (1/5) concerning birth control of the Council of the Academy of Muslim Law (Fiqh) of Jeddah (Fifth Session, Kuwait City, 10–15 December 1988) stated: (1) "general laws" that limit the procreative freedom of the spouses are unlawful. (2) It is prohibited to deprive a human being of the capacity to procreate (sterilisation) except in cases of necessity approved by the *Shari'a*. (3) It is lawful to temporarily control procreation to space out the periods of pregnancy or interrupt them for established periods in case of necessity recognised by the *Shari'a*; however, this should be done at the discretion of the married couple according to their mutual agreement and after consultation as to whether the instruments are lawful.[31]

In 1981, a *fatwa* by the Malaysian National Council for Islamic Affairs deemed contraception prohibited in order to limit the number of children but allowed in personal circumstances. Non-permanent contraception is lawful if the wife is weak or ill, if the couple suffers from hereditary pathologies or if pregnancies are too frequent. It is legitimate to space out births for health reasons (of both the parents and children), education and family affluence.

When he was still Grand *mufti* of the Republic of Egypt (from 1986 to 1996), M.S. Tantawi had taken up the main arguments maintained by the advocates of population control in the light of Muslim law but placed special

---

[29] Borrmans M., Fatwa-s algériennes, in Scarcia Amoretti B. and Rostagno L. (eds.), *Yad-Nama*, Roma, Bardi, 1991, Vol. I, 83–107.

[30] Nazer, Karmi, et al. (eds.), op. cit., Vol. II, 483.

[31] www.islamibankbd.com/page/ oicres.htm#39(1/5).

emphasis on the protection of the strength of the community.[32] The *mufti* advised parents to separate each new conception by at least three years in order to allow the mother to fully recover from the previous effort. Planning is admitted, but it cannot be "legally or morally" imposed; in any case, God orders us to avoid disorder in our lives hence the need for population planning without forgetting that the creation of a new life nevertheless depends on God. Contraception does not oppose faith in God, similarly to the case of the patient who goes to the doctor to follow the therapy recommended, as healing is attributable to direct divine action. Again according to Tantawi, the problems that make recourse to contraception lawful (the method ought to be advised by a doctor) vary according to the possibilities of the individual states and families. Economically strong countries and with advanced social systems can increase their population whilst the weak states, not having the means to cope, induce their inhabitants to a life totally dependent on others. In these circumstances, it is useful to plan births: "Let us welcome a numerous, strong, responsible and creative population. As far as a weak, ignorant and dissolute population is concerned, it would be preferable for itself if it were to remain a minority."

The *Islamic Manual of Family Planning* of the Egyptian University of Al-Azhar (1998) confirmed: "Large Muslim communities should be strong: An Islamic state should be strong so that it may be reckoned with in times of war and peace. Large Muslim communities should maintain their solidarity and unity."[33]

In January 2004 in the Philippines, a Catholic country where Muslims are a small minority (about 5%), the Filipino Muslim Ulamas issued a *fatwa* allowing the use of all contraceptive means (if safe and religiously lawful) except vasectomy and tubal ligation (as these are considered self-mutilation); these two methods are lawful only in special circumstances (e.g. risks for the mother's life or for the foetus). Filipino Muslims have one of the highest rates of population growth in the country and generally firmly believe that contraception is against the principles of Islam.[34]

On 10 March 2004, in Davao City, the *Dar al-Fatwa* of the Philippines, after having consulted the Egyptian religious authorities, issued a new fatwa confirming the previous one and stating that improved reproductive health conditions of the Muslim people benefit individual Muslims and strengthen the Muslim nation socially, economically and in all other aspects of human life. The main aim of the document was to protect the well-being of the mother and child so that

---

[32] Tantawi M.S., La protezione dell'infanzia e la legge dell'Islam, *Dolentium Hominum*, 1994, 25, 261–266.

[33] International Islamic Centre for Population Studies and Research (IICPSR), *Islamic Manual of Family Planning*, Cairo, IICPSR, 1998, 31.

[34] Mugas N.V., Muslim Decree for family Planning endorsed by Grand mufti, *The Manila Times*, 22 January 2004, in www.manilatimes.net

"the couple [could] raise . . . . children who are pious, health, educated, useful and well-behaved citizens".[35]

In general, the more influential *ulama* authorise family planning when it is the result of decisions taken freely by the couple, deeming it a problem of conscience, whilst any government interference is considered with suspicion. For example, Mahmud Shaltut, the highest authority in the 1950s and 1960s of the University of Al-Azhar in Cairo asserted that it was illegitimate for the state to establish by law the maximum number of children a couple should be allowed to have. According to Shaltut, however, state intervention becomes acceptable only with regard to women with too many children or parents who are the carriers of hereditary pathologies.[36] The Islamic Research Academy of Al-Azhar, mentioned above, reconfirmed (in 1965) the need for the free choice of the couple in accordance with their conscience and religious feeling, whilst the *Shari'a* prohibits any obligation with the aim of stopping procreation. The same position was agreed with by the Council of the Academy of Muslim Law (Fiqh) of Jedday (December 1988). The Syrian *sheikh* Said Ramadan al-Buti considers the use of modern contraceptives lawful by analogy with coitus interruptus; parents are authorised to space out and limit births; society has no rights over semen before fecundation nor has it the right to oblige parents not to procreate.[37]

Amongst *Shi'ite* opinions, *the ayatollah* Bahaeddin Mahallati, in a *fatwa* of 12 November 1964, to the question of whether a doctor can prescribe contraception with a temporary effect to avoid overpopulation, answered that it is lawful if it does not harm female fertility, causing sterility.[38]

Another *Shi'ite* opinion is that of *sheikh* Mohammad Mahdi Shamsuddin[39] who deemed recourse to *'azl* lawful as it represents a right of the husband but it is reprehensible to practise it without the wife's agreement. By analogy with coitus interruptus, modern contraception may be used, divided into methods used by the husband (condom, barriers that prevent the sperm reaching the uterus, drugs that inhibit the fertilising capacity of the spermatozoids) and methods used by the wife (drugs, barriers). According to Shamsuddin, the recommendation to have numerous children must remain subordinate to the possibility of fulfilling their needs.

It is clear that with the multiplication of contraceptive methods, jurists are unlikely to take up a position on which to choose, leaving the decision to doctors and limiting themselves to approving general criteria such as the analogy

---

[35] Jimenez-David R., *A Fatwah on Family Planning*, 30 March 2004 in www.inq7.net; Santos D.J.C. and Allpala J.S., *Muslim Religious Leaders support Family Planning*, 12 March 2004 in www.inq7.net

[36] Rispler-Chaim V., Islamic Medical Ethics in the 20th Century, *Journal of Medical Ethics*, 1989, 15, 203–208.

[37] Omran, op. cit., 211.

[38] Schieffelin, op. cit., 1.

[39] Shamsuddin M.M., Birth Control - Its Lawfulness and Methods, in Nazer I., Karmi H.S., et al. (eds.), op. cit., Vol. II, 259–262.

of the method with coitus interruptus, their harmlessness for the mother and possible foetus and lawfulness with respect to the *Shari'a*. From this perspective, the doctors of the law are often consulted on the impact that recourse to a given contraceptive method may have on correctly following some of the five pillars of the faith, e.g. the fast of Ramadan. In this regard, on 5 April 1991, the Committee of *Fatwa* of Al-Azhar stated that taking the contraceptive pill – even without water – during Ramadan interrupts the fast as it is taken by mouth and reaches the stomach. The Committee advised these women to make up later for the fast days interrupted by taking the pill.[40]

A cautious attitude towards contraception, except in extreme cases, was expressed in the early 1990s by the EMRO (Eastern Mediterranean Regional Office of the WHO) in a document on the influence of religions (Islam in the first place) in fighting the spread of AIDS and other sexually transmitted diseases. The recommendations drawn up by eight doctors of Muslim Law and two Orthodox Coptic priests from the Patriarchy of Alexandria, also deals with contraception. Point 7 states that to protect young people from sexual transgression, early marriage should be encouraged, therefore socio-economic problems should preliminarily be solved. The request for early marriage must however be accompanied by family planning to delay any possible pregnancy.[41] Point 9 accepts the use of the condom for the protection of the spouse of a person infected with AIDS. This use must be limited to spouses whilst it is unacceptable as protection against sexually transmitted diseases in sexual intercourse outside marriage.

The constant reference to methods that are analogous with coitus interruptus is not without ambiguity due to the differences existing between the many contraceptive methods and their action on the female body. For example, amongst the real contraceptives (first group), those that prevent the release of the egg-cells from the ovaries (e.g. the progestogen pill) can be distinguished from those that prevent the male and female gametes meeting (e.g. condom, diaphragm, spermicides, coitus interruptus). From the first group, a second group has to be distinguished that includes methods with abortifacient properties, the "interceptive" means capable of preventing the implantation of the egg that is already fertilised, and in their turn these can be divided into mechanical interceptive devices (e.g. IUD or coil) and hormonal devices (e.g. the day-before-pill or the day-after-pill). Another group includes the "contragestives" which provoke the detachment of an embryo which is already implanted (e.g. RU 486, prostaglandin). Lastly, these methods should be distinguished from the "natural methods" of regulating fertility (basal temperature, the ovulation or Billings method, symptothermal methods, etc.). The limited interest of jurists in all the differences connected with each method is a consequence of the traditional approach which preferred to analyse the motives

---

[40] Rispler-Chaim V., *Islamic Medical Ethics in the Twentieth Century*, Leiden, Brill, 1993, 59–60.
[41] World Health Organization, *The Role of Religion and Ethics in the Prevention and Control of AIDS*, Regional Office for the Eastern Mediterranean, 1992, 9 and 32.

that legitimise contraception, neglecting the analysis of the methods or adding them to those analogous with coitus interruptus.

On the new contraceptive techniques, the International Islamic Center for Population Studies and Research (IICPSR) of the University of Al-Azhar does not refuse any method, limiting itself to advising one or the other in accordance with the health of the mother and suggesting their use in three particular periods of a woman's life: postponing the first pregnancy (if the girl is too young), spacing out subsequent pregnancies and at the end of breastfeeding.[42] According to the IICPSR, in these three phases and depending on the cases, alongside natural methods (breastfeeding, evaluation of the fertile periods and coitus interruptus) use may be made of the condom, diaphragm, spermicides, inter-uterine devices (IUDs or the coil), hormonal contraceptives (oestroprogestin pill), injections of progesterone and the subcutaneous implant, Norplant. Gynaecologists specify that all modern methods may be harmful for the woman to a certain extent. As a consequence, if any of these methods risk causing greater harm than that caused by another pregnancy, then they should not be allowed as these methods have above all the aim of protecting the mother's life and health.

The tolerance shown by tradition to coitus interruptus and other artificial methods considered analogous has led to the neglect of natural ways of regulating fertility even if growing interest has recently been shown in these methods.

STERILISATION

The distinction between temporary and permanent sterilisation is now classic. In 1936 the Egyptian *sheikh* Ahmad Ibrahim stated the lawfulness of contraception and permanent sterilisation as it is not prohibited (*haram*) for a man to have recourse to any means to destroy spermatozoids and similarly women can use any means to prevent conception. This is lawful because sterilisation does not act on creatures that already exist. The opinion of another esteemed Egyptian doctor of the law, *sheikh* Mahmud Shaltut, is more rigid, for whom the permanent prevention of procreation is allowed only if the married partners or one of them is affected by hereditary pathologies, therefore preventing procreation satisfies the principle of avoiding a greater evil (the birth of incurable children) which has the priority over doing good (the birth).[43]

In the Congress held in Rabat (Morocco) in 1971 on "Islam and family planning", three trends emerged: (a) those who prohibit both forms of sterilisation as they can be assimilated with the crime of abortion; (b) those who authorise temporary sterilisation, deemed analogous with '*azl*, but not permanent sterilisation; (c) those for whom permanent sterilisation may be authorised according to strict limits. M.S. Madkur, Professor of *Shari'a* at Cairo University, who deems

---

[42] IICPSR, *Islamic Manual of Family Planning*, op. cit., 103–119.
[43] Madkur M.S., Sterilization and Abortion from the Point of View of Islam, in Nazer and Karmi, op. cit., Vol. II, 268. Omran, op. cit., 189–190 and 230.

permanent sterilisation lawful in the face of a grave need such as the presence of incurable mental illnesses or sexual diseases that can be transmitted to offspring, is in group c). Some maintain that the *Shari'a* does not condemn this provision but, on the contrary, requires it to avoid deformed or weakened offspring whose lives are destined to be full of psychological complexes and suffering.[44] The *Shi'ite Imamite sheikh* M. Shamsuddin also pronounced himself in favour of permanent sterilisation and personally deemed non-existent the juridical sources that oblige man to preserve the faculty of procreating.[45]

A recent document by the University of Al-Azhar (IICPSR) (1998), underlines the general prohibition of permanent sterilisation independently of the technique that causes it for one or for both the spouses; however, it is allowed when one or both of the spouses are affected by a serious and incurable hereditary pathology, which weakens progeny, becoming a burden for the community. *Sheikh* Jad al-Haq made an even more explicit pronouncement in a juridical opinion of 11 February 1979: if one or both partners suffer from hereditary diseases, a child would damage the nation, create useless citizens as well as representing a burden for society; sterilisation of the diseased parents is lawful or even compulsory in the light of the following principle of the *Shari'a*: avoiding evil has the precedence over doing good.[46]

Point 6 of the Conclusive Recommendations of the Seminar on "Human Reproduction in Islam" (IOMS, Kuwait, 1983) accepts surgical contraception (sterilisation) on the individual level in case of necessity but, it continues, "on the level of the Muslim nation . . .. it is unlawful, and the seminar denounces turning sterilisation into a general campaign and warns against its exploitation in demographic wars that aim at turning Muslims into minorities in their own countries or in the world as a whole".[47]

In these juridical opinions and documents, the political priority given to defending the force and health of the Community is clear.

THE OPPONENTS OF CONTRACEPTION

Over the course of history there has always been an essentially minority faction of jurists opposing contraception.[48] Two *ahadith* (quoted above) act as a premise to this interpretation: the Prophet is said to have defined '*azl* a "minor" or "hidden infanticide"; in another *hadith* he will be proud, on the Day of Judgement, of the number reached by the Muslim faithful compared to other faiths.

The Andalusian ("Spain") jurist of the *Zahirite* school, Ibn Hazm (d. 1063) may be considered the most representative "scholar" of this orientation. According to

---

[44] Madkur, op. cit.

[45] Shamsuddin, art. cit., 260.

[46] Arab Republic of Egypt, op. cit., 55–56.

[47] *The Seminar on Reproduction in Islam. Minutes of the Recommendation Committee*, in IOMS, *Human Reproduction in Islam (1983)*, (ed. al-Gindi A.R.) Kuwait, IOMS, 1989, 274–277.

[48] On this subject, see what has already been said in the first paragraph of this chapter.

his interpretation, the *ahadith* tolerating coitus interruptus were the oldest and were abrogated by the subsequent *hadith* defining *'azl* a "hidden infanticide" or a "minor infanticide"; since the Koran condemns infanticide and the Prophet considered coitus interruptus as equivalent to a hidden infanticide, Ibn Hazm deduces from this the prohibition of contraception in all circumstances.[49] The *Hanbalite* jurist Ibn Qayyim (d. 1351) challenged the position of Ibn Hazm maintaining that a *hadith* may abrogate a preceding one only if it can be accurately dated, which is impossible. Qayyim added that the prohibition of infanticide by the *Shari'a* does not concern *'azl.*[50]

The opponents of contraception judge procreation as the fundamental purpose of marriage (Koran 2.223: "Your wives are as a tilth unto you; so approach your tilth when or how ye will"), and this is nevertheless agreed with by the advocates of population control. The comment of Maududi, a contemporary Pakistani author of a "radical" orientation, on the Koranic passage, was as follows: Describing woman as a field, an important element is underlined. Biologically, man is a farmer and woman is a field and the main purpose of the relationship between the two is the procreation of human race. The farmer cultivates the land to produce crops. "Take away this purpose, and the entire pursuit becomes meaningless."[51] On several occasions, authoritative Pakistani *ulama* (in addition to Maududi, Muhammad Shafi, Haq Thanvi, etc.) have intervened publicly against the government programmes launched in the country to limit the birth rate. If their primary concern seemed to lie in the refusal of programmes imposed on a nationwide scale and therefore not very respectful of the choice of the couple (to procreate or not and with the preferred methods), in actual fact, their position embraces a set of reasons that always rejects contraception: Family Planning is judged similar to infanticide; it is against nature; it is based on distrust in Providence; it ignores the desire of the Prophet for Muslims to increase their numbers; it leads to social disaster; it represents a Western conspiracy against Islam.

Abd al-Halim Mahmud, Grand *mufti* of Egypt (1973–1978) and the Egyptian scholar M. Abu Zohra are close to these positions; the latter opposed population control with reasons that are very common today at popular level and almost identical to those of the *ulama* mentioned: (1) procreation is the first aim of marriage; (2) contraceptives harm public morals, stimulating promiscuity; (3) they are harmful for the health; (4) the imperialists spread it to control the growth of Islam; (5) there is no need to limit procreation because in the world there is room for many other Muslims.[52] Birth control is lawful only if the woman is too weak for new pregnancies and if there is the danger of transmitting hereditary diseases.

---

[49] Musallam B.F., *Sex and Society in Islam*, Cambridge, Cambridge University Press, 1989, 18–19. In the West, the books by Musallam and Omran (already quoted) are the historical works with the greatest wealth of data on the subject of population control in Islam.

[50] Ibid., op. cit., 19.

[51] Maududi A., *Birth Control*, Lahore, Islamic Publications, 1993, 83.

[52] Montague J., Islam and Family Planning, *Cultures et Développement*, 1975, 7 (1), 131–140.

The decision of the Council of Islamic Law (*Majma al-Fiqh al-Islami*) of Mec ca (Saudi Arabia) of 1987 also comes into this frame of ideas and presents very restrictive conditions for contraception. In fact, the practice is legitimate only if a possible pregnancy is not normal, endangering the woman's health. In all other circumstances, contraception is a "grave sin" including having recourse to it for economic reasons (poverty) and as an effect of national demographic policy.[53] The different position of this Saudi Academy of Muslim law (Mecca) should be noted with respect to Resolution 39 (1/5) issued in 1988 on birth control by the other Saudi body, namely the Council of the Academy of Muslim Law (Fiqh) of Jeddah, and quoted above.

Very strong pressure by the Hindu majority has contributed to the opposition of population control by the Islamic Fiqh (Muslim Law) Academy of India, with the aim of numerically reinforcing the Muslim community (approximately 12% of the population). However, instruments to temporarily delay procreation are allowed if a new pregnancy would be a hindrance to looking after an exist- ing child or if the woman is too weak for a pregnancy. Permanent birth control devices (female sterilisation) are lawful for the woman only if her life would be put at risk due to pregnancy or her health seriously damaged.[54]

At present, there is a great fear that population control plans may be "imposed" by Western organisations such as the International Monetary Fund as an indis- pensable condition to grant loans to certain Muslim countries.

The radical or "fundamentalist" Muslim component is inclined towards an increase of the population as an anti-Western weapon, based on the Koranic verse 11.6: "There is no moving creature on earth but its sustenance dependeth on God." On the contrary, in the opinion of "moderate" *mufti* (e.g. Tantawi) children, as a gift of God, have the right to grow up in decorous conditions, otherwise it is desirable to limit their number. In other words, a contrast emerg- es between the advocates of the quality of life (greater existential quality also thanks to a limitation of the number) and the advocates of population growth as a condition for improvement in all other fields.

As shown in the first chapter, a religious authority can change his opinion over time on a certain issue. This is the case of Jad al-Haq, the former Grand *mufti* of the Republic of Egypt and former *sheikh* of Al-Azhar, who, in a fatwa in 1979, deemed population control coherent with the *Shari'a*; any method "analogous" with coitus interruptus (application of the principle of analogy) is lawful if it does not cause permanent damage to fertility and independently of the fact that it is used by a man or a woman: indeed, the new methods, such as the pill or IUD (coil) are believed to be better than coitus interruptus as they allow the sexual act to take its natural course.[55] To the question of whether birth control is incompatible with faith in God and His help in the sustenance of creatures

---

[53] Omran, op. cit., 215–216.
[54] See www.ifa-india.org/english/
[55] Arab Republic of Egypt, op. cit., 53–56.

(Koran 11.6), the *sheikh* went back to Ghazali recalling that divine intervention does not excuse parents from the indispensable actions of providing for the sustenance of their children.

On the contrary, in 1994 the same *sheikh* – probably prompted by the controversies in view of the Cairo Conference – distinguished between regulation and limitation of the birth rate. The former is lawful in specific cases (e.g. spacing out births to raise the children) whilst limitation is prohibited. He added that overpopulation does not represent an obstacle to economic development.

All in all, at present, the majority of jurists seem to tolerate contraception if the wife's consent exists and if it is the result of a free choice. The most quoted justifications used currently, on condition that they do not cause a significant drop in the Muslim population, include:

1. the need to space out births to protect the mother's health;
2. prevention of transmitting hereditary and infectious diseases;
3. protecting a woman who is already ill from the risks caused by pregnancy;
4. avoiding economic hardship;
5. protecting the breastfed baby from the negative impact on the quality of the milk caused by a new pregnancy.

THE SOCIO-POLITICAL CONTEXT

Birth control policies adopted in Muslim countries are also diversified. National legislations nevertheless generally refer to the legitimisation of a national religious authority or body.

Alongside the main attitude of acceptance of contraceptive techniques and population control by doctors of the law and governments, the attitude of the public opinion in the "emerging countries" must also be noted and which, not infrequently, assimilates the idea of population control as the challenge of a West fearful of the numerical force reached by these countries: this is a preconceived idea that risks thwarting all the attempts at limiting the rate of growth of the populations. In Muslim countries, this possible prejudice is added to the false belief, often common at popular level, of the prohibition –by the *Shari'a*- of all contraceptive techniques.

Similar convictions must be associated with current reproductive conduct, giving rise to a picture that overall is somewhat problematic. In this regard,[56] it should be recalled that, in Muslim societies, there exists a high number of marriages of very young women and with consequent early pregnancies (together with health risks for the mother and the foetus); at least one-third of married women get married between 15 and 19 years of age. At the same time, many women continue to procreate after the ages of 35–40; many others have continuous pregnancies without intervals between them and it is not a coincidence that the average number of children per Muslim mother is six. The birth rate in

---

[56] IICPSR, 28–30.

Muslim countries is of 42 per thousand, the highest in the world with respect to other developing and industrialised countries. It has also been calculated that the Muslim world population will double in the next twenty years, amounting to more than two and a half billion.

Pragmatically, almost all the more populated Muslim states such as Egypt (the first to move in this direction), Iran, Pakistan, Indonesia, etc., have for years approved policies of population control; the only exception appears to be some desert countries such as Saudi Arabia which, for geopolitical reasons, have to increase their political weight also by increasing their numerical force.[57]

In Indonesia in 1967 the Government signed the "Declaration on Population", a document in which it is stated that family planning is a basic human right. Since 1969, Family Planning has been part of the development plans of the various governments. An inquiry in 1997 noted that 55% of married Indonesian women use contraceptives.

The conclusive Declaration of the International Conference on "Islam and Population Policy" held in 1990 in Indonesia (Aceh, 19–24 February) stated first that only some Muslim countries can cope with a rapid population increase whilst the majority of them will see their development and well-being threatened if beneficial counter-measures are not implemented. To this end, Muslim governments are invited to implement specific and targeted population policies for each national context, using contraceptive methods that are effective, not harmful and not prohibited by the *Shari'a*.[58]

After the International Conference of Aceh in 1990, in 1992 the Indonesian government approved law no. 10 concerning "Population and Family Planning."[59] Chapter VI art. 16,2 specifies: "The policies. . . . shall be carried out by efforts to improve the integration and participation of society, family guidance and birth control with due regard to religious values, harmony, proportion, and balance

---

[57] In 1981, a survey was made of the attitude of 510 married Saudi women Towards contraception in the Khobar and Dammam areas. The absolute majority (81.2%) used contraceptives only to space out births; the others for medical reasons. Spacing out births had the primary aim of protecting the health of mother and child rather than reduce the number of children. In fact, the majority of women did not put a limit on the number of children, wishing to follow divine will. Contraceptives were not understood as obstacles to the creative power of God. Oral contraceptives were the most widely used (63.2%). See Farrag O.A., Rahman M.S., et al., Attitude Towards Fertility Control in the Eastern Province of Saudi Arabia, *Saudi Medical Journal*, 1983, 4 (2), 111–116.

In a later survey on a sample of 30 Saudi families, it emerged that half of the wives used the contraceptive pill, 14 for the purpose of spacing out pregnancies and one to avoid having her period during the pilgrimage to Mecca; another woman used breastfeeding as a contraceptive instrument. The other women considered contraception prohibited by religion and harmful for the health or their husbands were opposed to it. See Panter-Brick C., Parental Responses to Consanguinity and Genetic Disease in Saudi Arabia, *Social Science and Medicine*, 1991, 33 (11), 1295–1302.

[58] IICPSR, op. cit., 8–11; Omran, op. cit., 219–222.

[59] See the law in: www.unescap.org/esid/psis/population/ database/poplaws/

between the size of the population and what the environment can support and accommodate, conditions of socio-economic and socio-cultural development." Article 17.1 states that birth control will be practised using methods "which can be accepted by husband and wife couples in accordance with their choice"; in addition, these methods "can be accounted for with regards to health, ethics, and the religion adhered to by the persons concerned", 17.2. Article 19 says: "The husband and wife have equal rights and responsibilities as well as equal status in determining the method of birth control."

Despite the opposition of many national religious authorities, Moroccan governments have supported family planning since independence in 1956. Royal Decree no. 181–66 of 1 July 1967 abrogated the previous French law of 10 July 1939 which prohibited the sale and advertising of contraceptives.[60] In the Kingdom, eroticism has been rehabilitated as an object of study thanks to the rediscovery of the Muslim tradition that gives an important place to pleasure in itself, dissociating it from procreation. This tradition, repressed until the present, has been resumed with the start of the country's population control policies.[61] Contraceptive pills are practically freely on sale in every chemist's without a prescription being necessary. The official statistics on population control do not include the consumption of contraceptives by unmarried women, being limited only to married ones. This falsifies the reality of sexual permissiveness.[62]

In 1972 Mali[63] was the first French-speaking West African country to abolish the French anti-contraception Law of 31 July 1920. The French law prohibited any use, propaganda, import and sale of contraceptives. Mali was also the first country in the Francophone West African area to legalise family planning. However, a population policy drawn up by the government was not adopted until 1991 (Declaration of the National Population Policy of Mali of 8 May 1991). On an average each woman has 6.6 children (1995–2000) and the increase in procured abortions (and abandoned children) is considered an effect of the shortage of contraceptives. There is strong opposition to their distribution. The obstacles include the government policy authorising the distribution of oral contraceptives only by doctors and midwives; the prescription of contraceptives to married women if they have their husband's consent; the low cultural level; the inclination to procreate; the reluctance to discuss sexual problems between partners; the low status of women.

In Lebanon in 1983, art. 537 and 538 of the 1943 Penal Code were abrogated. The two articles were based on the French anti-contraception law of 1920;

---

[60] United Nations, Population Division, *Abortion Policies: A Global Review*, New York, 2002, in www.un.org/esa/population/unpop.htm

[61] Dialmy A., Sexualité, émigration et SIDA au Maroc, in Observatoire Marocain des Mouvements Sociaux, *Emigration et identité*, Actes du colloque international, 24–25 Novembre 1995, Fez, 155–200.

[62] Ibid., art. cit., 165.

[63] United Nations, op. cit., 1999, 123–124 in www.unpopulation.org

similarly, Lebanese law imposed fines and imprisonment for these offences. Nevertheless, the trade in contraceptives was common. Today, many types of contraceptives are sold in chemists' without a prescription.

In Burkina Faso[64] the anti-abortion position, based on the French Napoleonic Code of 1810, was accompanied by the anti-contraceptive position based on the French Law of 31 July 1920 and which remained in force until the mid-1980s. In 1986 this rule was abrogated and a policy of National Family Planning was launched. In the period 1995–2000, the average number of children per woman was 6.6.

In Niger until the 1980s, the French law of 31 July 1920 was applied. The law was finally abrogated in 1988 (Ordinance no. 88–19 of 7 April 1988, art. 6). The new law[65] allows the use of modern contraceptive methods throughout the country. Recourse to reversible contraceptive methods (art. 2) is legitimate for: every adult individual; minors authorised by whoever exercises parental authority or legal guardianship; the mentally handicapped on medical advice. Permanent sterilisation may be practised only on women older than 35 with at least four living children; women whose lives are at risk in the case of pregnancy; men with at least six living children (art. 3). Despite the existence of a national Family Planning Programme, the knowledge and diffusion of contraceptives and contraceptive practices remains low due to social and economic reasons.

On 24 December 1980, with Law no. 80–49 Senegal abolished the French anti-contraception Law of 31 July 1920 which had been in force in Senegal since 30 May 1933. The National Population Council was established in the same year. In 1982, the first seminar held in the country on "Islam and family planning" recognised the lawfulness of population control in Islam as it is better to have children that can be protected and looked after than an unlimited number of children living in poor conditions.

In Sudan, despite the many population control plans, the traditionalist political trends against birth control (especially the use of the condom) as they are believed to go against the interests of the country, are still active.[66] In November 1998, several representatives of the Sudanese parliament asked the government to expel the agencies that operate voluntarily in the population control programmes in the country. The Minister for Health, Mahdi Babo Nimir, nevertheless expressed himself in favour of these activities as they reduce maternal mortality rates.

In Kuwait the government offers economic and administrative help to Kuwaiti couples who procreate, as they increase the number of citizens but, at the same time, the use of contraceptives is free.

Bahrain was the first country of the Gulf Cooperation Council (GCC) to have a population control plan; contraceptives are free and sterilisation is available.

---

[64] Ibid., 1999, 75–76.

[65] *Journal Officiel de la République du Niger*, 15 April 1988, 8, 370–371 and 384.

[66] Deutsche Press Agentur (DPA), 19 November 1998.

Since the 1980s, Algerian governments have officially promoted a population control programme. On 2 October 2002 Decree no. 02–312 established the National Population Committee and the Wilaya Population Committees. Contraceptives are free in public health facilities; oral contraceptives are available at chemists' on medical prescription.

In Libya until the mid-1980s, it was prohibited to import and sell modern contraceptives. Since then, the desire to plan births has been growing; however, there is no national plan of population control. To reinforce its geopolitical weight, Libya prefers to increase the number of its citizens. To this end, it provides allowances for dependent children, free education and health care, subsidies to find housing, social security etc.

In Mauritania, the French anti-contraception law of 31 July 1920 is still in force which prohibits the import, manufacture, sale, advertising and transport of contraceptives. Until the mid-1980s, the governments wanted to increase the population because the country was strongly under-populated.[67] In the 1980s the idea began to grow that one of the reasons for the gradual desertification could be attributed to the pressure of the population. In July 1986, at the Meeting on Family Health, the Director of Islamic Orientation declared that Islam allows population control with natural methods as well as the use of modern contraceptives for therapeutic purposes. Even if contraceptives are available in chemists, they are very rarely used.

THE CASE OF IRAN[68]

The first government plan for population control dates back to 1967 under the government of the Shah, Reza Pahlevi. In 1976 the population was 33.7 million, by 1986 it was 49.4 million, in 1991 it was 55.8 million, in 2001 64.5 million and today it is in the region of 70 million inhabitants.

Family Planning (FP) in 1967 was accompanied by laws which had the objective of changing the social status of women. The political value of these measures compromised support from the religious authorities.

Before the 1979 Revolution, the best known juridical opinions concerning contraception were those of the *ayatollah* Mahallati in 1964 (quoted above) and of the scholars M.M. Shamsuddin (already quoted) and M.H. Bahishti,[69] all in favour of contraception. They were subsequently allies of *ayatollah* Khomeini in his political struggle.

---

[67] United Nations, op. cit., 1999, 130–132;ibid., 1993, 136–138.

[68] See, inter alia, Aghajanian A., A New Direction in Population Policy and Family Planning in the Islamic Republic of Iran, *Asia-Pacific Population Journal*, 1995, 10 (1), 3–20. Ibid., Family Planning and Contraceptive Use in Iran, 1967–1992, *International Family Planning Perspectives*, 1994, 20 (2), 66–69.

[69] Bahishti M.H., Rules of Abortion and Sterilization in Islamic Law, in Nazer I., Karmi H.S., et al. (eds.), op. cit., Vol. II, 407–425.

The 1967 Family Planning programme did not produce valid results. With the Revolution, nobody knew if population control would have been allowed. The new political forces regarded the modernisation policy of the preceding monarchy with suspicion. For many, population control was an American idea to control Iran and developing countries; for others, population growth was not a real problem for an oil-producing country. Lastly, for some, development was the best contraceptive. All this, however, did not coincide with the positions of the highest religious authorities, as shown by the documents in favour of contraception issued by Khomeini and other *ayatollahs* a few months after the Revolution.

From as early as June 1979, Khomeini stated he was in favour of family planning. In September of the same year, the Ministry of Health gave the *ayatollah* a written report which, after having described the modern methods of contraception, highlighted the need to keep the family planning services. The *ayatollah* wrote in the margin of the report: "If the use of these methods does not expose women to any health problem (or harm) and it is also approved by her husband, [their use] to solve the problem [mentioned in the report] is religiously permissible."[70]

In the early 1980s, the then Deputy Minister of Health in charge of Family Planning presented to the Imam Khomeini the request for advice on the lawfulness of the IUD and sterilisation. The *ayatollah* answered "Prevention of pregnancy is not forbidden. As long as it is done with the consent of the couple, does not expose them to any harm, or require action inconsistent with religion, it is permissible."[71]

A circular dated 14 October 1980 from the Deputy Minister of Health for Public Health, Population and Family Planning stated that the methods, instruments and drugs provided by Family Planning clinics (the pill, condom, IUD, etc.) were free of charge and available to couples on request; sterilisation was not included amongst the free methods.

A subsequent letter (31 January 1981) from the Director General of Health of Isfahan Province asking for the opinion of Khomeini on the IUD, tubal ligation and vasectomy, received the following answer: "it is not permissible if it entails physical damage or sterility".[72]

In order not to completely suspend the activity of the population control services, the *ayatollahs* Gulpayegani, Najafi Marashi and Shirazi were also consulted on the use of different contraceptive methods. Golpaygani answered that "if they are not harmful and only temporarily prevent pregnancy nothing is wrong with their use".[73]

With the war against Iraq in September 1981, interest seemed to be lost in the problem of population control. The government promoted population growth

---

[70] Mehryar A.H., Delavar B., et al., *Iranian Miracle: How to Raise Contraceptive Prevalence Rate to above 70% and Cut TFR by Two-thirds in less than a Decade?* Institute for Research on Planning & Development, Ministry of Health & Medical Education, Tehran, 2002, 7–8.

[71] Ibid., 8.

[72] Ibid., 9.

[73] Ibid., 10.

by lowering the age limit for marriage, making it easier for large families to be given a home and consumer goods, etc. This did not mean the end of assistance for population control or an end to the supply of contraceptives to public and private structures. Unquestionable evidence is that concerning the number of people examined and of contraceptives given to couples in those years. This data was provided by the Statistical Centre of Iran in the *Statistical Yearbook of Iran* published each year or from the annual data provided by the MOHME (Ministry of Health and Medical Education). An overview of this data[74] shows that the number of people attended to in Family Planning services, pills and condoms distributed, IUDs fitted, grew constantly in Iran from 1978 to 1990, except in 1979–1980 and 1988.

In the mid-1980s, the growth rate of the population touched on 4% per annum. It was influenced by the increase in fertility of women over 30; the decline in infant mortality and the presence of almost 2 million Afghan refugees.

The census of 1986 showed that the Iranian population had reached 50 million; the news was received with great enthusiasm (even being called a "divine gift") by many political leaders. Nevertheless, the national experts were aware of the serious socio-economic risks linked with the impossibility of managing this growth with an economy that had been destroyed by the war with Iraq; a new population control policy was started. Influential religious authorities, not convinced of the urgency of the "population" problem or opposed to the legitimisation of family planning for Islam, resisted.

In December 1988, the High Judicial Council officially declared that there were no objections to Family Planning on the part of Islam.

The main elements of the new demographic policy were passed in 1989 with the approval in Parliament of the First Socio-economic Development Plan. Family Planning had three purposes[75]: to encourage spacing out new pregnancies every 3–4 years; to discourage pregnancies in women aged under 18 and over 35 and to limit the family to 3 children. Overpopulation is considered an obstacle to social development and the duty to increase the "quality" of children in place of the "quantity" is underlined.

In 1990, the High Judicial Council had declared male and female sterilisation as coherent with the principles of Islam.

On 22 April 1993, the Iranian Parliament approved a law imposing strong social obstacles for families from the fourth child: no government subsidies, no maternity leave, no help in finding a home, etc.

Condoms were easily available even before the official programme at groceries, chemists' and supermarkets. The Pill was free of charge in public clinics. Both male (vasectomy) and female sterilisation was encouraged.[76] All the methods and services are free of charge, including sterilisation. At present, Iran can claim

---

[74] Ibid., 11–12.
[75] Ibid., 17.
[76] Aghajanian, art. cit., 68.

to be the largest manufacturer of condoms in the Middle East, satisfying 90% of the domestic demand.

From the middle of the 1980s to the end of the 1990s, the number of children per woman in Iran dropped from 6.4 to 2.5 children. The Republic is the Muslim country where population control programmes have produced the best results.

At the end of the 20th century, only a dozen countries had reached a contraceptive prevalence rate higher than 70% of eligible people and the Islamic Republic of Iran is one of them.

It is worthwhile underlining the fact that both Saudi Arabia and Iran are committed to increasing the strength of the nation and the Muslim community. However, their demographic strategies are opposed but both legitimized by abundant juridical and religious sources. For Saudi Arabia, the aforementioned aim is reached by increasing the number of the faithful.

For Iran, the aim is achieved in the first place by limiting the number of the faithful-otherwise the survival of the community is at risk - and improving the quality of life of individuals.

CONCLUSION

The wealth of data and the variety of opinions of Muslim jurists and experts produce a highly articulated general picture. In addition there is the wide variety of popular attitudes which are frequently based on a social prejudice regarding contraception understood as a practice aimed at weakening Islam.

The political-religious dimension in population control policies of Muslim countries is all too present in the idea of reinforcing the Muslim community on all levels: numerical, economic, cultural, military, spiritual etc.

Nevertheless, the realism which is very often present in the authorities contrary to contraception but almost always willing to accept it to face up to difficulties of all kinds, whether individual (e.g. too many pregnancies, hereditary diseases, poverty) or social (economic crisis), is impressive. This juridical realism is very present in all issues of Muslim bioethics.

ABORTION

INTRODUCTION   91
THE SOURCES OF TRADITION   92
BEFORE ENSOULMENT   95
AFTER ENSOULMENT   98
THE PENAL SYSTEM   100
THE DURATION OF PREGNANCY   105
THREE BIOETHICAL PROBLEMS   106
THE DEBATE AND CONTEMPORARY OPINIONS   111
RAPE, ADULTERY AND FORNICATION   115
LEGISLATION IN SOME COUNTRIES   119
ABORTION AND THE DEFENCE OF HONOUR   131
CONCLUSION   133

INTRODUCTION

The Koran and the "sayings" of the Prophet contain many references to the development of the embryo and today these passages are kindling the interest of Muslim doctors and experts involved in highlighting the elements of modernity present in the Holy Texts.

The great variety of opinions that has been produced historically on the issue has not impaired the recognition of some common values by Muslim "scholars".

The first value has its roots in pure Muslim monotheism and can be summarised in the formula "everything belongs to God", including wealth, property, the family, the body, life and health; man enjoys the ownership of all these things on condition that he is responsible for them towards the Creator, who assigned them to him. The abuse of health, of the body, whether one's own or of someone else, is a sin. The second value concerns the value of human life, which can be inferred from many passages in the Koran,[1] and which is to be protected within the limits of the exceptions laid down by the *Shari'a*. The third value derives from the first two and is the Koranic prohibition of infanticide (e.g. 6.137; 6.151;

---

[1] "We have honoured the sons of Adam; provided them with transport on land and sea; given them for sustenance things good and pure; and conferred on them special favours, above a great part of Our creation" (Koran 17.70); "If any one saved a life, it would be as if he saved the life of the whole people" (Koran 5.32); "Nor take life – which God has made sacred – except for just cause. And if anyone is slain wrongfully, We have given his heir authority (to demand *Qisas* or to forgive): but let him not exceed bounds in the matter of taking life; for he is helped (by the Law)" (Koran 17.33); "Kill not your children for fear of want: We shall provide sustenance for them as well as for you. Verily the killing of them is a great sin" (Koran 17.31), see also 81.8; 4.29; 5.36.

Dariusch Atighetchi, Islamic Bioethics: Problems and Perspectives.
© Springer Science+Business Media B.V. 2009

6.140); the custom of burying alive newborn daughters, unwanted for economic, social and other reasons was a fairly widespread custom in pre-Islamic Arabia. The Koran and the *Shari'a* always prohibit this act. Some of these Koranic passages, whilst referring explicitly to the killing of a child that has already been born, are at times interpreted as a prohibition in the Koran of abortion.

The monotheistic religions agree on the protection and defence of nascent life but this protective attitude is articulated, enunciated and implemented according to characteristics, conceptions and beliefs specific to each faith, thus leaving room for differences on abortion which are at times important in the first phases of embryonic development and the individual aspects of abortion.

THE SOURCES OF TRADITION

It is obligatory to begin with the main passages in which God and Muhammad speak of the embryo. The most exhaustive is Koran 23.12–14 (as well as 40.67, 75.36–40, 35.11, etc.) in which seven stages of the development of the embryo from creation in the maternal womb are listed: "12. Man We did create from a quintessence (of clay) [*min sulalah min tin*]; 13. then We placed him as (a drop of) sperm [*nutfa*] in a place of rest, firmly fixed; 14. then We made the sperm into a clot of congealed blood [*alaqa*]; then of that clot We made a (foetus) lump [*mudgha*]; then We made out of that lump bones and clothed the bones with flesh; then We developed out of it another creature [khalqan akharan]."[2]

The last phrase of 23.12–14, "a new creation", introduces the most characteristic element of Muslim thought, i.e. the ensoulment of the foetus by God as confirmed, e.g. in Koran 38.71–72 "Behold, thy Lord said to the angels: 'I am about to create man from clay: when I have fashioned him (in due proportion) and breathed into him of My spirit, fall ye down in obeisance unto him'."

Ensoulment takes on fundamental importance for two reasons. Firstly, it underlines the superiority of man (who has a soul despite his role as "servant of God") compared to all other living creatures. Moreover, two phases in the development

---

[2] The term *alaqa* may be translated by "something hanging from the womb". The term in question has sparked off a controversy begun by M. Bucaille (a French surgeon) in the book *La Bible, le Coran et la science* (Paris, Seghers, 1976, 199–209), where he stated that *quelque chose qui s'accroche* ("something hanging from the womb") agrees with the phase of adherence shown by contemporary embryology, representing evidence of the modernity and scientific miracle of the Koran. On the contrary, the translation "blood clot" (and similar) is not confirmed scientifically and expresses the desire of non-Muslim translators to ignore the unsurpassable quality of the Word of God.
In fact, Bucaille's translation has only spread recently amongst Muslims whilst both versions, i.e. "blood clot" and "something which clings" (and similar) are present in Muslim authors who appear to prefer the term "clot" as, for example, in the translations of the Koran by A. Yusuf Ali, Maulana Muhammad Ali (Lahore, 1951), al-Hilali (Riyadh, 1993) and N.J. Dawood (Penguin Books, 2000). However, many authors dispute the literal use of Koranic terms such as *alaqa* and *mudgha*, as the eternity of the Koran is involved in the continuously new and contrasting discoveries in the medical and scientific field. In this regard see the last chapter on "Concordism."

of the foetus are distinguished, the first without a soul and the second with a soul:
historically this distinction has taken on great importance in the attempts to determine the juridical value of the different abortive acts (i.e. if it is an act that is permitted, reprehensible or prohibited according to the *Shari'a*).

The distinction between a foetus without a soul and after ensoulment opens
up many other problems, from those concerning the lawfulness of the funeral
and religious burial for foetuses without a soul to the question of whether foetuses without souls will be resuscitated on the Day of Judgement.

Another important passage on the subject is Koran 22.5: "O mankind! If ye
have a doubt about the Resurrection, (consider) that We created you out of dust,
then out of sperm, then out of a leech-like clot, then out of a morsel of flesh,
partly formed and partly unformed, in order that We may manifest (Our power)
to you; and We cause whom We will to rest in the wombs for an appointed term,
then do We bring you out as babes."[3]

The authentic "sayings" of Prophet Muhammad represent the "authentic" comment of the word of God for the Islamic tradition, as they develop and complete
the indications and precepts in it. As the Koran does not specify when ensoulment
takes place, such information is provided by various *ahadith* present (amongst others) in Bukhari, Muslim, and Nawawi. Muhammad says: "The germ of every one of
you is concentrated in his mother's womb in the form of a drop for forty days; then
he becomes a clot of blood for the same period; then he becomes a piece of flesh
for the same period; then the angel is sent to him to ensoul him"[4], 40+40+40=120
days. Another "saying" in Muslim indicates how on the 42nd night from ejaculation in the uterus, an angel sent by God begins to differentiate the organs of the
foetus but the text does not make any mention of ensoulment.[5] Lastly, two *ahadith*
in Nawawi quote the descent of the angel after the *nutfa* has been in the uterus for
between 40 and 45 nights (*hadith* no. 6392), or the descent of the angel in order to
give it shape after the *nutfa* has spent 40 nights in the uterus.[6]

---

[3] See also "It is He Who has created you from dust, then from a sperm-drop, then from a leech-
like clot; then does He get you out (into the light) as a child: then lets you (grow and) reach your
age of full strength; then lets you become old, though of you there are some who die before;
and lets you reach a term appointed" (Koran 40.67) or 75.36–40; 35.11. Koran 32.7–9 appears
to follow a different order: ensoulment seems to take place before the creation of hearing, sight
and the senses.

[4] Glossary of Hadiths, in Nazer I., Karmi H.S., et al. (eds.), *Islam and Family Planning: A Faithful Translation of the Arabic Edition of the Proceedings of the International Islamic Conference
held in Rabat (Morocco), December 1971*, Beirut, International Planned Parenthood Federation, 1974, Vol. II, 539–540.

[5] "When forty-two nights have passed over the sperm drops, Allah sends an angel to it, who shapes
it and makes its ears, eyes, skin, flesh and bones. Then, he says, 'O Lord! Is it a male or female?
And your Lord decides what He wishes and the angel records it' " in Ebrahim A.F.M., *Biomedical Issues-Islamic Perspective*, Mobeni, Islamic Medical Association of South Africa (IMASA),
1988, 115–116.

[6] Aksoy S., Can the Time of Ensoulment be the Beginning of an Individual Person?, *Journal of
the Islamic Medical Association of South Africa (JIMASA)*, 1998, 4 (1).

In conclusion, the *ahadith* indicate different times. The commonest limit coincides with 120 days from fecundation. Other less common times indicate 40 days, 40 days and a few nights, 40 days and 40 nights, 42 nights, 40 or 45 nights.

Less authoritative from the juridical point of view but significant is the exchange of opinions, revived by tradition, between the second Caliph Omar and the fourth Caliph Ali commenting on an opinion deeming that coitus interruptus is analogous with the burial of a live child, i.e. it is tantamount to infanticide and therefore should be prohibited. Ali refused this analogy: "This is not so before the completion of seven stages: being a product of the earth, then a drop of semen, then a clot, then a little lump of tissue, then bones, then bones clothed with flesh, which then become like another creation." "You are right", said Omar "may Allah prolong your life."[7] The passage shows the two figures refusing the equivalence of abortion and infanticide before the "new creation", i.e. before ensoulment.

Due to the absence of a supreme authority and a Magisterium capable of establishing the correct interpretation of the Holy Texts valid for all the faithful, the passages quoted have raised widely varying evaluations over the centuries.

Historically there has always been substantial agreement in prohibiting (*haram*) a procured abortion after ensoulment (120 days after fecundation or other previous date), except in the case of therapeutic abortion (i.e. to save the mother's life) as with ensoulment the foetus acquires features that require greater protection. On the contrary, before ensoulment the opinions of jurists on procured abortion are historically different, oscillating from authorisation and dissuasion to condemnation, whilst abortion to save the mother's life has not caused strong objections.

Two authoritative contemporary jurists, al-Qaradawi and the Egyptian Mahmud Shaltut, summarise the traditionally prevailing position on abortion. The former deems that pregnancy is to be protected from the very start and reasserts the prohibition (*haram*) of abortion after the full formation of the foetus and after ensoulment, except in the case of therapeutic abortion. After ensoulment, the *diyya* (i.e. the full blood price) is paid if the foetus is aborted alive and then dies, while the *ghurra* is paid if it is aborted dead.[8] *Sheikh* Shaltut expressed himself in similar terms: "Scholars are agreed that after quickening takes place abortion is prohibited to all Muslims, for it is a crime against a living being. Hence blood money becomes due if the foetus is delivered alive, and the *ghurra* if delivered dead. The jurists were not, on the other hand, agreed as to whether to sanction or prohibit abortion if performed prior to animation. Some felt it was possible on the grounds that there was no life at that stage and, therefore, no crime could be committed. Others held that it was unlawful, maintaining that it was already possessed of inviolable life, of growth and development. Among the

---

[7] Qaradawi (Al) Y., *The Lawful and the Prohibited in Islam*, Plainfield, American Trust Publications, 1994, 198–199.

[8] Qaradawi, op. cit., 201–202.

latter was . . . al-Ghazali. He dealt with this question, in contradistinction from coitus interruptus."[9]

BEFORE ENSOULMENT

Before ensoulment, the *Hanafite*[10] school (the most widespread in recent centuries) is historically the most open to tolerating abortion, which could be allowed or disapproved with little severity; however, abortion was suggested when there was a valid justification, the most frequent of which was the existence of a child to breastfeed whilst the new pregnancy was beginning: this situation was believed to dry up or impoverish the maternal milk to the detriment of the suckling infant.[11] Kamal Ibn al-Humam (15th century) in his *Sharh Fatth al-Qadir* says: abortion after conception is allowed unless the foetus has begun to form; according to Ibn Nujaim al-Masri (d. 1562) a woman who intends to abort does not commit a sin unless the foetus has begun to take shape.[12] The *Hanafite* erudite Haskafi (d. 1677) granted the pregnant woman the possibility of aborting within the first 4 months even without her husband's permission. According to Ibn Abidin (d. 1836 or 1842), the most important *Hanafite* jurist in the 18th to 19th centuries, causing an abortion is lawful "unless an embryo is already formed and this does not happen except after one hundred and twenty days"[13] and with the support of a valid reason such as avoiding a new pregnancy in the presence of a suckling infant. Similarly, Ibn Wahban,[14] a jurist from Damascus, specified that in the presence of a justification, abortion is authorised before the end of the fourth month, for example when, following a new pregnancy, the mother is no longer capable of breastfeeding whilst her husband does not have the means to pay for the services of a wet nurse. Wahban specified that abortion before ensoulment is reprehensible because the seed, once in the uterus, develops into a living organism; abortion remains lawful if there is a valid reason.

In conclusion, amongst the *Hanafites* there is an opinion that accepts abortion before ensoulment (four months) with or without a valid justification. Others maintain that abortion is reprehensible but becomes lawful if there is a valid reason.

---

[9] Hathout H., Induced Abortion, in Nazer I., Karmi H.S., et al. (eds.), *Islam and Family Planning*, op. cit., Vol. II, 287–325.

[10] Bouzidi M., L'Islam et la société marocaine face à la contraception, *Annuaire de l'Afrique du Nord*, 1979, XVIII, 285–303.

[11] The Prophet said: "Do not kill your children under false pretences, for the suckling of the child while the mother is pregnant has the same effect as when a horseman is overtaken (by an opponent) and thrown off the horse" (Abu Dawud) in Glossary of Hadiths, in Nazer I., Karmi H.S., et al. (eds.), *Islam and Family Planning*, op. cit., Vol. II, 538 and 543.

[12] Wa'ii (Al) T., Islamic Verdict on Abortion, in Mazkur (Al) K., Saif (Al) A., et al. (eds.), *Human Life: Its Inception and End as Viewed by Islam*, Kuwait, Islamic Organisation for Medical Sciences (IOMS), Vol. II, 1989, 210–221.

[13] Madkur M.S., Sterilization and Abortion from the Point of View of Islam, in Nazer, Karmi, et al. (eds.), *Islam and Family Planning*, op. cit., Vol. II, 263–285.

[14] Madkur, op. cit., 273.

Amongst the most significant *Shafi'ite* opinions, there emerges that of Ghazali (d. 1111), representing a synthesis between the different juridical theories, interpreting the seven phases listed in Koran 23.12–14. Ghazali states[15]: the first degree of existence is that in which sperm flows into the uterus and is mixed with the woman's sperm thus becoming receptive to life; destroying this is (already) an attempt on the life of an existing being; if the drop of mixture becomes *mudgha* and *alaqa* (clot), the attempt is even more serious; it is even more so when the soul has been infused and it has taken on human form; but the maximum of atrocity is reached when the crime is committed after the foetus has been delivered alive.

The passage indicates that biological life begins with fecundation and that interruption of foetal development in this phase is already an offence, a crime that progressively worsens (e.g. after ensoulment) up to murdering a child that has already been born.

Ghazali's opinion lends itself to two opposing interpretations in Muslim law on the phenomenon of abortion. The first prohibits abortion from fecundation with the exception of therapeutic abortion. The second, and most widespread, interpretation puts the emphasis on the progression of the gravity of abortion as time passes up to infanticide, but maintains a varyingly tolerant attitude before ensoulment.

Like many other jurists (including Ghazali), al-Bijarmi (d. 1806) distinguishes abortion from coitus interruptus as at the moment of ejaculation the seed is without life, which is not the case of its "state" in the uterus.[16] There is pronounced disagreement on the lawfulness of abortion before the term of 40 days (e.g. al-Bijarmi is contrary, Abu Ishaq al-Marwazi is in favour). Shabramalassi shows these differences amongst the *Shafi'ites* saying[17]: "They differed as to whether or not it is permissible to cause the removal or expulsion of the zygote, or the clot, after it settles in the womb."

Apart from Ghazali, however, the majority of *Shafi'ite* jurists allowed abortion within 40 or 42 days with the permission of the spouses, although this abortion nevertheless remained a legally detestable act (*makruh*).

Several *Hanbalites* tolerated the practice of abortion within the first 40 or 120 days from fecundation. Amongst the stricter positions, the medieval jurist Rajab (14th century) opposed those jurists who equated the abortion of a foetus without a soul to coitus interruptus because the embryo is a child that has already been conceived whilst there is not yet any child in coitus interruptus. Other *Hanbalites* say that when the embryo is already solidified it becomes unlawful to abort it, unlike the drop of sperm that has not yet solidified and cannot produce a child.[18]

---

[15] Ghazali, *Le Livre des Bons Usages en Matière de Mariage* (Extrait del *l'Ihiya Ouloum ed-Din*, or *Rivivification of the religious sciences*), Paris and Oxford, Maisonneuve-Thornton, 1953, 90.

[16] Madkur, op. cit., 276–277.

[17] Ibid., 276.

[18] Ashqar (Al) O.S., The Inception and End of Life, in Mazkur (Al) K., Saif (Al) A., et al. (eds.), *Human Life: Its Inception and End as Viewed by Islam*, Kuwait, IOMS, 1989, 123–146.

According to Ibn Abi Bakr (17th century) in the *Ghayat al-Muntaha* (The Final Purpose), a woman can take a lawful medicine to expel the sperm but not a clot of blood.[19] In one of his works, Ibn Qudama stated that anyone who strikes the belly of a pregnant woman making her miscarry owes expiation or the blood price; the same thing happens if the pregnant woman takes medicines to abort.[20] In his opinion, the recourse to expiation or fines indicates that a sin exists. Several *Hanbalites* allow the use of drugs to obtain an abortion, but only in the first 4 months as the embryo is not yet considered alive. For Ibn al-Jawzi (d. 1201), abortion is prohibited from conception onwards but the seriousness of the crime is proportional to the term of the pregnancy.

The last "canonical" *Sunni* school, the *Malikite* school, is the strictest on abortion in the first 4 months, the majority prohibiting it even in the first 40 days; simultaneously the *Malikites* agree in prohibiting it after 4 months, but they are strongly open towards therapeutic abortion. Al-Dardir (d. 1786) expressed himself in these terms: "It is not allowed to expel the zygote formed in the womb, even before the expiry of the first forty days. After animation takes place, abortion is prohibited in all cases."[21] The reason for this strictness derives from the concept of the embryo understood as a creature waiting to receive its soul from God. To this end, the seed should never be manipulated once in the uterus in the form of a clot adhering to the wall (the jurist Muhammad al-Qurtubi, d. 1273). Al-Rahwani said: Our sages establish the prohibition of anything that provokes the expulsion of the embryo from the uterus or makes the sperm come out; in the subsequent phase when the foetus has been formed but is still without a soul, the prohibition becomes stronger.[22] Al-'Izz Ibn Abd al Salam says, in coherence with this approach, but in more radical terms: no woman is allowed to use any means whatsoever that reduces the possibility of procreating.[23] The *Malikites* consequently impose the blood price on anyone who causes abortion even at the stage of the clot of blood.

A minority tolerated abortion within the first 40 days and in that case the act was deemed reprehensible (but not prohibited); a similar position was claimed by al-Dasuqi (d. 1815) who specified "it is also said that abortion before the expiry of the first forty days is an object of disfavour".

The fluctuation between those in favour and those against abortion before ensoulment is also present in the "non canonical"[24] juridical schools. For the *Zahirites*, abortion is unlawful before and after ensoulment.

The *Zaydite* position is the opposite in *Al-bahr al-zakhkhar*. They maintain that eliminating the zygote or the gastrula or the clot is like eliminating something without a soul, such as sperm, therefore it is lawful.

The *Ja'farite Shi'ites* and the *Ibadite* sect tend to refuse abortion at any stage.

---

[19] Wa'ii (Al) T., Islamic Verdict on Abortion, in Mazkur (Al)., Saif (Al), et al., op. cit., 216.
[20] Madkur, op. cit., 277–278. Wa'ii (Al), art. cit., 216.
[21] Madkur, op. cit., 274.
[22] Wa'ii (Al), art. cit., 214–215.
[23] Ibid.
[24] Madkur, op. cit., 278–280.

AFTER ENSOULMENT

After ensoulment, voluntary interruption of pregnancy is traditionally prohibited by the doctors of law as the foetus acquires characteristics that means it needs greater protection.

This prohibition after ensoulment does not apply in general to "therapeutic" abortion, a flexible phrase which can include different cases, three of which occur most frequently in Muslim law: (a) the continuation of the pregnancy can kill the mother; (b) it will have serious consequences on her physical and/or mental health; (c) the drying up of her milk caused by the new pregnancy threatens the life of her suckling infant whilst the parents do not have the resources to pay for a wet nurse.

It is intuitable that the acceptance of therapeutic abortion to save the woman, sacrificing the foetus, comes from a specific judgement on the value of the two lives. For the majority of doctors of the Law, the mother is recognised as having a greater value than the foetus as a form of life that has already developed and is possibly still the source of a new life; moreover, to help the mother there is recourse to the juridical principle of the "lesser ill" and that of the tree and the branch where the latter (the foetus or branch) can be sacrificed for the survival of the former (the mother or tree). One of the most highly appreciated Egyptian jurists of the 1950s and 1960s, *sheikh* Mahmud Shaltut, incarnates the "median" opinion prevailing on abortion as he deemed that Muslim jurists agree on the prohibition of aborting after four months of pregnancy and the infusion of life into the foetus considering it a crime against a living breathing creature. But if, after ensoulment, the new life risks causing the death of the mother, then Muslim law allows having recourse to the lesser of two ills, i.e. abortion. According to Shaltut, jurists diverge on their opinions before ensoulment (four months): some deem it permitted, others consider it a prohibited or a reprehensible act (*makruh*).[25] As far as therapeutic abortion is concerned, Shaltut specified: "The mother is the origin of the foetus; moreover, she is established in life, with duties and responsibilities, and she is also a pillar of the family. It would not be possible to sacrifice her life for the life of a foetus which has not yet acquired a personality and which has no responsibilities or obligations to fulfil."[26]

The range of the "valid reasons" to legitimise the interruption of pregnancy varies greatly. According to the *Hanafite* version developed by the contemporary Mustafa al-Zarqa (1904–1999), these valid reasons also include the mother's weak health or the conviction that the infant's health will be damaged (due to the deterioration of its mother's milk) if the mother is pregnant again and the father cannot afford to pay a wet nurse. In fact, modern medicine has shown that breastfeeding by a woman who becomes pregnant does not cause any harm for

---

[25] Arab Republic of Egypt, Ministry of Waqfs, *Islam's Attitude Towards Family Planning*, Cairo, 1994, 56–57.
[26] Qaradawi (Al), op. cit., 202.

the infant therefore this justification is scientifically superseded, even if it continues to be accepted by jurists such as *sheikh* Abdel-Aziz Icsa (Al-Azhar, 1987) and Mahmud Shaltut, Grand Imam of Al-Azhar in 1959.

Alongside the dominant trend in favour of therapeutic abortion, there has always been a minority of jurists (belonging to highly differing schools of law) against it after ensoulment, on the grounds of at least two considerations: (a) It is not certain that the mother will die if the foetus is not aborted, therefore this doubt is a sufficient motive not to kill it; Ibn Abidin (d. 1836), the advocate of this idea maintains: "If the foetus is alive, and the life of its mother is endangered by the pregnancy, it is not permissible to abort it [the pregnancy], because the mother's death is assumed, and it is not permissible to terminate it . . . and it is not permissible to kill a human being on the basis of an assumption."[27] (b) After ensoulment the foetus is a human being and, as such, must always be protected (e.g. *fatwa* of the Grand *mufti* of Egypt, Abdul-Majid Salim in 1937).

The distinction between animated and unanimated also affects the lawfulness of recourse to a funeral and "religious" burial. According to M.A. Ibn Abidin (d. 1836) the foetus that does not cry on birth (i.e. is stillborn) should be bathed ritually (*ghusl*), given a name, put in a piece of cloth and buried without prayers; this is independent of whether it has a soul or not.[28] Other authors think that it should not be given a name or washed and prayers should not be recited but it should only be buried in a shroud. Others yet believe that these operations should be carried out only if the foetus has a soul.

Muhammad Salam Madkur maintains that when a non-Muslim wife becomes pregnant by her Muslim husband, if the woman dies before the ensoulment of the foetus, the non-Muslim mother should be buried amongst non-Muslims because the foetus is not yet a human being. If the pregnant woman dies after the ensoulment of the foetus, the latter is a Muslim under the law and the woman should be buried in a Muslim cemetery.[29]

The eschatological aspect linked to ensoulment should not be overlooked. Some have wondered whether the foetus that dies without a soul will be resuscitated by God on the Day of Judgement. A trace of this problem can be found in the gloss by Ibrahim Baguri (1783–1861) to the theological poem *Gawharat al-Tawhid* by al-Laqani in which Baguri asserts that a foetus rejected after ensoulment will be resuscitated whilst the foetus without a soul, being in a state similar to that of an unanimated body (e.g. a stone) will become dust.[30]

[27] Yasin M.N., Abortion between Islamic Regulations and Medical Findings, in Gindi (Al) A.R. (ed.), *Human Reproduction in Islam*, Kuwait, IOMS, 1989, 202–209.

[28] Ebrahim, op. cit., 117.

[29] Abu Sahlieh A.A., Le Statut juridique du foetus chez les musulmans hier et aujourd'hui, in AA.VV., *Éthique, Religion, Droit et Reproduction*, Paris, GREF, 1998, 63–80.

[30] Borrmans M., *Statut Personnel et Famille au Maghreb de 1940 à nos Jours*, Mouton-Paris-La Haye, 1977, 358, note 121.

THE PENAL SYSTEM

The penal regulation of abortion has its origin in some passages of the Koran and the *Sunna*, the interpretation of which has given rise to different solutions amongst jurists. It is worth remembering that the juridical discussion deals mainly with the abortion of a free and Muslim foetus, namely with a free and Muslim father or the result of a relationship between a Muslim and his slave.

The mainpassage is Koran 4.92 which prescribes, for manslaughter, in addition to the payment of the blood price, *kaffara* (the penitence, reparation or expiation for a sin) consisting of the emancipation of slaves or the distribution of goods to the poor or prayer and fasting.[31] This Text, which is also used against involuntary abortion, says: "Never should a Believer kill a Believer; but (if it so happens) by mistake (compensation is due): if one (so) kills a Believer, it is ordained that he should free a believing slave, and pay compensation to the deceased's family, unless they remit it freely. If the deceased belonged to a people at war with you, and he was a Believer, the freeing of a believing slave (is enough). If he belonged to a people with whom ye have a treaty of mutual alliance, compensation should be paid to his family, and a believing slave be freed. For those who find this beyond their means, (is prescribed) a fast for two months running: by way of repentance to God." In the passage, the *kaffara* consists of the emancipation of a Muslim slave or fasting for two months. Today, as slavery is obsolete, *kaffara* – when laid down in a penal system – can be performed as a fast for two consecutive months.

One of the most important *ahadith* on abortion is in the collection by Bukhari: "Two women of Hudhail fought together, and one of them hurled a stone at the other and killed her and killed what was inside her. The case was referred to the Apostle of God. . . . The Apostle adjudicated that the blood money for the foetus was that of a slave or that of a new born female child; and he adjudicated that the blood money for the woman was to be exacted from her male kin, and that she was to be inherited by her children and those who go with them; whereupon Hamal b. al-Nabigha al-Hudhali expostulated saying: O Apostle of God, how is it that I should be penalized for someone who never drank, ate, uttered a word or cried, such a one should go unavenged. The Apostle of God said (in reply): He is only one of the fellow-ship of seers, because of his rhymed speech which he spoke."[32]

---

[31] See the introduction to the section "The Problem of Penal Mutilation" in Chapter 3.

[32] Glossary of Hadiths, in *Islam and Family Planning*, op. cit., Vol. II, 555. The *hadith* in question resembles the Biblical passage in Exodus 21, 22–23. Another *hadith* says: "The Apostle of God, in the case of a stillborn foetus of a woman from Bani Lihyan, awarded her compensation equal to a blood money payable for a slave or a slave-girl. Later, the (culprit) woman who was adjudged to pay the blood money died, and the Apostle of God adjudicated that her inheritance should go to the children and husband, and that the blood money (due for the stillborn foetus) was a liability of her descendants" (535).

On the basis of the elaborations of traditions, Muslim law will prevalently punish the procured abortion of an animated foetus (i.e. after 120 days or a shorter period) that is born alive (having shown signs of life such as breathing, coughing, sneezing, etc.)[33] but dies immediately afterwards, with the *diyya* or full blood price similar to the murder of an adult; this corresponds on average to 100 camels, 200 cattle, 2,000 sheep, 1,000 dinar or 12,000 dirham. On the other hand, if after ensoulment the foetus is stillborn, there is recourse to the *ghurra*, a lesser penalty than the *diyya*. In short, the *ghurra* is equivalent to one-twentieth (or one-tenth) of the *diyya* and can take the form of the emancipation of a male or female slave of excellent quality or the corresponding monetary value (or sheep, camels, etc.).[34] Lastly, in the case of multiple abortion, the blood price is to be multiplied by the number of foetuses aborted.

After ensoulment, some "scholars" asked for the application of the law of retaliation on the person responsible.

Here are some examples of the complicated juridical formulations on the subject, starting with that of the *Malikite* Khalil Ibn Ishaq[35] (d. 1365). Like Malik, he maintained that for a voluntary abortion, even of a simple clot, the guilty party owes one-tenth of the penalty laid down for the murder of the mother (even if a slave); this means that the settlement (or composition) is in money or a male or female slave (under seven years of age) of a value equal to one-tenth of the mother. If the foetus is delivered alive, even if it dies immediately afterwards, the full settlement (*diyya*) is due if the relatives swear that the death was involuntary. If the abortion is voluntary by means of blows on the stomach, back or head of the mother, there is uncertainty amongst the scholars on whether to apply the law of retaliation (as for murder) or full monetary settlement or composition (*diyya*). The dominant *Malikite* opinions[36] distinguished between blows to the back and

---

[33] According to Muslim juridical doctrine, the foetus becomes a subject of law as a "man" (*ragul*) only from birth. The foetus is understood as born when:
  (a) It has left the maternal cavity and has been detached from it.
  (b) It comes out alive. This takes place when, having left the cavity, it has given off a distinct cry (*istihlal*). For the *Malikite* school of law, it is the main sign of life together with the newborn's feeding as well as vigorous and repeated movements even if it has not cried or wailed (Dardir; Dimashqi).
  (c) It has emerged vital, i.e. formed and mature, namely capable of continuing to live outside the maternal cavity. The expulsion of a clot of blood is not considered delivery.
  See Santillana D., *Istituzioni di diritto musulmano malichita con riguardo anche al sistema sciafiita*, Roma, Istituto per l'Oriente, 1926–1938, Vol. I, 119–120.

[34] In the *ahadith* the value of the aborted foetus is equivalent to that of a male/female slave. The blood price for the killing of a slave was 5 camels (or 100 sheep) or 50 dinars, that is about 5% or 1/20 of the *diyya* or full monetary amount due for the murder of a free Muslim man, which is established at 100 camels or 1,000 dinars, etc; Samerra'i (el) F., The Rights of the Embryo, in AA.VV., *The Rights and Education of Children in Islam and Christianity*, Acts of a Muslim-Christian Colloquium, Amman 13–15 December 1990, Rome, Vatican Press, 33–59.

[35] Halil Ibn Ishaq, *Muhtasar* (*Sommario del diritto malechita*), Milano, Ulrico Hoepli, 1919, Vol. II, 680–681.

[36] Ibn Ishaq, op. cit., 681, note 224.

the stomach and those to the head; in the first case the law of retaliation is applied, in the second the *diyya*.

The *Hanbalite* Ibn Qudama specified[37]: "If abortion occurs when the organism is still in the gastrula stage and if expert midwives testify that it has the dim appearance of human shape, then the expiation will be in the form of *ghurra*, or an indemnity. If they testify that it would have assumed the early beginnings of a human creature and would have taken shape if it had remained, then there would be two possibilities: the more proper of which is that there is nothing in the organism because it has not yet acquired shape, and hence abortion at this stage, as at the stage of the clot, is not liable for indemnity." According to the *Hanbalite* Ibn Taymiyya (d. 1328), if an individual provokes a miscarriage by mistake or without intention, he must pay the blood price in the form of a male or female slave according to the *hadith*, the price equivalent to one-tenth of the blood price due for the murder of the mother (i.e. of an adult). In addition, the guilty party must expiate the manslaughter as shown in Koran 4.92.[38] If the mother intentionally interrupts her pregnancy with medicines or blows, etc. she must give a male or female slave as the blood price to the heirs of the foetus, herself excluded but the father included who may, however, waive it. If she does not own slaves, she can fast for 2 months or feed 60 poor people. The blood price for the foetus corresponds to one-tenth of the ordinary one.

The *Hanafites*[39] maintain that if a person strikes the stomach of a pregnant woman, causing her to miscarry, the male relatives of the offending party have to pay the *ghurra*. If the pregnant woman voluntarily causes the miscarriage (without her husband's permission), her male relatives must pay the *ghurra*; if the miscarriage is involuntary or her husband has given his permission, then there is no obligation of *ghurra* because there has been no aggression.

As far as the *Zahirites* are concerned, Ibn Hazm in the *Kitab al-Muhalla* says: if a woman voluntarily causes miscarriage "with the intention of killing the embryo, then she is liable for expiation and to the payment of a *ghurra*, if the embryo is less than four months old; but if it is more than four months old, then she lays herself open to reprisal or is liable for indemnity as a ransom".[40] If the miscarriage is provoked involuntarily within the first 4 months, a *ghurra* is paid, afterwards both *ghurra* and expiation are laid down.

According to one of the most important *Shi'ite* jurists, Ja'far Ibn al-Hasan al-Hilli (d. 1277), in his work *Shara'i al-Islam* (*The Laws of Islam*), anyone who voluntarily strikes a pregnant woman causing the miscarriage of the foetus is responsible for murder, which is punishable with the lex talionis; if the person has acted "quasi-voluntarily", he is responsible for the blood price; if he has

[37] Madkur, op. cit., 277–278. Wa'ii (Al), art. cit., 216.
[38] Ashqar (Al), art. cit., 138–140.
[39] Madkur, op. cit., 281.
[40] Ibid., 278–280.

acted involuntarily, the payment of the blood price is due to the parents. In all three cases, the person responsible for the miscarriage must also do expiation.[41]

Regarding the blood price for the abortion of an unformed foetus, the prevailing opinion amongst the *Shi'ites* indicates a blood price of 20 dinars for the abortion of sperm; 40 dinars for a clot of blood; 60 dinars for a gelatinous foetus; and 80 dinars if the bones are formed.

The blood price to pay for a miscarried foetus of free Muslim parents is 200 dinars (i.e. 720 g of gold) if the foetus is formed but the presence of life is not evident. The person responsible for the miscarriage must pay the blood price but expiation is not foreseen.

If the aborted foetus is alive, the blood price of a male foetus is the same as the blood price of an adult man, whilst the price for a female foetus will be half. In addition, the person responsible is subject to expiation.[42]

A different penal pattern has been suggested, in *Hanafite* circles, by the contemporary Syrian Mustafa al-Zarqa (d. 1999), who divides the first 4 months into two phases.[43] The first is that in which no organ is formed and the foetus does not have human qualities therefore some jurists justify abortion. Zarqa deems that abortion without justification is forbidden although without implying, in this first phase, penal or civil responsibilities (but only a religious one) except if it is performed unknown to the husband or if it is the consequence of aggression by a third party (in which case responsibility implies castigation at the discretion of the judge). The second phase prior to ensoulment sees some organs emerging with the start of the formation of the foetus. With effect from this development of the organs, abortion is prohibited, corresponding to an abortion performed after 120 days but the guilty party must pay the ghurra. After four months – Zarqa continues – abortion is more serious than in all the previous stages; furthermore, it is lawful only to save the mother's life or if the woman is breastfeeding an earlier child and the husband cannot afford to pay a wet nurse. At this stage, if the foetus is aborted live but dies immediately afterwards, the offender must expiate a penalty equivalent to that for an accidental death: the expiation consists of the emancipation of a slave or charity for 60 people or fasting for 60 days; the offender must also pay the full blood price as for an adult. If the foetus is stillborn after the first 4 months, the guilty party must pay the *ghurra* as when abortion takes place within the first 4 months but after the organs have begun to be formed.

If ensoulment and/or the formation of the foetus represent the fundamental criteria to qualify abortion and to establish the penalties, there are also other criteria. From the *Hanbalite* perspective, for example, the blood price was also established by the religion of the foetus, by its social status but not by its sex.

Again in the *Hanbalite* context, there is also the particular position according to which the full blood price (*diyya kamila*) is due exclusively if the procured

[41] Querry A. (Par), *Recueil de Lois concernant les Musulmans Schyites*, Paris, Imprimerie Nationale, MDCCCLXXII, Vol. II, 663–669.

[42] Ibid., see in particular art. 329–337, 353 and 355.

[43] Hathout, Induced Abortion, op. cit., 317–318.

abortion takes place from when the foetus is considered alive and viable, i.e. from the 24th week, privileging this criteria over that of ensoulment, although the latter is not ignored (Ibn Qudama). For many *Hanbalites*, this term of six months is imposed as a sufficient period of time for the foetus to survive alone[44] outside the mother's womb; the term coincides with the modern criterion of viability. If after this period a man beats a pregnant woman who miscarries a living foetus which then immediately dies, the complete *diyya* is paid for a free foetus. The *ghurra* is paid for a stillborn foetus. However, the *diyya* is to be paid if the cause of the miscarriage is violence. Lastly, if the foetus is aborted alive within the first 6 months, the *ghurra* is paid as for a stillborn aborted foetus (Ibn Qudama).[45]

Returning to ensoulment, jurists showed a greater tolerance towards abortion before it took place, due also to the lack of agreement between the "scholars" even on the obligation of paying the *ghurra* (for a procured abortion) within that period.

According to Muslim Law, in the case of a procured abortion, monetary compensation (*ghurra*) is due to the mother (unless she herself caused the abortion) either as the heir of the foetus to which, by *fictio iuris* (legal fiction) the sum is due or *iure proprio* (own law) due to the destruction of the foetus considered as part of the mother's body.[46]

When one of the partners has voluntarily or unintentionally caused the abortion, the *ghurra* (blood price) will be paid by the paternal family of the guilty parent who is excluded from the inheritance according to the juridical principle that prohibits the assassin from inheriting from the victim (principally a *Hanafite* and *Shafi'ite* orientation).[47] A widespread conception held that the *ghurra* was not due when both the parents provoked the abortion as there would not be prevarication of one party over the other. Indeed, voluntary, but unlawful, abortion, desired by both parents does not contemplate the payment of the *ghurra* although it can be qualified as a "haram" (prohibited) act. For some jurists, abortion in these circumstances is not even a crime.[48]

The *Hanbalites* maintain that if the foetus dies with the mother when the aggression on the latter is the result of an error or semi-intentional, then the *diyya* is paid for the mother and the *ghurra* for the foetus, to be paid by the family of the guilty party. However, for an intentional aggression on the woman or if only the foetus dies, only the aggressor must pay, excluding his family group.[49]

---

[44] Cilardo A., Il prezzo del sangue p er l'aborto secondo la scuola hanbalita, in Atti della V Settimana di Studi, *Sangue e Antropologia. Riti e culto*, Roma, 1984, 485–494.

[45] Cilardo, op. cit., 492–493.

[46] Castro F., *Diritto musulmano e dei paesi musulmani*, in *Enciclopedia Giuridica Treccani*, Roma, Istituto Poligrafico e Zecca di Stato, 1989, Vol. 11, 1–17; Ghanem I., *Islamic Medical Jurisprudence*, London, Probsthain, 1982, 58–59; D'Emilia A., *Scritti di diritto islamico*, Roma, Istituto per l'Oriente, 1976, 522.

[47] Vallaro, M., Cenni sulla valutazione dell'aborto nella religione islamica, *Sociologia del Diritto*, 1980, 3.

[48] Vallaro, art. cit., 98–99.

[49] Ebrahim, op. cit., 141; Cilardo, op. cit., 491.

Taking drugs is prohibited if the purpose is to cause an abortion (e.g. Ibn Qudama) whereas if they are taken out of necessity then, according to Zarakshi, the mother is not responsible for any possible subsequent abortion.[50] If in the month of Ramadan the mother fears that fasting may cause an abortion, then it is not necessary for her to fast. However, according to the *Shafi'ite* jurist al-Mawardi (d. 1058), if she fasts and then miscarries, then she becomes responsible and she cannot inherit the *ghurra* as she is guilty of manslaughter.[51]

THE DURATION OF PREGNANCY

Whilst for some *Hanbalites* the full blood price is paid for the abortion of a live foetus from the period of viability (six months), the other juridical schools identify six months only as the minimum period for the formation of a child capable of surviving.

However, according to the *Shari'a*, the first 6 months correspond to the minimum period of gestation, a period of time calculated by comparing three Koranic passages, namely 46.15, 31.14 and 2.233.[52] A man can legitimately maintain the non-paternity of a child born in less than six months to his wife, on condition that the baby does not shown any signs of being premature.

If the *Shari'a* fixes the minimum period of pregnancy,[53] it does not lay down the maximum period; therefore the different opinions on the subject should be of no surprise, from the origins, where many jurists deemed that a pregnancy could last from two years to an indefinite period of time, or the varying periods of two, four or five years. The term of gestation could even be calculated from the time of death of the husband or of the final repudiation, i.e. from when lawful sexual intercourse between the spouses was impossible. Paradoxically the idea that the pregnancy could last only nine months was held by a minority school, the *Zahirites*, whilst amongst the canonical schools there was often recourse to the theory of the "sleeping foetus" which allowed extending the pregnancy

---

[50] Ebrahim, op. cit., 142.

[51] Ibid.

[52] Koran 46.15 says "In pain did his mother bear him, and in pain did she give him birth. The carrying of the (child) to his weaning is (a period of) thirty months"; Koran 31.14 "We have enjoined on man (to be good) to his parents: in travail upon travail did his mother bear him, and in years twain was his weaning"; Koran 2.233 "The mothers shall give suck to their offspring for the whole years, if the father desires to complete the term. But he shall bear the cost of their food and clothing on equitable terms. No soul shall have a burden laid on it greater than it can bear. No mother shall be treated unfairly on account of her child. Nor father on account of his child."

[53] The determination of the moment of the start of the period of gestation is uncertain. For the *Hanafite* school, it is calculated from the conclusion of the marriage contract and the sexual union of the partners is not indispensable; the *Zahirite* school "admits explicitly that the conclusion of the contract is sufficient for the *nasab* to be established". On the contrary, for the other schools of law, the presumption of paternity is calculated from the sexual union of the spouses. See Aluffi Beck-Peccoz R., *La modernizzazione del diritto di famiglia nei paesi arabi*, Milano, Giuffrè, 1990, 157–158.

to well beyond this limit.[54] In this regard, the *Hanafites* based their opinion on a *hadith* of 'A'isha, one of the Prophet's wives, according to whom a child cannot remain in the mother's uterus for more than two years. Amongst the *Malikites*, the period can even be as long as seven years, although Malik was personally inclined to accept the period of five years. The theory of the sleeping foetus, the non-scientific nature of which was well known to the jurists, nevertheless performed two functions. In the first place these were legal terms that allowed, in the case of a child that was the result of adultery or fornication, "automatically establishing the paternal *nasab*"[55] (the *nasab* is the juridical relationship that binds the parent to the child); at the same time, it had a function of social protection for the mother and future child as the accusation of adultery for the former and the role of illegitimate child for the latter could cost the women her life and certain marginalisation for the baby. Today the belief continues in some areas of North Africa, in particular, because of its social utility, e.g. the case of the immigrant who returns home after several years and finds his wife pregnant. The doctor, if the husband appears to accept the belief in the sleeping foetus, could corroborate this belief to avoid serious social consequences for the woman and the child. There are many legal cases where a woman, who has given birth years after separation from her husband, has had recourse to the theory of the sleeping foetus to obtain economic benefits from the presumed father, her former husband.[56]

An attempt to reinterpret the Koranic text in the light of modern evidence to confirm the validity of the former was made at the beginning of the 20th century by *sheikh* Muhammad Rashid Rida (d. 1935), who reached the conclusion that if the duration of the pregnancy followed by breastfeeding is a total of 30 months (Koran 46.15) whilst breastfeeding alone takes 2 full years (Koran 2.233), after subtraction the pregnancy must last between 6 and 9 months.[57]

Modern national legislation, at least in Egypt, Syria, Sudan, Tunisia and Morocco, has established the maximum duration of pregnancy as one year, whilst the Personal Status Code of Kuwait (1984) and Libya (art. 53, Law no 10 of 1984) fix the minimum and maximum duration of the pregnancy at 6 months and 365 days, respectively.

THREE BIOETHICAL PROBLEMS

I. Speaking today of ensoulment, in the light of the enormous progress made in the field of embryology in the West, raises the question of whether there exists a significant correspondence between the periods indicated for ensoulment (120 or

[54] Colin J., *L'enfant endormi dans le ventre de sa mère*, Presses Universitaires de Perpignan, 1998; Colin J., L'enfant endormi dans le ventre de sa mère, *Médecine/Sciences*, 1999, 15, 260–263; Aluffi Beck-Peccoz, op. cit., 158–159. Ghanem, op. cit., 36.

[55] Aluffi, op. cit., 158.

[56] A similar case that took place in Libya is described by Benomran F.A., Sleeping Foetus? – Medicolegal Consideration of an Incredibly Prolonged Gestational Period, *Medicine Science and the Law*, 1995, 35 (1), 75–78.

[57] Rispler-Chaim V., *Islamic Medical Ethics in the Twentieth Century*, Leiden, Brill, 1993, 8.

40–42–45 days) and the phases identified today as most important in the development of the embryo and of the foetus. There still appears to be a significant tendency to reconcile certain scientific data with religious information (Koran and *Sunna*) according to a "concordist" approach (see chapter on "Concordism"), aimed at finding anticipations and evidence of contemporary science in the Holy Text.

According to M. Shaltut, when we speak of life that begins in the fourth month, we refer today to the life that the mother perceives through the movements of the foetus, a phase indicated by the term "animation".[58]

Different interpretations on the importance of animation are provided by contemporary and past jurists. Amongst the latter, Ibn Qayyim agreed with the existence of the harmony between science and the Muslim religion on the phases of development of the foetus but, he specified, this coincidence is not valid for ensoulment because "this can be known only through revelation, for there is nothing in nature as such which requires it. As a result, the best physicians and philosophers have been perplexed by this issue and say that it cannot be known except by conjecture (*al-zann al-ba'id*)".[59]

The prevalent position in Islam nevertheless considers that the beginning of biological-animal life coincides with fecundation whilst human life takes place only with animation.

It is worthwhile quoting some of the interpretations that can be found today on ensoulment. The neurosurgeon Mukhtar al-Mahdi refuses the idea that human life takes place with the fecundation of the egg as the case of identical twins or a triplet birth would show, where the twins or triplets develop from the same egg, sharing the same chromosomes and genes but not having the same soul or identity or life.[60] On the contrary, the start of human existence takes place with the birth of the brain, i.e. in the 12th week (about the 90th day).[61]

The opinion of M.N. Yasin is more traditional, according to whom the *ahadith* do not make explicit reference to when embryonic life starts; they refer to when the foetus acquires human characteristics therefore the embryo without a soul[62] cannot be called human. This doctor, like Ibn Qayyim a few centuries earlier, deems ensoulment a fact of faith that is not scientifically demonstrable. The consequences: before ensoulment, abortion remains a prohibited act but less serious than that performed after ensoulment, which is lawful only to save the mother. Before ensoulment, the interruption of pregnancy is lawful in the

---

[58] Hathout, Induced Abortion, op. cit., 316.

[59] Musallam B.F., *Sex and Society in Islam*, Cambridge, Cambridge University Press, 1989, 56–57.

[60] Mahdi (Al) M., The Beginning of Human Life, in Mazkur (Al) K., Saif (Al) A., et al. (eds.), *Human Life: Its Inception and End as Viewed by Islam*, op. cit., 59–68.

[61] In fact, the neural plate which will be followed by the structures of the central nervous system takes place from the 18th day. However, how can this theory be reconciled with ensoulment on the 120th day (40+40+40)? Al-Mahdi reinterprets the Koran maintaining that the three phases can be partially overlapping, lasting less than 120 days; this would correspond to the information of modern science that suggests three phases for a total of 84 days after which human life begins.

[62] Yasin, op. cit., 204–207.

presence of a valid reason such as a serious foetal pathology or fears for the mother's health.

Equally traditional is the interpretation of Sulaiman al-Ashqar, for whom life is the consequence of ensoulment whilst death coincides with the departure of the soul; indeed, the author refers to a double birth and a double death quoting Koran 40.11 "Our Lord! Twice hast Thou made us without life, and twice hast Thou given us Life!" and 2.28.[63] Following the interpretation in the Commentary to the Koran by al-Tabari, the first death coincides with the phase that goes from fecundation to ensoulment (the end of the fourth month); the first life coincides with animation. The second death coincides with the departure of the soul from the body whilst the second life coincides with the Resurrection and re-ensoulment.[64]

Isam Ghanem, as well as accepting the criterion of 120 days, endeavours to establish its scientific basis with reference to the experience of modern intensive care units which allow the survival of increasingly premature babies, therefore it is legitimate to hope in the future that babies born prematurely at only 120 days will be able to survive.[65]

A Turkish expert, Dr. S. Aksoy, shows that the traditional interpretation of animation on the 120th day is not valid in the light of contemporary embryology. Therefore the criterion of animation is updated: on the basis of the *ahadith* by Nawawi no. 6392–6393, the angel arrives to give the embryo a soul (but the a*hadith* do not state this explicitly) after the *nutfa* (understood as the zygote) has been in the uterus for 40–45 days or nights after implantation which takes place between the 9th and 10th day from conception. According to Aksoy, summing 9–10 to 40–45 days, the date of animation is between the 49th and 55th day after conception.[66]

The debate on ensoulment continues. The authoritative Honorary Director at the CNRS of Paris, the Tunisian F. Ben Hamida, calculates that the embryo becomes a full person after 100 days.[67]

Whilst many Muslim authors try to demonstrate the modernity of the references on the embryonic development present in the Sacred Sources, it must not be overlooked that the terms around which the debates revolve in the West on the beginning of the person generally do not coincide with the references to Muslim Law.[68]

---

[63] Koran 2.28 "How can ye reject the faith in God? Seeing that ye were without life, and He gave you life; then will He cause you to die, and will again bring you to life; and again to Him will ye return."

[64] Ashqar (Al), art. cit., 126–127.

[65] Ghanem, op. cit., 61.

[66] Aksoy, art. cit.

[67] Ben Hamida F., L'Islam e la bioetica, in Massué J.P., and Gerin G. (eds.), *Diritti umani e bioetica*, Roma, Sapere, 2000, 79–91.

[68] An overview of some of these positions will make the statement clear. In the West, after the Warnock Report (1984), several scholars maintained the impossibility of recognising a human character for the embryo before the 14th day from conception, i.e. from the completion of the implantation

II. If ensoulment is real, it imposes a transformation in the nascent life with repercussions on its "right" to life. Opinions on this have also always been diversified but there remains the perception that ensoulment confirms a situation of "weakness" characterising the embryo-foetus before animation with regard to the "status" of the more protected animated foetus. For example, the Egyptian *sheikh* Sha'rawi has maintained that on the 120th day the foetus passes from a state of a potential human being with few rights (*insan bi-l-quwwa*) to that of a real human being (*insan bi-l-fi'l*) against whom any aggression is punishable (in *Al-Liwa al-Islami*, 6 April 1989, 55).[69] According to Fazlur Rahman, the animated foetus "legally" becomes a person with greater rights including hereditary rights, therefore aborting it is equivalent to murder; furthermore, thanks to ensoulment, the foetus becomes independent of the mother.[70] A similar position is expressed by M. Ebrahim for whom the foetus is of human constitution but is not yet a person as it is dependent on its mother. Nevertheless, any aggression against a foetus (foeticide), whilst not equivalent to murder, is still a crime punishable with the blood price (*ghurra* or *diyya*) and *kaffara*.[71]

In fact, according to Muslim juridical doctrine, the unborn child in the uterus is a "person" for whom existence is hoped, therefore the *Shari'a* protects some of its rights and interests even before birth. Consequently, the unborn child can receive freedom by emancipation even separately from its mother; it can acquire a gift (donation or legacy); it can establish a foundation (*hubus* or *waqf*) in its favour (the *Shafi'ites* do not admit this because the unborn child cannot take possession of it and this is an essential condition for the validity of the donation),

---

and from the formation of the primitive streak; the 14th day also corresponds to term beyond which the twin division or hybridisation are no longer possible. According to other authors, human life takes place only after the implantation of the blastocysts (sixth to nineth day) when the cells pass from the state of totipotency to that of unipotency. Others yet maintain that the start of human life can only be talked of with the formation of the nervous system and the start of cerebral activity in the uterus, i.e. at the end of the second month of life since fecundation. Other schools of thought, again in the West, deem it lawful to speak of a potentially human being only when it has a self-consciousness and intentional and free activity (e.g. H.T. Engelhardt). Whilst the position of the Catholic Church is well known, less known is the debate on ensoulment that developed in particular with the Fathers of the Church and in the Middle Ages. For St. Thomas, there initially exist vegetable and animal souls whilst the human soul was infused between the 30th and 40th day. For the Greek Fathers (e.g. Gregory of Nazianzum), on the other hand, ensoulment coincided with fecundation. However, both the teachings of the Church and Canonical law always considered abortion a crime independently of when it was performed to be punished very severely. The only real convergence can be identified between the more rigid current of Muslim law on abortion (historically a minority position) and the Catholic position, but bearing in mind that Islam generally accepts therapeutic abortion on principle.

[69] Rispler-Chaim V., The Right not to be Born, in Brockopp J.E. (ed.), *Islamic Ethics of Life*, Columbia, University of South Carolina Press, 2003, 81–95.

[70] Rahman F., Islam and Health/Medicine: A Historical Perspective, in Sullivan L.E. (ed.), *Healing and Restoring*, New York, Macmillan, 1989, 149–172; Rahman, *Health and Medicine in the Islamic Tradition*, New York, Crossroad, 1989, 113.

[71] Ebrahim, op. cit., 138 and 145–146.

In addition, if there is an inheritance and one of the heirs is an unborn child, the division must be suspended until delivery or, according to other jurists (Ibn Asim; Khalil; Dasuqi), the share must be reserved for the unborn child. However, the provisions quoted are subordinate to the birth of the foetus. If it is not born or is stillborn, then every right reserved to it lapses.[72]

The value of the foetus is shown by the Envoy of God when he postponed corporal punishment on an adulterous pregnant woman until after the birth of her child or after its weaning, precisely for the purpose of protecting the unborn child. This postponement is made with all the more reason in the event that a mother commits a crime punishable by death whilst she is pregnant.

Alongside those who protect the embryo in the light of the traditional criterion of ensoulment, there are numerous affirmations that criticise or ignore ensoulment. Today, these are often made by doctors tending towards a radical refusal of abortion. Ensoulment can be seen as an element that weakens, to varying degrees, the defence of the embryo. Dr. al-Abbadi condemns abortion from the moment the embryo adheres to the uterus (not from fecundation), the real time when new life is alleged to begin.[73] The role of ensoulment becomes of little importance. Hassan Hathout objects to ensoulment. In his opinion, the *fatwas* of the jurists of the past were conditioned by the mistaken medical conceptions of previous centuries. According to those conceptions, ensoulment was shown by the movements of the foetus from the fourth month onwards. On the contrary, modern embryology does not accept frontiers that clearly separate embryonic development which appears to be continuous, as the foetus moves even long before the fourth month. The author observes that a juridical opinion must adapt to new discoveries.[74] The start of life precedes both the 4th month and the 40th day as it takes place with the fecundation of the gametes and the formation of a zygote containing human genetic makeup and the characteristics of the individual that will develop subsequently. According to Hathout, in Islam the foetus is a human being with the right to life and its protection – even before ensoulment – is shown by the prohibitions, fines and punishments traditionally laid down against abortion in the first stages according to the *ahadith* of the Prophet which forbid punishing a pregnant woman until after the end of her pregnancy.[75]

III. Hathout's observations are also significant because they challenge the idea that the foetus is "part" of the mother's body, an issue that may have repercussions in the bioethical debate. In the Muslim world there is full consensus on the idea that the woman does not create the new life in her womb but she receives it from God; the foetus, says Hathout, is not part of the mother's body and can have a different blood group, rhesus, HLA and biological and

---

[72] Santillana, op. cit., 119.

[73] Abbadi (El) A.S., The Rights of the Embryo: Muslim Response, in AA.VV., op. cit., 23–32.

[74] Hathout, op. cit., 314–315.

[75] Ibid., 321 and 324.

genetic attributes.[76] In the *Encyclopaedia of Bioethics* Abdul-Rauf underlined the autonomy of the foetus with regard to the mother of whom it does not represent a mere part; even if it were part of the mother, abortion would remain prohibited as the *Shari'a* forbids the mutilation of the body given to us by God.[77] Fazlur Rahman interprets the new creature indicated in Koran 23.14 as a new being independent of the mother.

As mentioned, there are also opposing positions, for example the Egyptian Law of Inheritance of 1943 when it laid down that the mother is entitled to the *ghurra* for miscarriage caused by aggression as the death of the embryo is evaluated as the destruction of a part of the mother's body. On the occasion of the heated debate for the opening of the Cairo Conference on "Population and Development" in 1994, the Islamic Research Academy of Al-Azhar chaired by *sheikh* Jad al-Haq, on 4 August, examined the project of action by the Conference publishing a document stating that the mother cannot be sacrificed to save an embryo which does not yet have an independent life and that is a mere organ of the mother.[78] In fact, the origin of the idea of the foetus as a part of the mother can be interpreted as a juridical "device". Indeed, according to the juridical doctrine, in the case of a procured abortion, the monetary compensation (*ghurra*) is due to the mother (unless she herself provoked the abortion) either as the heir of the foetus or because the foetus is considered as part of the mother's body.[79]

THE DEBATE AND CONTEMPORARY OPINIONS

Some representative opinions from recent decades expressed by individual authorities or by Muslim organisations which follow the different historical positions are presented in this section.

The Grand mufti of Egypt, Abdul Majid Salim, in a *fatwa* of January 1937, declared that *Hanafite* jurists, although against abortion, accepted it only for a good reason such as that of preventing the mother's milk drying up after the start of a new pregnancy; on the other hand, after ensoulment, abortion was "completely prohibited".[80]

---

[76] Hathout H., Islamic Concepts and Bioethics, in Various Authors, *Bioethics Yearbook. Theological Developments in Bioethics: 1988–1990*, Dordrecht, Kluwer Academic Publishers, 1991, Vol. I, 103–117.

[77] Abdul-Rauf M., Contemporary Muslim Perspective, in Reich W. (ed.), *Encyclopaedia of Bioethics*, New York, Free Press, 1978, 894.

[78] Statement of the Islamic Research Academy on the Occasion of the United Nations Conference on Population and Development in Cairo, in Al-Azhar, *Views on the Draft Programme of Action of the International Conference for Population and Development held in Cairo*, Cairo, 1994, 5; International Islamic Center for Population Studies and Research (IICPSR), *Islamic Manual of Family Planning*, Cairo, IICPSR, 1998, 18.

[79] Castro, art. cit., 13.

[80] Arab Republic of Egypt, op. cit., 50–51.

In a *fatwa* of 1964, the Grand *mufti* of the Kingdom of Jordan, *sheikh* Abd Allah al-Qalqili, maintained that the four "orthodox" juridical schools accept coitus interruptus and contraception. By analogy, recourse to abortive drugs is acceptable within the first 120 days as the foetus has not yet taken on a human form; the *Hanafites* allow it in the presence of some justifications such as the presence of a child to be weaned.[81]

The justifications for abortion do not include abortion as a means of birth control "on a wide scale", precisely because Islam tolerates contraception. An example was given in 1965 by the Second Congress of the Egyptian Islamic Research Academy in a document which confirmed the prohibition of the *Shari'a* to have recourse to abortion and sterilisation to limit births.[82] A fatwa by the President of the Court of Appeal of the Republic of Yemen (1968) underlined the lawfulness of abortion in the *Shari'a* for a woman with many children if there is the consent of her husband and within the first 120 days.[83]

In a *fatwa* of 11 February 1979, the Egyptian *sheikh* Jad al-Haq summarised the traditional positions of the juridical schools: for the *Hanafites* abortion without valid justification and within the first 120 days is a reprehensible act (not prohibited); the justifications include the case of a mother who is again pregnant and without milk to breastfeed a previous child whilst the husband cannot afford a wet nurse. This opinion is accepted by some *Shafi'ites*. Some *Malikites* prohibit it in general, others deem it only reprehensible whilst the *Zaydite Shi'ites* always allow it. After animation (120 days), however, jurists of all tendencies agree on the prohibition of abortion except when a difficult birth is feared or if the pregnancy appears harmful for the woman; this is based on the principle of the "lesser ill".[84]

In February 1987, the Committee of Senior Ulama of Saudi Arabia in Resolution no. 140 declared abortion possible within very narrow limits[85]: (1) within the first 40 days of fecundation if necessary to promote legitimate benefits (it is not specified what these are) or to avoid foreseen damage (not specified) but not out of the fear of having too many children or the future economic burden. (2) Between 40 and 120 days, it is lawful only if the continuation of the pregnancy damages the health of the mother, possibly also causing her death; this answer is to be expressed only by a special medical committee. (3) After four months it is lawful only if the mother's life is in danger. It should be noted that both the Fatwa Committee of Kuwait[86] and the Saudi Committee simultaneously refer to the terms of 40 and 120 days.

---

[81] Schieffelin O. (ed.), *Muslim Attitudes toward Family Planning*, New York, The Population Council, 1974, 3–5; Bouzidi, art. cit., 302.
[82] Nazer I., and Karmi H.S., *Islam and Family Planning*, op. cit., Vol. II, 532.
[83] Bouzidi, art. cit., 303.
[84] Arab Republic of Egypt, op. cit., 56–57.
[85] See "*International Digest of Health Legislation*", 1992, 43 (1), 29.
[86] See Kuwaiti law under section "Legislation in Some Countries".

A contemporary *Hanafite*, the Syrian Muhammad al-Buti, allows abortion within the first 40 days as the *nutfa* is a drop or cell of life without a soul and form; within this period consent of both parents is required. Within the first 120 days, it may be lawful if the life of a suckling baby is at risk due to the shortage of milk caused by a new pregnancy. Furthermore, after 120 days, abortion is lawful if the doctor fears for the mother's life or there are dangers for the unweaned child.[87]

The Ethics Committee of the Islamic Medical Association of North America (IMANA) has recently maintained the lawfulness of abortion in the first 120 days to save the mother's life and/or her health from serious psycho-physical damage.[88] Abortion is permitted in the first 120 days in the case of lethal malformations for the foetus but also of non-lethal malformations on the consent of experts. Abortion is also lawful in cases of rape, incest and war crimes.

In 2004 the last version of the Islamic Code for Medical Ethics by the IOMS prohibits abortion unless the mother's health and life are threatened.[89] In these cases, abortion is permitted before the end of the fourth month of pregnancy. The risk of serious injuries to the mother must be confirmed by no fewer than three doctors. When abortion is recommended, written consent must be signed by the husband and wife (or guardian) in order to abort.

Amongst the *Imamite Shi'ites* of the second half of the 20th century, *sheikh* M.H. Bahishti based his opinion on the Koranic verses to maintain that the new creature becomes human only after ensoulment, a fact that distinguishes man from all other animals. Adam became worthy of the angels prostrating themselves before him only after ensoulment. The abortion of a foetus before ensoulment is not the equivalent of the murder of a human being; if anything, it is the equivalent of killing an animal.[90] Abortion is prohibited only after ensoulment because it is a mortal sin. Before that, it may be considered an aggression against the parents or one of them, if performed without their consent.

Many of the recent documents appear more restrictive as they agree with the traditional *Malikite* approach which is essentially against abortion even before animation. These include the opinion of *sheikh* Ahmad al-Sharabassi, professor at Al-Azhar, in the book *Islam and Population Control* (1965), for whom the *Shari'a* prohibits abortion at any stage of motherhood except in the case of necessity. In his words, the different opinions on the lawfulness of abortion before ensoulment are to be attributed to a lack of information by doctors of

---

[87] Omran A.R., *Family Planning in the Legacy of Islam*, London and New York, Routledge, 1994, 192–193. Abu-Salieh, op. cit., 203 and 205.

[88] Imana Ethics Committee, *Islamic Medical Ethics: The IMANA Perspective* (PDF format), in www.imana.org, 1–12.

[89] IOMS, *The Islamic Code for Medical and Health* Ethics, 2004, in www.islamset.com/ioms/Code2004/index.html, 1–61.

[90] Bahishti M.H., Rules of Abortion and Sterilization in Islamic Law, in Nazer I., Karmi H.S., et al. (eds.), *Islam and Family Planning*, op. cit., Vol. II, 407–425.

the law on the development of the embryo. At that time, however (1965), it was known that the embryo is alive from conception therefore from this moment onwards abortion is prohibited.[91] In 1969, the Committee of the *Fatwas* of Al-Azhar maintained an equally rigorous opinion according to which the embryo has a human nature and consequently abortion is unlawful even in the first months, independently of the soul.

In 1971 the final statement of the Conference of Islam regarding birth control held in Rabat (Morocco) in the presence of experts from more than 20 Muslim countries clearly summarised the position of Islam with reference to the value of "opinions" held by these documents: "As regards abortion , with the intention of disposing of it, the conference reviewed the opinions of Muslim scholars in jurisprudence and what came to light was that it was forbidden after the fourth month except for a pressing personal necessity, that is if the mother's life is at stake. As for the period before the end of the fourth month, diverse are the opinions in jurisprudence on the matter, the correct view favours forbidding it at any stage of pregnancy except for the most extreme personal necessity, in order to safeguard the life of the mother or in the case of there being no hope for the life of the foetus."[92]

The opinion (15 February 1973) of *sheikh* Ahmad Hammani, at the time President of the Higher Muslim Council of Algeria was very rigid, with abortion being a crime for him[93] and the abortionist a murderer who exposes himself to revenge and the law of retaliation.

The conclusive statement of the "International Conference on Islamic Medicine" held in Kuwait in 1981 stated that the sacred principle of life includes all its stages, including inter-uterine life. This must not be damaged by the doctor, except in the cases of medical necessity recognised by Muslim law.[94]

The final document of the IOMS at its 1983 conference on "Human Reproduction in Islam" once again confirmed the prohibition of abortion especially after 120 days but also recalled that "some participants, however, disagreed, and believe that abortion before the fortieth day, particularly when there is justification, is lawful".[95]

In August 1994, at the Cairo Conference on "Population and Development", the Islamic Research Academy (chaired by Jad al-Haq) of the University of Al-Azhar issued an opinion. The document, influenced by the political and religious

---

[91] Bronsveld J., The Official Position of the Egyptian Muslim Community on the Subject of Birth Control, in *Arab Culture 1977*, Beirut, Dar el-Mashreq, Vol. 5, 163–185.

[92] Conclusive document of the conference on *Islam and Family Planning*, Beirut, IPPF, 1974, Vol. II, 483.

[93] Borrmans M., Fatwa-s algériennes, in Scarcia Amoretti B. and Rostagno L. (eds.), *Yad-Nama*, Roma, Bardi, 1991, 83–107.

[94] *Islamic Code of Medical Ethics*, Chapter 8, The Sanctity of Human Life, Kuwait, 1981 in www.islamset.com/ethics/code/.

[95] Recommendations in Gindi (Al) A.R. (ed.), *Human Reproduction in Islam*, op. cit., 276.

tension prior to the Conference, defined abortion unlawful even in cases of adultery and rape, except when the doctors certified that the mother's life was in danger. In addition, confirmation was given for the option of saving the mother's life instead of the foetus, deemed an organ of the mother and not independent of her, therefore it may be logically sacrificed instead of the mother.[96]

RAPE, ADULTERY AND FORNICATION

For the *Shari'a*, sexual relations are lawful only between spouses (with up to a maximum of four wives) and/or between a man and his slaves or concubines. Any carnal relationship between the two sexes outside these limits (i.e. fornication and adultery) comes under the forbidden acts of *zina* for which penalties may be very harsh, including death. It is spontaneous to ask whether it is lawful to abort when the foetus is the result of pre-marital relations or the rape of a Muslim woman by a Muslim, an infidel (*kafir*) or an atheist. There is no unanimous approach to these last issues.

In a *hadith* in Muslim, Prophet Muhammad ordered a woman pregnant due to adultery to carry through her pregnancy and weaning – to protect the child – and then to undergo the due punishment.

For many doctors of the Law, a pregnancy that is the result of adultery or fornication cannot terminate with killing the foetus which would end up by paying for the sin of the couple with its life; in this case, foeticide would be added to the crime of adultery.[97] To denounce such conduct, reference is made to Koran 17.15 "No bearer of burdens can bear the burden of another." Moreover, to abort, the consent of the legal parents would be necessary who here are obviously non-existent. Lastly, the right to life is believed to be greater and independent of the circumstances and conditions of its start. Any recourse to the "principle of necessity" appears invalid according to Koran 5.3 which allows unlawful acts only if performed for serious necessity and without inclination towards sinning, a condition that is absent in the case of adultery and fornication. On 4 August 1994, the aforementioned Islamic Academy for Research confirmed once again the unlawfulness of abortion both due to rape and adultery, except when the mother's life is in danger.

The countless cases of the rape of Muslim women by non-Muslims in Bosnia and Kosovo reopened the question of the lawfulness of abortion. It should be remembered that the underlying intention of the *Shari'a* is always to increase the spiritual, material and numerical development of the Muslim community which can also be supported by the obligation of marriage between Muslim partners or between a Muslim man and a woman of another religion of the book (*Judaism and Christianity*) as the child must take the father's religion. In

---

[96] *Statement of the Islamic Research Academy on the Occasion of the United Nations Conference on Population and Development in Cairo*, op. cit., 5; IICPSR, op. cit., 18 and 23.

[97] Samerra'i (El), art. cit., 56.

the case of a non-Muslim man (whether he is of the People of the Book or not) and a Muslim woman, marriage is prohibited and the *umma* (the Muslim community) would be weakened; this is all the more true in the case of a Muslim woman raped by a non-Muslim. Furthermore, Koran 2.221 explicitly prohibits Muslim women marrying a polytheist. Rape, however, always compromises the social condition of the Muslim woman who becomes the victim of great ostracism by society, with the risk of not finding a husband. The illegitimate child (*walad zina*) or "bastard" is marginalised because, due to social status and rights, it is considered the lowest member of society.

These problems have induced many jurists to considerable tolerance towards the abortion by Muslim women raped by infidels in Bosnia: the Saudi *sheikh* Bin Baz declared abortion within the first 40 days lawful whilst *sheikh* Tantawi, at the time Grand *mufti* of Egypt, maintained that these pregnant women had to abort as soon as possible.[98] Yusuf al-Qaradawi justifies this abortion especially in the first few days according to the principle of necessity ascertained by competent individuals.[99] The Chief *mufti* of the Palestinian Authority, Ikrima Sabri, in 1999 allowed abortive drugs for women raped by Serbs in Kosovo in order not to strengthen the latter group.[100]

According to some jurists, the best conduct by a raped woman consists of immediate recourse to medical care to prevent the pregnancy before adherence to the uterus, as in these phases – resuming the opinion of the *Malikite* Qurtubi – the seed is still something undefined (*yaqinan*) therefore it can be eliminated.[101] If this does not happen or fails, Ebrahim, while recognising the innocence of the foetus and its right to live, nevertheless deems the abortion of a woman who has been raped justified. Similarly, the Islamic Medical Association of South Africa in a comment[102] on the new South African law of 27 September 1996 with which abortion was liberalised within the first 12 weeks, limited to 3 situations the possibility of abortion in Islam before 120 days: (1) when the mother's health would be seriously damaged by the continuation of the pregnancy; (2) in the presence of serious foetal anomalies which make life impossible and (3) in the presence of incest and rape. After 120 days abortion is lawful only if the mother's life is threatened by the pregnancy. The positions of juridical-religious bodies[103] in the same country, such as the Islamic Council of South Africa's Judicial Committee (which permitted abortion on grounds of poverty) and the Muslim Judicial Council's Fatwa Committee, were similar but more articulate.

---

[98] *La Repubblica, 3 March 1993.*

[99] Qaradawi (Al) Y., *Aborting a Fetus resulting from Rape*, in www.IslamOnline.net, Fatwa Bank, 12 May 2005.

[100] Rispler-Chaim, art. cit., 88.

[101] Ebrahim, op. cit., 123–124.

[102] Islamic Medical Association of South Africa, *Objection to the Enactment of Termination of Pregnancy Legislation*, 15 October 1997, in www.ima.org.za/abortion.html.

[103] Moosa N., A Descriptive Analysis of South African and Islamic Abortion Legislation and Local Muslim Community Responses, *Medicine and Law*, 2002, 21 (2), 257–279.

The condition required by jurists who tolerate the abortion of a foetus resulting from a rape is that it is performed before the period of ensoulment. In
October 1998, a *fatwa* by Nasr Farid Wassel, Grand *mufti* of Egypt, stated that
in the case of rape it is lawful (*halal*) for a woman to abort within the first 120
days of pregnancy as rape is equivalent to psychological murder in a cultural and
religious context where female virginity is an essential condition for marriage.[104]
Wassel declared that a reparatory marriage, from the Muslim point of view, is
not valid. Amongst those in favour of the *fatwa*, S.M. Tantawi, the *sheikh* of Al-
Azhar, added that the raped woman is not obliged to inform her future husband
of the abortion following rape: both are entitled to keep their pre-marital past
to themselves.

In 2004 Tantawi approved a draft law (that has not yet been discussed in the
Egyptian parliament) allowing the abortion of a foetus which was the result
of rape. On the contrary, the present Grand *mufti* of Egypt, Ali Gumaa, has
defined killing an innocent foetus which is the result of incest or rape as sacri-
lege; abortion is lawful to save the mother's life but only before infusion of the
soul on the 120th day.[105]

Due to the harsh social consequences for the woman, it is not infrequent for
the raped woman to marry her rapist to save her honour and that of the family,
according to art. 291 of the Egyptian Penal Code. In this, as in other cases, the
protection of the honour of the group has priority over the rights of the individual. However, this legal loophole was abrogated on 20 April 1999, by the People's
Assembly, whilst the death penalty remains in force for the proven perpetrator
of a rape (art. 290 Penal Code). Other Middle Eastern countries maintain the
annulment of the penalty for rape if the rapist marries his victim; this is the case
of art. 522 of the Penal Code of Lebanon and art. 162 of the Turkish Penal
Code Draft Law.

In several areas of the Arab world, a girl discovered not to be a virgin on her
wedding night risks being killed by her own family in order to cleanse the offence.
For these reasons Nasr Farid Wassel, when he was the Grand *mufti* of Egypt,
also deemed recourse to the surgical reconstruction of the hymen of a raped girl
lawful. This operation is illegal but frequently performed in Egyptian hospitals
to remedy rape and one which arouses considerable ethical problems. Previously,
many religious figures leader of Al-Azhar had condemned the operation, deeming it
deception as the Koran requires the bride to be a virgin; if she is no longer a virgin,
it is pointless lying to the husband. At the end of the 1980s, E.A. al-Tamimi, Grand
mufti of the Kingdom of Jordan, recalled that the laceration of the hymen may
be caused by reasons other than fornication or adultery. However, whatever the
reason of the laceration, artificial reconstruction (hymenorraphy or hymen repair

---

[104] Tadros M., The Shame of It, *Al-Ahram Weekly On-line*, 3–9 December 1998, no. 406. Mufti
backs Hymen Restorations for Rape Victims, *Cairo Times*, 29 October to 11 November 1999,
Vol. 2, no 18. Legal Loophole for Rapists Closed, *Cairo Times*, 28 April to 12 May 1999, 3 (5).

[105] Said S., *Abortion Issue in Egyptian Spotlight*, 16 March 2005 in www.arabnews.com

surgery) is not beneficial. According to Tamimi, in the event of a girl having been raped, the doctor (in agreement with an official) can give her an "official certificate" which absolves the girl of any fault before her family and society; even if the girl is not to blame, surgical reconstruction of the hymen does not protect her interest or that of her family. The operation may be discovered by the husband on their wedding night (or afterwards) and he will realise that he has been deceived; or "he may divorce her at the wedding night, kill her, and an unwarranted scandal may erupt".[106]

In any case, the operation consists of the insertion, during suture, of a capsule of red gelatine which breaks during intercourse on the first night of the wedding, staining the white silk handkerchief with "blood", the proof of virginity to be shown to the families. The cost of the operation varies according to the social condition of the woman and, as well as in clinics, it is performed in private homes. There are health risks associated with the operation which is often performed by inexperienced nurses. To avoid risks and due to the associated ethical problems, the Egyptian Medical Association prohibited its members from performing the operation.[107]

Ahmed Aroua,[108] an Algerian doctor and former Rector of the University of Islamic Sciences of Constantine, considers that life is an absolute right of the child from the very beginning; however, it is possible to abort if procreation is a risk for the mother, the child or society. Termination of pregnancy is a lesser ill if the child presents a genetic risk that will prevent it from leading a normal life or when one or both of the parents are affected by serious hereditary or transmittable pathologies or when the foetus is the result of irresponsible sexual relations as in the case of mental illness, rape and incest, which are prejudicial for the rights of the child, the mother and society. In these cases, abortion may be performed in the first weeks of life of the foetus.

Sexual ethics are intrinsically associated with social ones and this explains why the field of sexual relations becomes one of the privileged subjects for Muslim apologetics against the moral corruption of the West. This opinion is very widespread in the Muslim world and the solution proposed for the risk of rape is simple: eliminate all forms of indecent exposure of the human body in public; ban any hint of pornography in films, literature and songs; reduce free mingling of the sexes (promiscuity) and bring to an end the use of attractive women in advertising. It is also essential that those guilty of rape be punished in public.[109] These laws against fornication and adultery are a deterrent to sexual violence but are currently seldom applied by Muslim countries.[110]

---

[106] Tamimi (Al) E.A., Hymenorraphy from an Islamic Perspective, in Mazkur (Al) K., Saif (Al) A., et al. (eds.), *The Islamic Vision of Some Medical Practices*, Kuwait, IOMS, 1989, 517–530.

[107] Kandela P., Egypt's Trade in Hymen Repair, *Lancet*, 8 June 1996, 347, 1615.

[108] Aroua A., *Islam et contraception*, Alger, OPU, 1993, 66–68.

[109] Ebrahim, op. cit., 123.

[110] Ibid., 124.

Lastly, what should be done if the illegitimate relationship took place between consenting minors? Islam prohibits adoption (Koran 33.4–5) as a possible solution for the child (see section "Legal Adoption" in Chapter 6) avoiding abortion; abortion allows the girl to avoid being a single mother and to get married in the future, but the problem remains as to whether it is morally correct not to tell the husband of the previous pregnancy. According to Mohsin Ebrahim, the best solution is abortion in the first phases of the pregnancy.[111]

LEGISLATION IN SOME COUNTRIES

From a general analysis of the legislation on abortion in countries with a Muslim majority,[112] some evident points emerge. Only a handful of countries make explicit reference to multiple criteria of the *Shari'a* to deal with abortion: ensoulment, blood price, etc. (e.g. Iran, Saudi Arabia and Pakistan); in these cases too, however, there are great differences between the individual countries. Other countries refer only to some Muslim characteristics (generally only the term of 120 days to abort) but without adapting the penal system to the classic criteria. Most countries do not apply any Islamic criteria.

The widespread influence of the Penal Codes of colonising countries (especially France, Great Britain and Holland) is to be mentioned and this gave rise – for a certain period – in the colonised areas to much harsher laws on abortion (refusing even therapeutic abortion) than those based on the criteria of the *Shari'a*. In addition, many countries have also applied the laws of the colonising countries to deal with contraception, imposing harsh positions on population control which are extraneous to the moderate approaches of the Muslim tradition on this issue. Lastly, the enormous ex-Soviet Muslim area has generally resumed the liberal positions in force on abortion in those same republics before their independence.

The laws based on the embryonic phases of Koranic origin seem to offer less protection for the embryo compared to those that ignore the criteria of the *Shari'a* but, at the same time, prohibit abortion from fecundation (e.g. Egypt); this is perhaps because the presence of intermediate stages in the development of the embryo favours diversified responses to abortion at different stages.

Iran

The Islamic Republic of Iran has attempted to re-Islamise every aspect of social life; therefore it has also reintroduced monetary compensation for many offences including abortion. The Iranian Penal Code passed on 15 December 1982 dedicated a certain amount of attention to abortion in the section on *diyat* (*diyya*)[113] and to the related penal measures which were reconfirmed by the Penal Code

---

[111] Ibid., 121–122.

[112] See also Atighetchi D., Le leggi sull'aborto in alcuni Stati musulmani: tra diritto islamico e diritto positivo, *Medicina e Morale*, 1998, 5, 969–988.

[113] Islamic Penal Code of Iran (trans. A.R. Naqvi), *Islamic Studies*, 1986, 25 (2), 243–273.

of 30 July 1991 (arts. 487–497). The *diyat* is the sum of money to be paid to the person who has suffered injury or to the heirs in the event of death. The *diyat* section of the Code of 1982 resumes the phases of embryonic and foetal development listed in Koran 23.12–14: (1) the *diyat* to be paid for the termination of pregnancy when the sperm is still deposited in the uterus is 20 dinars (section 194); (2) the *diyat* for the abortion of a clot of blood (Arabic *alaqa*)[114] is 40 dinars; (3) the *diyat* for the abortion of a small shapeless mass of flesh (*mudgha*) is 60 dinars; (4) the *diyat* for the abortion of a foetus with bones that are already formed but not covered with flesh is 80 dinars; (5) the *diyat* for the abortion of a foetus in which the flesh and bones are formed but is without a soul is 100 dinars. In these first five phases, the law does not distinguish between abortions of male and female foetuses; (6) For an aborted foetus after ensoulment, if it is a male the full *diyat* will be paid (corresponding to 100 camels, 200 cows, 1,000 sheep, 200 garments of Yemenite material, 1,000 gold dinars, or 2,000 silver dinars) whilst if it is female the *diyat* is halved. If the sex is not established, three-quarters of the full *diyat* is paid. If the abortion follows the murder of the mother, the *diyat* for the foetus (independently of its stage of development) will be added to that of the mother (section 195). If it is the mother herself who provoked her abortion, she will have to pay the *diyat*, independently of the stage of development of the foetus, without receiving any of the sum (section 196). When there are several foetuses in the mother's womb, the culprit will pay the amount owed for each one (section 197). Each lesion suffered by the foetus (whether animated or not) has a corresponding amount of the *diyat* (section 198). Lastly, the payment of the *diyat* for a procured abortion is paid by the aggressor whilst if the abortion is the result of an accident it is paid by his *aqila* (clan) independently of whether the foetus is animated or not (section 199). Expiation (*kaffara*)[115] is not foreseen for an abortion procured before animation whilst for an abortion procured after animation both expiation and the full payment of the *diyat* (section 200) are contemplated.

To legitimise abortion to save the mother's life, reference must be made to other sections of the Penal Code which exonerate those who commit a crime to save the life of another person from punishment.

As well as the payment of blood money, art. 622 of the Penal Code (Punishment Law) establishes that anyone who causes a pregnant woman to miscarry through violence risks from one to three years' imprisonment. Article 624 lays down that if a doctor, midwife, chemist, surgeon, etc. supplies the instruments or takes part in the abortion, they risk from two to five years' imprisonment as well as blood money.[116]

---

[114] The term "clot" is present in the translation by Naqvi.

[115] *Kaffara* or expiation by the repentant culprit in Muslim law consists of donating goods to the poor, in the emancipation of slaves or praying and fasting.

[116] Jahani F., *Abortion in Iranian Law*, 15/1/2004 in www.NETIRAN.com.

In January 2004, the Coroner's office indicated the cases in which abortion is lawful. These are clinical situations which may be fatal for the mother if she continues the pregnancy; diseases that cause the death of the foetus in the uterus; incurable pathologies of the foetus and that cause its death after birth.[117]

Iranian law differs from the Westernised laws of almost all other countries in the world. Its characteristics include:

(a) Punishment does not consist of fines to pay to the State but compensation to the victim or to the victim's family.

(b) Compensation is not calculated with the common paper money but with traditional means: gold and silver coins, flocks, clothes.

(c) Evaluating the gravity of the abortion is based on the stages of embryonic development according to the Koran, which do not correspond to modern medical conceptions. Lastly, according to the Iranian law, ensoulment takes place after 4 month (about 120 days).

On 31st May 2005 the Iranian Parliament approved a new law that allows abortion within the first 4 months of pregnancy to save the mother's life and if the foetus is mentally and/or physically handicapped. The law was approved by the Guardian Council on 15 June 2005 and finally ratified by Parliament on 21st June 2005. In the new law, the woman's consent is sufficient to carry out the abortion instead of the consent of both spouses. Three doctors must confirm that the foetus is disabled or the mother has a life-threatening condition.[118] Abortion for a woman who is pregnant after rape is not contemplated.

There is very little data on the practice of abortion in the country. However, It seems that each year there are at least 80,000 abortions, the majority of which are illegal.[119]

Saudi Arabia

Article 24 of Ministerial Resolution no. 288/17/L of 23 January 1990 promulgating Rules for the implementation of the Regulations on the practice of medicine ratified by Royal Decree no M/3 of 2 October 1988 simultaneously refers to the legal terms of 40 and 120 days for abortion. The article prohibits abortion except to save the mother's life; however, abortion within the first 4 months is possible when there is the certainty that continuation of the pregnancy will cause serious harm for the woman's health. To articulate this general principle, the text of the law includes Resolution no. 140 of the Saudi Committee of Senior Ulama of 19 February 1987 which prohibits abortion at any stage of the pregnancy except in the limited cases allowed by the *Shari'a* (see section on "The Debate and Contemporary Opinions"): in the first 40 days abortion is lawful to produce

---

[117] Ibid.

[118] Larijani B., Zahedi F., Changing Parameters for Abortion in Iran, *Indian Journal of Medical Ethics*, 2006, 3(4), 130–131; *Majilis Authorizes Abortion to Save Mother, Avoid Problematic Fetus*, in www.IRNA.com.

[119] See *UNFPA in the News*, 8–14 February 2003.

a legitimate benefit. From the 40th day to the end of the first 4 months abortion is unlawful unless the pregnancy harms the woman's health and life. After the fourth month, abortion is prohibited except in the event that the continuation of the pregnancy will be fatal to the mother. Abortion is allowed, within the limits quoted, in order to avoid the worse of two risks and to choose the better of two benefits (classic principles of Muslim law and Muslim medical ethics).

Article 24/2/L of Ministerial Resolution indicates the procedure to be followed to perform the abortion. The director of the hospital must form a committee of three specialists with the task of preparing a report (signed by the members of the committee and approved by the director of the hospital) illustrating the risks that the mother will incur if the pregnancy is continued. If abortion is recommended, the situation has to be explained to the woman (or her guardian) and her husband and their written consent is necessary for the operation.

The Islamic Jurisprudence Council (linked with the Islamic World League), 12th session, Mecca, 10–17 February 1990, approved abortion within the first 120 days of conception if the foetus is affected by incurable handicaps and its life, after birth, would be miserable for itself and for its family. The prior consent of both parents is necessary. On these grounds, abortions of foetuses with serious congenital diseases appear to be performable in Saudi hospitals.[120]

Pakistan

In Pakistan, the law on abortion was amended in 1990 to make it compliant with the *Shari'a* and to supersede the previous regulations in the Penal Code imported by the British to the Indian subcontinent in 1860.[121] The latter Code severely punished abortion except if performed "in good faith" to save the mother's life; the punishment varied according to the stage of the pregnancy. If the pregnancy was not very far advanced (the exact period is not specified), those guilty of procured abortion were fined or imprisoned for up to three years or both. In the case of a more advanced pregnancy, the culprit risked up to seven years' imprisonment plus a possible fine. If the woman was against aborting, the punishment for the abortionist consisted of life imprisonment or a 10-year sentence.

Since 1979 with the gradual re-application of the *Shari'a*, the laws of retaliation (*lex talionis*) and monetary compensation have formally been re-introduced: murder and permanent injuries are punishable by the retaliation which may be performed by doctors on the subject responsible (Text of Qisas and Diyat Ordinance, 337-P)[122]; the injured party or the heirs of the person killed may opt

---

[120] Albar M.A., Counselling about Genetic Disease: an Islamic Perspective, *Eastern Mediterranean Health Journal* 1999, 5 (6), 1129–1133. Ibid., Ethical Considerations in the Prevention and Management of Genetic Disorders with Special Emphasis on Religious Considerations, *Saudi Medical Journal*, 2002, 23 (6), 627–632.

[121] Boland R., Recent Developments in Abortion Law, *Law, Medicine and Health*, 1991, 19 (3/4), 267–277.

[122] Mehdi R., *The Islamization of the Law in Pakistan*, Surrey, Curzon Press, 1994, 313–314.

for monetary compensation to which the government may add imprisonment. Minor physical damage foresees the payment of a fine.

In 1989 a decision of the Pakistani Supreme Court stated that the rules of the Penal Code of 1860 concerning offences against physical integrity were incompatible with Muslim laws. Since 1990 voluntary abortion has been a crime, the gravity of which depends on the development of the foetus (Text of Qisas and Diyat Ordinance, 338, 338 A–B–C); if the organs are not formed (the law does not specify when this happens but it is presumed on the 120th day) the culprit risks three years' imprisonment which can be increased up to ten if the woman was not consenting. If any organs or limbs of the foetus are formed, there are three types of punishment, independently of maternal consent: (a) the culprit is given a fine equivalent to one-twentieth of the sum of money to be paid to the heir of someone killed (i.e. the *diyya*) if the child is stillborn; (b) the full sum (*diyya*) if the child is born alive but then dies; (c) up to seven years' imprisonment if the child is born alive but then dies for a reason independent of the aggression. Abortion is lawful only if performed "in good faith" in the first phase of foetal development to save the mother's life or to give her otherwise unspecified "necessary treatment" (338). These words seem to refer to the physical and mental health of the woman. This is more liberal compared to the Penal Code of 1860. After the first phase, abortion remains lawful if performed in good faith to save the mother's life (338-B). If a pregnant woman has been condemned to the law of retaliation or to death, the court may, in agreement with a medical officer, postpone the execution of the punishment until two years after the birth of the child (337-P-3).

The rate of maternal mortality is very high, also because only 5–10% of births take place in hospital. Illegal abortion is one of the main causes of maternal mortality.

Kuwait

Law no. 25 of 1981 and Ministerial Decree no. 55 of 1984 allow abortion within the first 120 days when serious damage for the mother's health is imminent or it is certain that the child will be affected by incurable physical or mental lesions. In addition, the 1984 decree requires the previous and unanimous consent of three Muslim doctors guided by an obstetrician or gynaecologist; the consent of the parents is also indispensable.

The Committee of Fatwa of Kuwait had pronounced itself in 1984 in a more "articulated" way.[123] The opinion prohibited abortion after 120 days following fecundation except to save the mother's life whilst if requested by both parents abortion is legal within the first 40 days. Between 40 and 120 days, abortion may be tolerated if the continuation of the pregnancy causes serious damage to the

---

[123] Ashqar (Al) M.S., The Beginning of Life, in Mazkur (Al) K., et al. (eds.), *Human Life: Its Inception and End as Viewed by Islam*, op. cit., 115–122.

mother's health or when it is certain that the child will be born with serious and incurable physical or mental handicaps.

For the penalties, the Penal Code of 23 November 1960 remains in force. It establishes that anyone who intentionally provokes the abortion of a pregnant woman with or without her consent – except in the lawful cases – risks up to 10 years' imprisonment to which a fine may be added. If the culprit is a health professional (doctor, chemist, midwife) he or she risks 15 years' imprisonment and a fine. Any pregnant woman who makes an attempt on the life of her foetus without therapeutic justifications risks five yeas' imprisonment and/or a fine.

Sudan[124]

Until 1983, abortion was regulated by the Penal Code of 1 August 1925 (sections 262–267). It was allowed only to save the mother's life. Anyone who performed it with the consent of the woman risked three years' imprisonment and/or a fine if the foetus had not reached the stage of "quickening". The woman who performed her own abortion is subject to the same penalties. The penalties are harsher if the woman does not give her consent, if the abortion is performed after the stage of "quickening" or if it causes the death of the pregnant woman. The pregnant woman who provokes her own abortion to cancel dishonour risks a reduced penalty of two years' imprisonment and/or a fine.

In 1983 the Penal Code was replaced by legislation closer to the principles of the *Shari'a*. Abortion is lawful only to save the mother's life. Blood money was reintroduced; anyone who aborts outside the law pays compensation and is imprisoned and/or has to pay a fine. Payment is made to the relatives of the foetus and/or to the mother depending on the circumstances of the abortion.

In 1991, the Penal Code was again amended and the law on abortion was softened. Abortion is lawful to save the mother's life and if a raped woman wants to abort within 90 days of the rape (art. 135). Illegal abortion within the first 90 days is punishable with up to three years' imprisonment and/or a fine. After 90 days the punishment is extended to five years and a fine. Both before and after 90 days the person responsible must pay compensation.

Since 1991 the new legislation has not applied to the Christian south of the country.

Qatar

The Penal Code of 28 August 1971 allows abortion only to save the mother's life. Law no 3/1983 of 22 February 1983 in section 17 allows abortion only to save the mother. However, within the first 4 months (120 days, a classic legacy), abortion may be performed even if the pregnancy is certain to damage the mother's health

---

[124] United Nations, Population Division, *Abortion Policies: A Global Review*, New York, 2002, 107–108, in www.un.org/esa/population/unpop.htm; United Nations, Population Division, *Abortion Policies: A Global Review*, New York, 1995, 105–107 in www.unpopulation.org.

(17-a) or cause incurable damage to the physical and mental health of the foetus (17-b). The consent of both parents is required.

Malaysia

Until 1989, section 312 of Act 727 of the Penal Code allowed abortion only to save the mother's life. In April of the same year, the section was amended allowing abortion even in the presence of risks for the physical and mental health of the pregnant woman. Abortion may be performed within the first 120 days of pregnancy (influence of the classic doctrine). The Obstetrical and Gynaecological Society of Malaysia supports the initiative by the Ministry of Health to amend the existing abortion laws (section 312, Penal Code, 1989) in order to allow abortion after rape or incest.

Tunisia

The two strictest laws on abortion copied the similar French laws: art. 214 of the Tunisian Penal Code of 1913, a copy of the French Penal Code of 1810 (art. 317), prohibited abortion in all cases. The decree of 18 September 1920, a copy of the French decree of 31 July 1920, also severely punished propaganda for contraceptives. The Bey's decree of 25 April 1940, a copy of the French decree of 29 July 1939, harshly prosecuted abortionist doctors; this last version allowed abortion to save the mother's life.

In the 1960s, there was great concern by the authorities over the numerous illegal abortions performed with "primitive" means and in precarious medical and hygienic conditions; these abortions often concluded in fatal "accidents" for the women.[125] To reduce the phenomenon, the legislator, with law no. 65–24 of 1 July 1965, which recast art. 214 of the Penal Code (of which the previous version of 25 April 1940 accepted abortion exclusively to save the mother's life), authorised social and therapeutic abortion, making Tunisia the most "open" Muslim country towards abortion. Voluntary interruption of pregnancy was authorised in the first 3 months in the presence of at least five living children. In addition, abortion was lawful if the continuation of the pregnancy could harm the mother's health. The operation had to be performed in all cases in a hospital and by authorised doctors. Outside these limits, the law laid down, for anyone who had attempted to perform an abortion on a woman – whether she was consenting or not – five years' imprisonment and/or a fine of 10,000 dinars. When the woman aborts alone (or tries to abort), two years' imprisonment and /or a fine of 2,000 dinars was laid down.

Subsequently, the law-decree of 26 September 1973 modified art. 214 of the Tunisian Penal Code more liberally and it still allows free abortion within the first 3 months of pregnancy in hospital or an authorised clinic.[126] After this period, abortion is possible when the health and the mental balance of the mother are

---

[125] Borrmans, op. cit., 356–358.
[126] République Tunisienne, *Code Pénal*, Tunis, Imprimerie Officielle de la République, 1997, 54–55.

at risk due to the pregnancy (the law does not specify the risks in question) or if the child will suffer from a serious illness or handicap (the mere possibility seems sufficient to terminate the pregnancy). The only possible obstacle is a report from the woman's family practitioner to the doctor who will perform the abortion. In other words, the law allows abortion throughout the period of pregnancy although the first 3 months are to be privileged.

The Tunisian Code of Medical Ethics of 17 May 1993 regulates the intervention of the doctor subordinating it to the respect of the laws in force (art. 40) and recognising the doctor as the only judge of the interests of the mother and child, in the case of prolonged deliveries, against any pressure by the family.[127]

The United Nations[128] reported the growing number of abortions from 1973 until their figure stabilised towards 1983; however, abortion remained the third most used method for population control after inter-uterine devices and oral contraceptives. The difficulty for women to have recourse to legal abortion due to ignorance of the laws in force was also reported. Abortion for unmarried women was particularly problematic, especially in traditional communities and rural areas where illegal abortion remains widespread.

Egypt

The reference on abortion is provided by arts. 260–264 of the Penal Code of 31 July 1937 and by arts. 60–61. Articles 260–264 prohibit abortion in all cases, articulated as follows. Article 260 lays down temporarily forced labour for anyone who intentionally causes an abortion through blows. Article 262 prescribes imprisonment for anyone who intentionally causes the interruption of a pregnancy by means of drugs and similar, with or without the consent of the pregnant woman; in addition, the same punishment is laid down for the pregnant woman who knowingly takes drugs to abort. Article 263 increases the punishment (forced labour for a defined period of time) for medical staff guilty of procured abortion. According to art. 264, attempted abortion is not prosecutable. These prohibitions are moderated by arts. 60–61. The former states that an act performed in good faith, in the exercise of a right recognised by Muslim law, does not come under the application of penal law; in other words, the article decriminalises the work of medical staff if they are authorised to perform, do not commit errors or negligence and there is the consent of the patient. Article 61 says "A person who commits a crime in case of necessity to prevent a grave and imminent danger which threatens him or another person shall not be punished, on condition that he has not caused it of his own volition or prevented it by other means."[129] This article decriminalises the actions performed out of necessity and it can be used to justify abortion not only when the pregnancy threatens

---

[127] Republique Tunisienne, Code de Déontologie Médicale, *Journal Officiel de la République Tunisienne*, 1993, 28 mai – 1er juin, no 40, 766.

[128] United Nations, op. cit., 1995, 138–139. Ibid., 2002, 139–140.

[129] Ibid., 135.

the life of the mother, but often also to protect the mother's health; the approval of two doctors and the consent of both parents is also required. In fact, the law does not appear to be particularly respected and a voluntary interruption of pregnancy is not difficult especially for higher class women who can afford to pay for the operation with adequate guarantees for their health. Nevertheless, the fear by abortionist doctors of being legally prosecuted has contributed to the development of illegal abortion which today appears to represent the main cause of mortality of mothers.[130]

There is an extensive production in Egypt of religious opinions on abortion. These oscillate between rigid and tolerant, especially regarding abortion during the first foetal stages. The state law appears more rigid than the positions prevalently taken by the religious authorities.

Bahrain

Abortion in Bahrain was regulated by the Penal Code of 1956 (sections 201–203) which prohibited abortion except[131]: (a) to save the mother's life, (b) to preserve her mental and physical health, (c) in the event of ascertained deformation of the foetus. Infringement of these limits lays down 10 years' imprisonment for the abortionist, up to five years' imprisonment for the woman causing her own abortion and up to three years' imprisonment for anyone who helps her.

The Penal Code of 20 March 1976 (sections 321–323) allows abortion to save the mother's life, to protect her physical and mental health, in the case of rape or incest, foetal malformations and socio-economic problems. It is illegal if performed by the pregnant woman herself without adequate medical knowledge or assistance or if performed without the consent of the pregnant woman. In the first case the punishment is of 6 months' imprisonment or a fine of up to 50 dinars. In the second case, the abortionist risks 10 years' imprisonment.

A clause of the Law-Decree no. 24 of 1977 limits the performance of abortion in the cases contemplated by the law to doctors only.

Algeria

In Algeria the Penal Code of 8 June 1966 in sections 304–313 prohibits abortion except to save the mother's life (art. 308). Article 304 prescribes for anyone – including medical staff – who provokes abortion, with or without the pregnant woman's consent, from one to five years' imprisonment and a fine. If the woman dies, the penalty increases from ten to twenty years. In any case the person found guilty of abortion also risks local banishment (*interdit de séjour*). Medical staff involved in the procured abortion may be prohibited from practising the profession (art. 306). A women who procures (or tries to

---

[130] Lane S.D., Madut Jok J., et al., Buying Safety: The Economics of Reproductive Risk and Abortion in Egypt, *Social Science and Medicine*, 1998, 47 (8), 1089–1099.
[131] United Nations, ibid., 1992, 37.

procure) an abortion risks from six months' to two years' imprisonment plus a fine (art. 309).

Subsequently, the Public Health Code, Ordinance no. 76–79 of 23 October 1976 in art. 414 and 28 allows therapeutic abortion to save the mother's life and her health if seriously threatened; this is allowed only before the foetus is viable. Article 72 of the Law on the Protection of Health no. 85–05 of 1985 adds "Abortion with a therapeutic purpose is considered as an indispensable measure to save the life of the mother or to preserve her seriously threatened physiological and mental balance. Abortion is performed by a doctor in a specialised structure, after a joint medical examination with a specialist." For doctors who perform abortions outside the law, imprisonment and a fine are laid down as well as the possible prohibition of the profession.

It is worthwhile comparing this with a *fatwa*, dated 15 February 1973, by the then President of the Higher Islamic Council of Algeria, *sheikh* Ahmad Hammani.[132] Discussing *ijhad* (abortion) in Muslim legislation, the religious figure qualified abortion as a crime and the person responsible for a voluntary abortion as a criminal. The latter, even if attenuating circumstances are available, is exposed to the "law of retaliation", whilst the monetary punishment (blood price) laid down for the foetus is equivalent to one-tenth of that to pay for the murder of an adult person. At this point, the *sheikh* resumes the criterion of ensoulment or of the "full human form" of the foetus that is believed to take place on the 120th day after fecundation. According to Hammani, Muslim jurists agreed on the compulsory nature of the "blood price" when a foetus that is already formed is aborted, whilst there is no agreement for the earlier phases. There are also exceptions for Hammani that moderate his severe initial statements of principle. Indeed, it is lawful to interrupt pregnancy due to the illness of the mother or the probable serious damage caused to her by the pregnancy. In other words, therapeutic abortion becomes legitimate when it is not exclusively the mother's life that is at risk but also her health. The latter solution is similar to the position of Algerian law. However, Hammani continues, voluntary abortion is no longer lawful after "full human configuration" (about 120 days).

Indonesia

The Indonesian Criminal Code, enacted on 1 January 1918, was modelled on the Dutch Criminal Code, with the aim of stemming the practice of abortion in the country.[133] Section 348 of the Code prohibited abortion in all cases, punishing it with at least five and a half years of imprisonment. According to section 346, the pregnant woman who provokes her own abortion risks up to four years in prison. Sections 349–350 prescribe harsher penalties and the revocation of practising the profession for medical staff involved in the abortion. Therapeutic abortion was not lawful.

---

[132] Borrmans, art. cit., 88–89.
[133] United Nations, op. cit., 2002, 59–61; ibid., 1993, 60–62 in www.unpopulation.org.

It was not until 1992 that Health Law no. 23 stated that abortion may be permitted "to save the life of a pregnant mother and/or her foetus" (art. 15.1). The abortion must be performed by obstetricians or gynaecologists. The decision is up to the pregnant woman; approval may be given by her husband or family only if the woman is unconscious or incapable of approving (art. 15.2).

The law is restrictive but not very respected; the number of illegal abortions is very high and there is often recourse to traditional practitioners despite the serious health risks for the woman. Complications caused by abortion are the main cause of death amongst women of reproductive age.

Senegal[134]

The Penal Code of the country imitates art. 317 of the French Penal Code of 1810, subsequently amended by the French Law decree of 1939. On these grounds, art. 305 of the Senegalese Penal Code prohibits abortion with imprisonment and a fine while therapeutic abortion is not mentioned. Health personnel involved voluntarily risk no longer being able to practise. However, there is recourse to the principle of necessity to save the mother's life, a fact explicitly laid down by the Code of Medical Ethics of Senegal. Abortion amongst adolescents is very widespread and surveys have shown that at least 12% of secondary school girls have had abortions.

Burkina Faso

Until 1994 the law against abortion in Burkina Faso[135] was based on the French Napoleonic Code of 1810. Taking the same position, the 1984 Penal Code prohibited abortion without any exceptions. However, there was recourse to the juridical principle of necessity to allow abortion to save the mother's life. The anti-abortion position was accompanied by the anti-contraception position based on the French Law of 31 July 1920 and which remained in force until the mid-1980s. In 1996 the law changed to improve protection for the mother's health as too many women were suffering from serious complications and dying due to dangerous illegal abortions. In the 1996 Penal Code, abortion remains a crime and anyone who procures one risks from one to five years' imprisonment and a fine. At the same time, abortion is lawful at any time if two doctors certify that the physical and mental health of the woman is at risk or the health of the child will be seriously compromised. In addition, in the case of rape or incest, abortion is lawful in the first 10 weeks.

Nigeria[136]

Abortion is regulated by two different laws. In the northern states with a Muslim majority, Penal Code Law no. 18 of 1959 is in force, which refers to British

---

[134] Centre pour le Droit et les Politiques en Matière de Santé et de la Reproduction (CRLP) – Groupe de Recherche Femmes et Lois au Sénégal (GREFELS), *Les droits des femmes en matière de santé reproductive au Sénégal*, 2001 août, 1–36 (PDF).

[135] United Nations, op. cit., 2002, 75–76.

[136] Op. cit., 2002, 169–171; 1993, 176–178.

criminal law adopted in the past from India and Pakistan. In the south where there is a Christian majority, the Criminal Code of 1916 is in force, modelled on the English Offences against the Person Act of 1861. Both prohibit abortion but divergences in interpretation have caused slightly different applications.

In the North, abortion is lawful only to save the mother's life; in other cases anyone who provokes abortion voluntarily risks up to 14 years' imprisonment and/or a fine. The same penalty applies to the pregnant woman who provokes her own abortion.

In the South, the Criminal Code with art. 297 allows abortion only to save the pregnant woman. Article 228 inflicts up to 14 years' imprisonment for the person guilty of voluntary abortion; art. 229 inflicts up to seven years' imprisonment for the pregnant woman who provokes her own abortion. However, the Criminal Code has been interpreted in such a way as to allow abortion in other situations applying the ruling of 1938, in the *Rex v. Bourne* case, which concerns permission to abort to protect health. The sentence acquitted the doctor of the offence of abortion on a woman who had been raped, the abortion having been performed to avoid the serious physical and mental suffering for the latter. The sentence became a precedent for similar sentences.

In Nigeria, subsequent attempts at legally liberalising abortion failed, especially due to the opposition by religious leaders; in the meantime, procured abortion is very common and performed privately. It represents one of the major causes of maternal mortality. The main reasons for abortion amongst young women are: the fear of interrupting their education, the fear of losing their jobs and the social stigma of having children outside marriage.

Mauritania

Article 293 of the 1983 Penal Code of Mauritania, which concerns abortion, derives directly from the French Napoleonic Code of 1810 and the subsequent French Code of 1939. Abortion is prohibited in all cases.[137] However, the criminal law of 1972 accepts recourse to the principle of necessity to save the mother's life. Although there is no law, in 1987 the Government legitimised abortion in the case of rape or incest. Abortion seems rare; the social and religious values persuade a woman to keep an unwanted child. At the same time, the number of unmarried women who abandon their children is on the increase.

Mali

Articles 170–171–172 of the Penal Code of Mali (Law no. 99, 3 August 1961) do not lay down exceptions to the prohibition of abortion. Attempted or successful voluntary abortion, by the woman or by another person, with or without the consent of the woman, is punished by a sentence of one to five years in prison and a fine (art. 171). Any doctor or paramedical employees involved in the practice, as well as the penalties of art. 171, risks being prohibited from practising for

---

[137] Ibid., 2002, 130–132; 1993, 136–138.

five years or forever. However the principle of necessity and legitimate defence allows saving the mother's life (art. 176). Article 13 of Law no. 02–044 of 24 June 2002 on Reproductive Health always prohibits abortion except to save the mother's life and if the pregnancy is the result of rape or incest.

Morocco

In Morocco, Royal Decree no. 181–66 of 1 July 1967 amended the previous version of art. 453 of the Penal Code which still prohibited abortion. The new version allows abortion to protect the physical and mental health of the mother if it is openly practised by a doctor with the consent of the husband. This consent is not required when the woman's life is in danger. If the husband is absent or if he refuses to give his consent to the abortion or is prevented from consenting, the doctor requires the written authorisation of the chief medical officer who certifies that the abortion is necessary to protect the mother's health.[138]

The legislation in Niger is based on the Napoleonic Code of 1810, amended by the French law of 1939. The Penal Code of 15 July 1961 (Law no. 61–27, art. 295) generally prohibits abortion. In any case, according to the principle of necessity abortion is lawful only to save the life of the pregnant woman.

The legislation on abortion in Bangladesh is based on the Indian Penal Code of 1960. Sections 312–316 allow abortion to be performed only to save the mother's life.

The laws of the Muslim-majority countries of the former Soviet Union, namely Azerbaijan, Kazakhstan, Kyrgyzstan, Tajikistan, Turkmenistan and Uzbekistan all lay down great freedom of abortion especially within the first 12 weeks, but sometimes even up to 28 weeks.

ABORTION AND THE DEFENCE OF HONOUR

The logic of protecting the honour of the group – the family, the clan, and ultimately the *umma* (the Muslim community) – may prevail over the rights of the individual. This is the case of abortion to save the honour of the woman or of her family.

Abortion is prohibited by art. 16 of the Jordanian Penal Code of 1960.[139] The person performing it risks one to three years' imprisonment; if it is a doctor the punishment increases. The woman who provokes her own abortion risks from six months to three years in prison. The punishments increase if the abortion is performed without the permission of the woman or if she dies. On the other hand, the punishments are reduced if the abortion is performed to save the honour of the woman by the woman herself or by a descendant or relative up to the third degree of kinship; these are rules 324 and 322–323 of the Penal Code which also

---

[138] *Le Code Pénal*, Publications de la Revue Marocaine d'Administration Locale et de Développement (REMALD), série "Textes et Documents", 1997, no 5, 101–102.

[139] United Nations, op. cit., 2002, 82.

allow abortion in the case of incest and rape.[140] According to Public Health Law no. 20 of 1971, abortion is lawful to save the mother's life and health. There are no time limits for the abortion. However, the laws are interpreted with considerable discretion.

The same approach aimed at protecting the honour of the pregnant woman exists in other Penal Codes of Muslim countries. The Lebanese Penal Code[141] of 1 March 1943 and that of 16 September 1983 prohibited abortion in all circumstances with arts. 539–546. Anyone performing an abortion to save the honour of a descendant or relative up to the second degree, as well as a woman who provokes her own abortion to save her own honour, incurs a reduced punishment (art. 545).[142] The subsequent Presidential Decree no. 13187 of 20 October 1969 allows abortion to save the mother's life. The consent of the pregnant woman is required unless she is unconscious; in this case the doctor may perform the abortion despite the opposition of her family.

The Syrian Penal Code of 22 June 1949 (Legislative Decree no. 148) prohibits abortion in all cases. Articles 527–528 lay down from six months to three years of prison for anyone performing an abortion, the pregnant woman included. Article 531 specifies that the penalty will be reduced for the woman who has had an abortion to save her honour or for the person that acts to save the honour of their descendants or of a relative to the second degree.[143] According to the principle of necessity, abortion is lawful to save the mother's life.

The Libyan Penal Code of 1953 (art. 390–395) prohibits abortion; the penalties are reduced by half if the abortion has been performed to save the honour of the person performing it or the honour of a relative (art. 394). Health Law no. 106 of 1973 accepts abortion to save the mother's life similarly to law no. 17 (paragraph 19) on the responsibility of the doctor of 3 November 1986.

---

[140] Some countries that have a reduced penalty for abortion performed to save honour also have other laws reducing the penalty for the "crime of honour" or of passion. This is the case of art. 340 of the Penal Code of Jordan concerning the crime committed to vindicate one's honour, which states: "Anyone catching his wife or one of his immediate family in a flagrant act of fornication with another person, and kills, injures or harms both or either of them, will benefit from the exculpating excuse." The article is inspired by similar articles of the Napoleonic Code of 1810 and the Ottoman Penal Code of 1858. Similar articles, with the same origin, can be found in the Penal Codes of other countries in the region. In the Jordanian case, art. 340 has been amended by Temporary Law no. 86 of 2001 which allows for the wife the same attenuating circumstance valid previously only for her husband, see HUMAN RIGHTS WATCH, *Honoring the Killers: Justice denied for "Honor" Crimes in Jordan*, April 2004, Vol. 16 (1E), 1–39.

In February 2002, the Jordanian government abolished the penal exemptions in favour of those who commit crimes of honour.

[141] United Nations, op. cit., 2002, 100–101.

[142] In February 1999, Lebanon modified the previous article of the Penal Code which tolerated the crime of honour with a new articles that made the penalty for man and women committing the crime equal.

[143] In Syria art. 548 of the Penal Code is very similar to Jordan's art. 340.

In Sudan, until 1983 abortion was regulated by the Penal Code of 1 August 1925 (sections 262–267). Abortion was permitted only to save the mother's life. The pregnant woman provoking her own abortion to cancel dishonour risked a reduced penalty of two years' imprisonment and/or a fine. This position appears superseded by the Penal Codes of 1983 and 1991.

Before the collapse of the central government of Somalia, the Penal Code of 16 December 1962 (art. 418–422 and 424) prohibited abortion in all cases except to save the mother's life. If the abortion was performed to save her honour or that of a close relative, the penalty could be reduced by two-thirds.

In Palestine, the prevailing law on abortion in the West Bank is based on Penal Code no. 16 of 1960 of Jordan and amended by Law no. 39 of 1963 of Jordan.[144] The rules prohibit the voluntary interruption of pregnancy with or without the consent of the pregnant woman. The penalty varies from 7 to 14 years' imprisonment. In the event that the pregnant women has an abortion for reasons of honour, the penalty will be significantly reduced. The penalty is increased if the abortion is performed by medical staff, chemists and obstetricians.

CONCLUSION

The Muslim views on abortion are historically complicated and diversified, especially due to the criterion of the infusion of the soul and the stages of development mentioned in the Koran and *Sunna*. These criteria form a "step" structure both on the level of juridical thought and on the penal level. It appears very difficult to summarise such a variety of positions on abortion in Islam with a single expression. Before the infusion of the soul, abortion may be prohibited, criticised or allowed. After animation, the attitude becomes more severe, generally allowing only therapeutic abortion: this expression allows, especially today, different possibilities of abortion according to the definition.

Regarding the laws of the contemporary nations, the laws based on the embryonic stages of Koranic derivation seem to offer less protection for the embryo compared to those that ignore the criteria of the *Shari'a* (the embryonic stages) but, at the same time, prohibit abortion from fecundation (e.g. Egypt); this is perhaps because the presence of intermediate stages in the development of the embryo favours diversified responses to abortion at the various stages.

It must not be forgotten that, whilst in previous centuries the faithful could follow the different positions on abortion of the individual jurists of the different juridical schools (at times present in the same territory), today the only legislator is the state, the law of which is valid for all citizens in the same territory.

---

[144] Rishmawi M., *On the Legislation relating to Palestinian Children*, Birzeit, The Law Centre of Birzeit University, 1996.

ASSISTED PROCREATION

INTRODUCTION   136
LEGAL ADOPTION   139
JURIDICAL-RELIGIOUS FORMULATIONS   140
OPINIONS IN SHI'ITE ISLAM   148
PROBLEMS RELATIVE TO THE EMBRYO   151
SOCIETY AND LEGISLATION   154
CONCLUSION   159

The issue of artificial procreation does not appear new in the history of Muslim civilisation. The two examples we will give show that some aspects of artificial procreation both in man and amongst animals had been broached in the distant past. It would seem that (a) artificial insemination was practised by the Arabs on horses, for the first time in history, as early as 1322, in order to guarantee the purity of their stallions.[1] (b) In addition, in pre-Islamic Arab history there existed a type of carnal union called *al-istibdaa* or "procreative union" which allowed a married woman to form a new couple with a famous man or a man with a good reputation other than her husband, in order to have a child by him. The goal was to give the infant the characteristics inherited from this "second father"[2] and, once pregnant, the woman would return to her original home. This custom, however, was abolished by the Prophet and his position was developed by Muslim law, leading to the prohibition of heterologous artificial fertilisation.

In the Muslim world, there was initially a preclusive or critical attitude towards assisted reproductive technologies dictated by several motives, the main one of which was to be found in the fear of its use outside the principles of the *Shari'a*. The problem of sterility was historically solved by recourse to polygamy or repudiation and marriage to another woman. Furthermore, fertility and sterility were believed to be the result of divine will and consequently techniques of conception, other than natural ones, were avoided. Lastly, this practice means that the doctor performing it observes the body of the woman in parts which normally can only lawfully be seen by her husband. On the basis of these presuppositions, even homologous artificial insemination was prohibited as prescribed by the Libyan Penal Code in 1972.

There are at present Muslim scholars who prohibit any type of artificial insemination on the basis of the observation that in the West steps have been

---

[1] Chafi M., L'Islam et l'Insémination artificielle, *Revue Juridique Politique Indépendance et Coopération*, 1987, 356–362.
[2] Ibid., 359.

Dariusch Atighetchi, Islamic Bioethics: Problems and Perspectives.
© Springer Science+Business Media B.V. 2009

taken that go far beyond insemination between husband and wife.[3] Abd al-Halim Mahmud, *sheikh* of Al-Azhar from 1973 to 1978, prohibited any form of in vitro fertilisation (IVF) because a child obtained this way does not have a father and mother and human warmth in growing up. In addition, he added, in consideration of the overpopulation of Egypt, there is no need to produce test-tube babies (*fatwa*, 1986). In the 1960s, the Lebanese *Shi'ite sheikh* M.G. Mughniyyah, in a *fatwa* on this subject, stated literally: "Artificial fecundation is *haram* [prohibited] and no Muslim may come out in its favour."[4]

Today, this opposition has been largely overcome. Particular resistance to Western juridical influence is offered by the rules of Muslim family law (marriage, divorce, legitimate filiation, etc.), which precisely due to the fact that they are derived from the Koran, have less probability of being modified by contemporary Muslim countries. The issue of assisted procreation refers directly to the principles of Muslim family law more than any other topic in bioethics. Nevertheless, we will see an increasingly complex debate emerge and even see positions which to some extent are in favour of heterologous procreative practices.

INTRODUCTION

How is consent on the new techniques of medically assisted conception articulated? Two aspects should be considered in the first place:

1. The numerical dimension of the phenomenon of infertility in the Muslim world, which is by no means negligible, requires methods of intervention beyond those traditionally considered (polygamy, repudiation, etc.). In 1992 the world population of the Muslim faith amounted to about 1 billion and 250 million individuals, 24% of whom were girls and women of reproductive age; the average rate of infertility hovers between 10% and 15%, mainly due to obstructions of the tubes (e.g. in Egypt) whilst in Tunisia sterility affects about 20% of married couples.

2. The second aspect regards the concept of marriage and the family. Muslim law considers marriage a duty, on condition that the man is capable of maintaining the family, can pay for the wedding gift to the woman and he fears, if he remains unmarried, being unable to resist incontinence. Some sayings of the Prophet exemplify this concept[5]: "Marriage is my way; he who deviates from my way is none of me", "Marry the affectionate and prolific woman",

---

[3] Rispler-Chaim V., *Islamic Medical Ethics in the Twentieth Century*, Leiden, Brill, 1993, 23.

[4] Castro F. (ed.), Illiceità della fecondazione artificiale in diritto musulmano secondo una recente pubblicazione dello Shaykh Muhammad Gawad Mughniyyah, *Oriente Moderno*, 1974, LIV, 222–225.

[5] Glossary of Hadiths, in Nazer I., Karmi H.S., et al. (eds.), *Islam and Family Planning: A Faithful Translation of the Arabic Edition of the Proceedings of the International Islamic Conference held in Rabat (Morocco), December 1971*, Beirut, International Planned Parenthood Federation, 1974, Vol. II, 547.

"No institution in Islam finds more favour with God than marriage."[6] In short, celibacy is not well thought of in Islam as it is considered a condition that goes against nature; moreover, the single man and, to an even greater extent, the single woman, are considered to be in a mutilated and socially marginalised position. In this context, in the Koran, offspring are considered a divine blessing and procreation is one of the main reasons for marriage.

In the wake of these values, the Muslim doctor has the duty of protecting the faculty of conception and curing infertility. This is in fact the most important duty if we consider that the social role of the woman has always been conditioned by her procreative capacity in marriage, therefore sterility is strongly detrimental to her position, damaging her self-esteem. It is well known that in the rules of the *Shari'a* in general, the juridical condition of the woman, within the family, appears very different from that of the man and confirmation is provided by the different opportunities offered to a husband or wife when the partner is sterile. In fact, if the wife is sterile, the husband can remarry, without repudiating his previous wife, as Muslim law allows polygamy with up to a maximum of four wives (Koran 4.3). Alternatively, the husband can choose to repudiate his wife. Unilateral repudiation by the husband has always led to abuse, to the extent that the excuse of supposed sterility of the wife was sufficient for the husband to repudiate her.[7]

On the contrary, when it is the husband that is sterile, the *Shari'a* does not grant the wife the faculty of repudiating him. However, in current legislation in many Muslim countries, as marriage maintains the character of a private contract, the woman can include in the contract a number of conditions in her favour (e.g. that the husband cannot marry another woman, that he is not sterile, etc.), breach of which by the husband gives her the right to obtain a divorce; or the wife can put an end to the marriage by paying compensation or redemption to the husband so that he repudiates her; lastly, in some countries, in the case of certain male illnesses, the wife is allowed to have recourse to judicial divorce.[8]

At the same time, the specific condition of the family in Islam must be borne in mind. The family is a divine institution. The Koran and the *hadith* condemn as fornication (*zina*) any intercourse between a man and a woman who is not his wife or his slave,[9] based on Koran 17.32 "Nor come nigh to adultery: for

---

[6] A *hadith*, again in the collection by Bukhari, lists the reasons for choosing a wife: "A woman is taken in marriage for four things: her wealth, her family, her beauty and her religiousness; so get hold of the religious woman, and you will prosper" in Glossary of Hadiths, in Nazer and Karmi, op. cit., 558.

[7] In general, see Santillana D., *Istituzioni di diritto musulmano malichita con riguardo anche al sistema sciafiita*, Roma, Istituto per l'Oriente, 1926–1938, Vol. I, 253 etc.

[8] Chafi, art. cit., 361; Prader J., Il diritto matrimoniale islamico, in AA.VV., *Migrazioni e diritto ecclesiale*, Padova, Edizioni Messaggero, 1992, 133–167.

[9] Santillana, op. cit., 198–199.

it is a shameful (deed) and an evil, opening the road (to other evils)". As slavery (here, concubinage) although accepted in the *Shari'a*, is an institution that is now obsolete in the Muslim world, sexuality remains permissible exclusively within marriage.

The term *zina* takes on a fundamental value in establishing the general attitude of Islam towards assisted reproductive technologies. Technically the term *zina* includes both fornication and adultery (both harshly punishable by the *Shari'a*)[10] and, at present, all techniques of heterologous artificial conception, where a stranger to the couple provides sperm, ovules, embryos, uterus, etc. to the married couple incapable of using their own to generate children, are assimilated to acts of *zina*. On the contrary, homologous practices are tolerated or accepted.

As the only legitimate filiation for Islam is that with respect to the paternal figure, children generated following intercourse between an individual and a woman who is not his wife (relationship of *zina*) do not and cannot belong to the paternal family; there is no recognition for them; they have no connection with the father, no right to his inheritance; the illegitimate child (*walad az-zina*) has no connections other than with the mother and maternal family. Fundamentally, contemporary *Sunni* jurists also apply these legal consequences to children generated through heterologous artificial procreation. From the penal point of view, heterologous procreation does not seem to be punishable by the penalties of the *Shari'a* contemplated for fornication and adultery as there is no direct physical contact between male and female.

However, the child remains attributed and bound to the mother both when it is the result of either a licit or an illicit relationship. In the latter case, the bond with the mother does not compensate for the exclusion from the bonds with the father and, through the father, with the agnates: these relations guarantee essential protection for the social life of the individual in traditional societies.[11]

The traditional approach has been modified in some contemporary legislations. For example, the previous Code de Statut Personnel (Moudawana) of Morocco, in art. 83 (1) recalled that legitimate filiation is that thanks to which "the child takes on the kinship of his father and follows the religion of the latter". The new version of the Moudawana (2004) with art. 142 also introduces the figure of the mother: "Filiation is through the procreation of the child by his/her parents".

---

[10] The punishment established by the Koran 24.2 consists of 100 whiplashes for fornication. The origin of the punishment of death by stoning, however, is not in the Koran.

[11] For example, in Muslim law, if an individual commits a murder, the agnates are responsible for him for the payment of the blood price (*diyya*). If the individual is killed, it is the duty of the closest agnate to apply the law of talion on the person responsible or claim the blood price (see the case of a procured abortion), etc. See Aluffi Beck-Peccoz R., *La modernizzazione del diritto di famiglia nei paesi arabi*, Milano, Giuffrè, 1990, 153–154.

LEGAL ADOPTION

Faced with the impossibility of having children, the couple cannot have recourse to adoption as it is prohibited by the *Shari'a* on the grounds of Koran 33.4–5.[12] Only "compensation" or "testamentary" adoption is accepted, according to which a family may raise, bring up and protect a child without considering it their own as it remains bound to its original family. In this way, the "protected" child does not enjoy the same inheritance rights as the other "true" children, is not bound to the marital rules laid down by the Koran between certain degrees of kinship, etc.

Article 46 of the Algerian Family Code (9 June 1984) states: "Adoption is prohibited by the *Shari'a* and by the law" and this prohibition is in force in other Arab countries with the exception of Tunisia (law 27 of 1958) and the Code of Personal Status of Somalia (art. 110–115).[13] In addition, art. 116 of the Algerian Code replaces adoption by *kafala*, consisting of caring, educating and protecting a minor in the same way as a father with his own children but, in the event that the real parents are unknown, the child cannot take on the surname of its "adoptive father"[14]; art. 60 of Law no. 10 of 1984 in Libya is very similar.[15] However, art. 123 of the Algerian Family Code and art. 60c of the aforementioned Libyan law establish that the "adopting" parent can leave one-third of his estate by means of his will to a child subject to *kafala*.

The previous *Moudawana* (Code of Personal Status) of Morocco stated in art. 83 (3): "Adoption has no legal value and does not have any of the effects of filiation. However, adoption called 'of gratification' (*jaza*) or testamentary (by which the adopted child is placed in the position of a first degree heir), does not establish any link of filiation and follows the rules of inheritance."[16] Under art. 149, the new *Moudawana* states: "Adoption does not produce any effect for legitimate filiation. Adoption called 'of gratification' (*jazaa*) or assimilation to the child by a will (*tanzil*) do not establish filiation and are subject to the rules of inheritance."[17]

---

[12] Koran 33.4–5 "4. God has not made for any man two hearts in his (one) body: nor has He made your wives whom ye divorce by *zihar* your mothers: nor has He made your adopted sons your sons. Such is (only) your (manner of) speech by your mouths. But God tells (you) the Truth, and He shows the (right) Way. 5. Call them by (the names of) their fathers: that is juster in the sight of God. But if ye know not their father's (names, call them) your Brothers in faith, or your *Maulas*." The term *zihar* indicates one of the ways of repudiation.

[13] See Aluffi Beck-Peccoz, op. cit., 152.

[14] Aluffi Beck-Peccoz R. (ed.), *Le leggi del diritto di famiglia negli stati arabi del Nord-Africa*, Torino, Edizioni della Fondazione Giovanni Agnelli, 1997, 40 and 48.

[15] Ibid., 118.

[16] Livre III, Chapitre I, in Blanc F.-P. and Zeidguy R. (eds.), *Moudawana. Code de statut personnel et des successions*, Casablanca, Sochepress-Université, 2000–2001, 87.

[17] Law no. 70.03 called the Family Code (3 February 2004), in Mounir O., *La Mudawana. Le nouveau Droit de la Famille au Maroc*, Rabat, Editions Marsam, 2005, 137–179, 154.

Whilst in the West adoption is often presented as a natural solution to sterility and the absence of children, in Islam, although the Koran prohibits legal
adoption, at the same time the "Word of God" encourages bringing up and
educating orphans. Despite this willingness, there is a strong cultural resistance
in Muslim countries to this humanitarian solution.[18] Amongst the "prejudices"
towards bringing up children other than one's own, the following can be mentioned: (a) the fear of having to do with illegitimate and therefore potentially
immoral children; (b) the original parents could reappear and claim back their
"protected" children; (c) the parents and the new arrival could fail to establish
bonds of affection; (d) a sexual attraction may arise between the parents and the
new arrival or between the latter and the natural children; (e) children who are
brought up by a family not their own are often stigmatised by the community
and by the other members of the family; (f) the "adoptive" parents, but especially the mother, are likely to risk stigmatisation due to the inability to procreate
"true" children. Due to these popular convictions, it is not surprising that only
very few sterile and affluent couples in Egypt have thought of bringing up an
orphan to whom they cannot, however, pass on their wealth according to the
current laws on inheritance.

JURIDICAL-RELIGIOUS FORMULATIONS

As far as recourse to artificial means for conception is concerned, a precedent
can be identified in the possibility given by the *Shari'a* of a pregnancy that is not
derived from direct carnal contact, i.e. penetration. For example, some medieval
jurists accepted that a woman could introduce the sperm that she considered to
be her husband's into her uterus on her own.[19]

Similarly to the contraceptive practices which succeed in blocking the procreative process only if God wants, becoming a divine instrument, in the same way
pregnancy achieved through means that are other than natural could be deemed
as not representing an obstacle to divine creation. In fact, God desires the happiness of Muslims and insemination succeeds only if God so wishes, independently of the technique (but always in the respect of the precepts of the *Shari'a*)
which thus becomes one of His instruments.

A document which follows the above indications is the *fatwa* by Jad al-Haq,
Grand *mufti* of Egypt, dated 23 March 1980,[20] i.e. six years before the opening
of the Egypt's first IVF centre and only two years after the first IVF baby in
England. According to this ruling, any technique is acceptable if the gametes
come from the married couple and the embryo is implanted in the wife without
the participation of third parties, which would make the practice similar to

---

[18] Inhorn M.C., Global Infertility and the Globalization of New Reproductive Technologies:
Illustrations from Egypt, *Social Science and Medicine*, 2003, 56, 1837–1851.

[19] Rispler-Chaim, op. cit., 20–21.

[20] International Islamic Center for Population Studies and Research, *Islamic Manual of Family
Planning*, Cairo, IICPSR, 1998, 98–99.

adultery (*zina*). IVF is also forbidden with mixed sperm from a donor and from a dead consort. For Islam, the offspring, regardless of the type of fertilisation, must be the fruit of the biological father and mother who are legally married. Adoption is unlawful. Recourse to sperm banks is judged a form of *zina* with the difference that the perpetrator is unknown; in addition, this possibility may induce a selection of gametes to produce a better race. The child born from heterologous artificial insemination is tantamount, for the jurist, to a foundling (*laqit*) or a *walad zina*, i.e. a "son of sin" (*ibn haram*) whose paternity cannot be attributed to a father but only to the mother who generated it, in the same way as a child that is the result of an illegitimate sexual relationship.[21] In addition, a man who accepts that his wife is inseminated with the semen of another man loses his honour, deserving the description of *dayyuth*, i.e. a despicable man in that he is too weak to protect his wife, allowing her dignity to be violated.[22]

In a *fatwa* by the *mufti* of the Arab Republic of Egypt, *sheikh* Abdullatif Ghani Hamza, published in the newspaper *Al-Muslimin* in 1982, it was reconfirmed that when the sperm that fertilises the egg does not come from the husband, then the insemination indirectly comes under the category of *zina* (illicit sexual relationship which also includes adultery).[23]

---

[21] One example of the serious situation in which women and any children they may have conceived outside marriage find themselves is shown by the 1984 Algerian Family Code influenced by Muslim law. The only legitimate filiation is that resulting from legal marriage thanks to which the child is related to the father (art. 41 of the Family Code). "Natural" filiation and adoption are prohibited (art. 40 F.C.). An extra-marital relationship with a subsequent pregnancy represents an offence that may be prosecuted under penal law. There is no place for a natural child in the law or in society. Article 339 of the Penal Code referring to adultery states: "Any married woman proven guilty of adultery is punished by imprisonment of one to two years. Any married man found guilty of adultery is punished by imprisonment of six months to one year." Due to social pressure, illegitimate pregnancy is deemed unacceptable. To save her honour and that of her family, the unmarried mother, often with the support of her family, gets rid of the child or abandons it near a mosque or in the street. At other times, the male members of the family ignore the affair, leaving the women the job of saving the honour of the family that they have infringed (the mother whose unmarried daughter becomes pregnant is responsible for not having educated her adequately to the social values). The girl has six months to recognise it as her own child (otherwise she can abandon it) but legal recognition does not exist as the child will never be able to bear its father's surname. The Registry office will automatically impose on it two names (e.g. Ahmed Mohamed, the last used as a patronym) but no surname except, sometimes, that of the mother. If the child is without known parents, the Family Code allows a family to take it in by means of *kafala* but without giving it their surname. The 1976 Code of Public Health had foreseen hostels for unmarried pregnant women in order to try and avoid children being abandoned. In addition, monthly aid and treatment for the immediate needs of the mother and child were also contemplated. The Code of Public Health was abrogated in 1985 without the measures to help unmarried women and their children ever being implemented and they remain unprotected by the law. See Aït Zaï N., L'Enfant illégitime dans la société musulmane, *Peuples Méditerranéens*, 1989, 48–49, 113–122. Aït-Zaï N., L'abandon d'enfant et la loi, *Revue Algérienne des Sciences Juridiques, Économiques et Politiques*, 1991, XXIX (3), 473–493.

[22] Rispler-Chaim, op. cit., 26.

[23] Ghanem I., The Response of Islamic Jurisprudence to Ectopic Pregnancies, Frozen Embryo Implantation and Euthanasia, *Med. Sci. Law*, 1987, 3, 187–190.

In 1982 the Malaysian National *Fatwa* Council approved IVF. The Council decreed that Muslims may use frozen embryos in assisted reproduction procedures. However, the embryo can only be used if a couple is legally married and both spouses are alive.

Sheikh Yusuf al-Qaradawi recalls that Islam protects descendants by prohibiting any illicit relationship of *zina* (adultery and fornication) and adoption and that the purpose of this is to preserve the offspring from any extraneous and ambiguous additions. His opinion coincides with that of Mahmud Shaltut, according to whom insemination by a donor other than the husband represents a crime that condenses both the negative result of adoption – i.e. the introduction of an alien element in descendants – and the sin of adultery. "By this action the human being is degraded to the level of an animal, who has no consciousness of the noble bonds (of morality and lineage) which exist among the members of a human society."[24]

In view of the recourse to the practice of IVF in 1984 the Farah Hospital in Amman (Jordan) had asked the *mufti* of Jordan, *sheikh* Ezzedine Al-Khatib, for an opinion and he replied with a *fatwa* on 2 January 1985. The *sheikh* recalled that the *Majlis al-Ifta* (Council of *Fatwa*) of the Kingdom had already dealt with the issue in depth, reaching the conclusion that in vitro fertilisation with embryo transfer (IVFET) is licit in the case of need, only if the insemination takes place between the gametes of a married couple; he also recalled that the surgeon performing the operation must take every precaution so that no gametes of the individual couples are exchanged, which would have serious effects for their descendants, dignity and honour. Furthermore, insemination between individuals who are not lawfully married is equivalent to "camouflaged zina" forbidden by the *Shari'a*.[25]

In 1985 in Saudi Arabia, the 8th session of the Fiqh (Muslim Law) Academy of Makkah, a body of the Muslim World League, one of the three Organisations that theoretically represent the *umma* (community of Muslim faithful) pronounced itself on artificial fertilisation,[26] in a document, the juridical value of which does not go far beyond that of a *fatwa* (juridical opinion).

The text of the Academy deems the techniques of internal and external homologous insemination (i.e. in vivo and in vitro) coherent with the *Shari'a*, whilst it considers all other systems of assisted conception forbidden (*haram*) as they come under the category of acts of *zina* (notion including both adultery and fornication) either because the gametes do not come from the couple or because the "volunteer" woman who brings the pregnancy to term is not part of the married couple, etc. In addition, both due to the uncertainties of the results

---

[24] Qaradawi (Al) Y., *The Lawful and the Prohibited in Islam*, ATP, 1994, 227–228.
[25] Botiveau B., Avortement, procréation médicalement assistée et éthique médicale: la cas de la Jordanie, *Journal International de Bioéthique*, 1998, 9 (1–2), 87–93.
[26] Borrmans M., Fécondation artificielle et éthique musulmane, *Lateranum*, 1987, 1, 88–103, in particular 91–98.

and to the risks of error which are constantly present (e.g. the exchange of sperm or embryos), recourse to assisted conception is tolerated only in cases of extreme necessity.

In order to procure the sperm for in vitro insemination, Islamic morals do not place any obstacles on masturbation. Apart from this case, the *Shari'a* does not generally consent free recourse to masturbation. Prophet Muhammad invited those who were unable to marry to control their sexual instinct through fasting (*hadith* in Bukhari). Imam Malik referred to Koran 23.5–7[27] to maintain that the emission of semen was lawful in intercourse in marriage and with one's own slaves ("(the captives) whom their right hands possess"); in the other cases it was forbidden except for necessity as masturbation allows the release of waste material from the body.[28] Ahmad Ibn Hanbal, like Ibn Hazm, considered semen in the same way as any other corporeal secretion, therefore the act was allowed. In particular, Hanbal considered it lawful for prisoners, travellers, soldiers and individuals on their own without a legal sexual partner. In general, the *Hanbalite* jurists placed two conditions, i.e. that the act was performed to avoid more serious sins (fornication or adultery) and the man did not have the resources to get married. Other *Hanbalite* jurists also tolerated female masturbation for women without a husband or alone. For the *Hanafite* school, it is allowed if, for example, the man is not married and does not intend to fornicate whilst masturbation done whilst thinking of one's own wife seems lawful.[29] Amongst contemporary jurists, Qaradawi considers masturbation possible in a context of excitement in order to avoid more serious sins. On the contrary, Abd al-Halim Mahmud judges it an ignoble action, on the risks of which many doctors have expressed their views (danger for the sight, the respiratory system, sexual activity, etc.). The remedy for the sexual drive leading to masturbation lies above all in marriage, fasting, performing various religious duties and lastly, sport and social activities that prevent thinking about masturbation.[30] In a *fatwa*, al-Sha'rawi mentions an indecent custom by young people which does not procure physical harm but of which it is better to be released by work, fasting and reading the Koran until marriage, the real cure for the problem.

During the 8th session of the Fiqh Academy of Makkah (1985) the decision adopted at the 7th session,[31] held in 1984, was also re-examined. The 7th session had admitted the last case which has not yet been discussed here, namely IVF where the gametes of husband and wife are used, after which the embryo is inserted into the uterus of another wife of the same husband (polygamous

---

[27] Koran 23.5–7: "5. Who abstain from sex; 6. except with those joined to them in the marriage bond, or (the captives) whom their right hands possess, for (in their case) they are free from blame, 7. but those whose desires exceed those limits are transgressors;"

[28] See Abd al-Halim Mahmud in Rispler-Chaim, op. cit., 132.

[29] Rispler-Chaim, op. cit., 24 and note 14–15.

[30] Ibid., 132–133.

[31] A Ruling of the Fiqh Academy of Makkah, in Gindi (Al) A.R. (ed.), *Human Reproduction in Islam (1983)*, Kuwait, IOMS, 1989, 395–399.

relationship) who voluntarily brings the pregnancy to its term as the woman who had provided the ovules does not have a uterus. In the case in question, the insertion of the embryo which is the result of the gametes of the husband and wife in another wife maintains the procreative relationship within the same family nucleus and for this reason it would be treated as a homologous technique of artificial insemination.

Furthermore, the 7th session (1984) had also clarified the juridical condition of the second mother, i.e. of the "carrier", who became the equivalent of a "milk mother" of the child. In Muslim law, a milk relationship comes about when a wet nurse feeds the child of another woman; the milk mother and father (i.e. the husband of the wet nurse as well as "owner" of the milk) become equivalent to the natural parents whilst the suckled child becomes the equivalent of a brother or sister of the other natural children of the couple "offering the service". In this case, the milk child is subject to the same rules and prohibitions with regard to marriage as for brothers and sisters based on Koran 4.23–24 which prohibits the marriage of a person with his milk sister or mother.[32] At the same time, however, milk relatives do not inherit one from another whilst milk parents do not have any obligations of protection or maintenance with regard to the child who has been breastfed.

In the Muslim world today, but depending on the area, polygamy concerns a very low number of families, between 1% and 5% of the total; this is therefore a very small minority also due to the preference shown by national authorities and jurists for monogamy, without this explicitly prohibiting polygamy, permitted by Koran 4.3 (polygamy is prohibited by the law in Tunisia and Turkey).

In order to dispute the resolution of the 7th session (1984), the 8th session (1985) analysed two extreme cases (A and B) which lead, for their juridical implications, to the refusal of embryo-transfer between wives of the same husband.[33] Therefore, (A) the Academy noted that the second wife, in whom the embryo from the first wife had been implanted, could remain spontaneously pregnant in turn following sexual intercourse in the period of the implantation. Twins would be born without it being possible to distinguish which one is the result of the implantation and which one is the result of natural intercourse, whilst it is equally improbable to determine the real mother. There may also be the case (B) where one of the two foetuses dies during the gestation and is expelled only at the birth of the other foetus: as it is impossible to recognise the implanted baby from the natural one, confusion and uncertainty would follow in recognising the filiation with respect to the real mother.

---

[32] Koran 4.23–24: "23. Prohibited to you (for marriage) are: Your mothers, daughters, sisters; father's sisters, mother's sisters; brother's daughters, sister's daughters; foster-mothers (who gave you suck), foster-sisters; your wives' mothers; your step-daughters under your guardianship, born of your wives to whom ye have gone in, no prohibition if ye have not gone in; (those who have been) wives of your sons proceeding from your loins; and two sisters in wedlock at one and the same time, except for what is past; for God is Oft-forgiving, Most Merciful; 24. also (prohibited are) women already married, except those whom your right hands possess,".

[33] Borrmans, art. cit., 96–97.

Bearing in mind that for the *Shari'a* the only valid filiation is that with respect to the paternal figure, Muslim law refuses any case of uncertainty on the determination of legitimate filiation and family relations. Refusing all promiscuity, the 8th session of the Fiqh Academy of Makkah (1985) refused "surrogate maternity" in a polygamous relationship in order to guarantee definite filiation for the child, also with regard to the figure of the mother.

The Fiqh (Muslim Law) Academy of Jeddah (which is part of the Organisation of Islamic Conference) expressed itself in the same way in 1986 at the 3rd session in Amman (11–16 October) with Resolution no. 5 (4–3) concerning test-tube babies. The First International Conference on "Bioethics in Human Reproduction Research in the Muslim World", held in Cairo in 1991, made the same pronouncement.

Naturally, in a context where there is no supreme juridical authority, it is likely that there is no shortage of specialists who nevertheless consider the practice of "surrogate mothers" within polygamy in coherence with the principles of the *Shari'a*.[34] The Fiqh Academy of Makkah at the 7th session in 1984 (see above) considered the "surrogate mother" equivalent to the "milk mother". Similarly, other *Sunni* doctors of the law consider the surrogate mother as a wet nurse and the woman who provided the ovule as the biological, real and legal mother.[35] On the contrary, according to another opinion, the real mother is the one who carries the embryo in her womb until the birth, based on the Koran "None can be their mothers except those who gave them birth" as well as the passages that underline that motherhood is sacrifice, patience, tolerance and suffering.[36]

The clear condemnation of heterologous artificial fertilisation represents the fundamental point of Muslim thought and this opinion is shared by the political, religious, juridical and scientific authorities[37] beyond the divergences on many other aspects, as well as corresponding with the sensitivity and mentality of the population. Ironically, the Lebanese *Shi'ite sheikh* M.H. Fadlallah, wondered[38]: "If we want the uterus of a bride to be fertilised by a man other than her husband,

---

[34] Murad H., Surrogate Mothers as a Solution to the Infertility of Spouses Unable to Procreate, in Serour G.I. (ed.), *Proceedings of a Seminar on Ethical Implications of Assisted Reproductive Technology for the Treatment of Human Infertility*, 25–27 August 1997, Cairo, 1997, 104–108, Massué J.-P. and Gerin G. (eds.), *Diritti umani e bioetica*, Roma, Sapere 2000, 86.

[35] Murad H., art. cit., 108.

[36] Jaber (Al) A.M.Y., Ethical Fundamentals of the Concept of Surrogate Mothers, in Serour G.I. (ed.), op. cit., 109–127.

[37] Amongst the numerous organisms that have confirmed this refusal, the Egyptian Association of Criminal Law can be mentioned, which had advocated the intervention of the legislator to discourage recourse to heteroinsemination as a prosecutable offence under criminal law on the basis of the texts of the Penal Code relative to immoral offences (Cfr. Association Égyptienne de Droit Criminel, Le Droit pénal et les Méthodes Biomédicales modernes, *Revue Internationale de Droit Pénal*, 1987, 58 (1–2), 253–256).

[38] Dupré la Tour A. and Nashabé H., *Questions de Bioéthique au regard de l'Islam et du Christianisme*, Beirut, Dar el-Machreq, 2000, 55.

why don't we allow unlawful sexual relations? It would be a preferable solution to any medically assisted reproduction."

Nevertheless, more open opinions are beginning to emerge on this rigid position in some countries (especially amongst the *Shi'ites*).[39] In March 2001 M.S. Tantawi, *sheikh* of Al-Azhar, issued a *fatwa* in which he confirmed as illicit the practice of "renting out a uterus". Fertilisation with the sperm of the deceased husband is also *"haram"* because a woman ceases to be the wife of that man when he dies. The *fatwa* was issued in reply to the opinion of the Member of Parliament Abdel Mo'ti Bayoumi, head of the Faculty of the Fundamentals of Religion at Cairo University, for whom women could rent out their uterus in the case of serious economic difficulties. These contrasting opinions were accompanied by a lively debate in the Egyptian mass media.[40]

In April 2001, the Islamic Research Council of the University of Al-Azhar (Cairo) condemned recourse to surrogate maternity which would have enabled a woman without a uterus, due to a tumour, to have a child from her own gamete and from that of her husband in the uterus of a carrier mother. However, this prohibition was not voted unanimously although it concerned a heterologous procreative technique.[41]

In 1999, a *fatwa* by *ayatollah* Ali Khamenei, the *Shi'ite* religious leader of the Islamic Republic of Iran, permitted some heterologous practices under certain conditions (see the next section).

Regarding heterologous procreation, the trend by North African and Middle Eastern couples with serious problems of sterility to go to Western countries where it is possible to have recourse to heterologous reproduction practices must not be overlooked. With an excuse (tourism, a trip to see relatives or friends, etc.) they go to hospitals where these practices are available in order to have a child; at home, their relatives do not know that the couple has problems of sterility and so few doubts are raised when the woman returns pregnant.

Amongst the techniques on which there is not yet agreement by the religious authorities is the possibility of making a woman pregnant after menopause or the possibility of using a cryopreserved embryo which is the result of the gametes of two legally married and still living spouses. In assessing the case, the risks and the complications connected with the woman's age must not be overlooked, nor must the problems involved in raising a child for a mother and father who are well on in years be ignored.[42]

---

[39] Concerning Turkey, a Sunnite country, see for example: Isikoglu M., Senol Y., et al., *Public Opinion regarding Oocyte Donation in Turkey: First Data from a Secular Population among the Islamic World*, "Human Reproduction," 2006, 21(1), 318–323, which mentions a new tolerant trend regarding oocyte donation.

[40] On this kind of debate, see Farag F., Legislating Morality, *Al-Ahram Weekly On-line*, 17–23 May 2001, no. 534; Farag F., Cost of Bearing, *Al-Ahram Weekly On-line*, 17–23 May 2001, no. 534.

[41] Serour G.I., *Attitudes and Cultural Perspectives on Infertility and its Alleviation in the Middle East Area*, www.who.int/reproductive-health/infertility: 46.

[42] Serour G.I., Reproductive Choice: A Muslim Perspective, in Harris J. and Holm S. (eds.), *The Future of Human Reproduction*, Oxford, Clarendon Press, 1998, 191–202.

In November 2000 the University of Al-Azhar in Cairo and the International Islamic Center for Population Studies and Research (IICPSR) of Al-Azhar organised an international Workshop on "Ethical Implications of ART". The conclusive Recommendations consider post-menopausal pregnancy lawful in exceptional cases, for example "by maintenance of the integrity of a child's genetic parentage".[43]

At the same Workshop, interest was shown in the lawfulness of the implantation of embryos in a widow during the period (idda) when widows are prohibited from remarriage.[44] Formally, the opinion prevailing amongst jurists, doctors and Muslim experts is that marriage is terminated by the death of one of the partners therefore it is religiously illegal to implant an embryo after the death of the husband. However, a second opinion was presented, according to which the widow's moral, psychological and economic interests would be better protected if she had a child by her late husband. The question was put to the Islamic Research Council for further study on the problem.[45]

It is to be underlined that a few years ago, Farid Wassel, the third last Grand mufti of Egypt, personally allowed a case of embryo implant in a legitimate wife after the death of her husband. This couple had decided to undergo IVFET. The husband had already provided the sperm sample and was accompanying his wife for the implant of the cryopreserved embryos. On the way, the man died in a car accident. On the request of the relatives, Wassel gave his consent to the implant of the embryos.

A similar case has been analysed in a fatwa, by the sheikh Abd Allah al-Mushidd,[46] head of the Fatwa Committee of Al-Azhar. Al-Mushidd had considered it licit, in the light of the Shari'a, to fertilise the gamete of a widow in the period of the idda, with the cryopreserved sperm of her late husband. This theory raised protests such as that of June 1989 by M.M. Shalabi, Professor at the Universities of Alexandria and Beirut according to whom al-Mushidd does not take into account that the mother could become pregnant on the last day of the idda or after the end of the idda. In these cases, the child would be born 365 days later (from the divorce or the death of the father) and would be illegitimate. This would lead to a state of anarchy. However, if the child is born within a year from the divorce or the death of the father, it still belongs to the bed (firash) of the father and there are no problems on its status as a legitimate child. There remains the fact that the absence of the husband due to death or divorce causes

[43] Serour G.I. (ed.), Proceedings of the Workshop on Ethical Implications of Assisted Reproductive Technology for the Treatment of Infertility, 22–25 November 2000, Cairo, Al-Ahram Press, 2002, 106.

[44] The idda is the period of legal withdrawal that a woman must respect after the marriage has been dissolved. The idda of a widow lasts for four months and ten days; that of a divorcee is of three monthly cycles of menstruation.

[45] Serour G.I. (ed.), op. cit., 98 and 106; Serour G.I. and Dickens B.M., Assisted Reproduction Developments in the Islamic World, International Journal of Gynaecology and Obstetrics, 2001, 74, 187–193.

[46] See Rispler-Chaim, op. cit., 23 and 129–132.

excessive risks both for the genealogy (*nasab*) of the child and the distribution of the property of the deceased. Besides, the jurists are unanimous in showing the stability of the genealogy when it is the result of a legitimate relationship between a couple in the marital bed (*firash*). In this regard, Prophet Muhammad said that "the child belongs to the bed [of a married couple] and the stone [stoning] belongs to the prostitute". For all these reasons, concludes Shalabi, artificial fertilisation remains licit only between living married couples.

In 1997, according to the Committee of Bioethics of Lebanon,[47] the IVFET of a widow or divorcee with the cryopreserved semen of the husband is unlawful as the contract of marriage terminates with the death of the husband or the divorce.

OPINIONS IN SHI'ITE ISLAM

The prohibition of the donation of gametes (sperm and eggs) from persons who are extraneous to the married couple also prevails in the *Shi'ite* world. However, the *Shi'ite* positions are more complex and differ from one another, as well as at times being distant from the *Sunni* positions. The reason is historical and theological. *Shi'ite* Islam (which accounts for some 10% of Muslims in the world and present mainly in Iran, Iraq, Pakistan and Lebanon) still considers open the possibility of a personal (but qualified) interpretation of the Sacred Sources whereas this possibility has formally been closed for the *Sunnite*s since about the 12th century. This possibility often makes the *Shi'ite* positions more liberal on many doctrinal and ethical questions compared to the *Sunnite* world.

For some *Shi'ite* jurists, only the child produced by direct adulterous sexual intercourse has no juridical relationship with its father or rights of inheritance; the child maintains a juridical relationship with its mother only. As a consequence, according to these jurists, heterologous artificial insemination and fertilisation are unlawful acts, but as the man does not play a direct part in the sexual-procreation act, the child remains legally related to the owner of the seed and all the rules concerning ascendance and inheritance valid for the other children of the man apply to the child who is the result of heterologous insemination (response 785).[48] On the contrary, for the *Sunnite*s, natural or artificial heterologous procreative sexual intercourse (both in vivo and in vitro) always leads to the birth of an illegitimate child that has no relationship with the father whilst it has a connection only with its mother and with her family.

For some *Shi'ite* jurists (response 786) homologous in vitro fecundation is lawful and the father of the child will be the owner of the sperm, however this child does not have a mother (probably as it has been fertilised in vitro instead of in the mother's womb).[49]

---

[47] Comité de Bioéthique – Ordre des Médecins du Liban, *Bioéthique et Jurisprudence Islamique, Lettre du Comité de Bioéthique*, January 1997, Beirut, 1–19.

[48] Bostani (al) A.A. (ed.), *Le Guide du Musulman*, Publication du Séminaire Islamique, 1991, 269–271.

[49] Ibid., 270.

In the opinion of some, artificially injecting sperm into the uterus of one's own wife is lawful but becomes unlawful if the person injecting the sperm is an outsider (someone other than the husband) and the injection entails touching and seeing the intimate parts of the woman (response 787). This motivation had a certain importance with those *Sunni* and *Shi'ite* Muslim "scholars" who were initially against every type of artificial technique of procreation.

For other jurists, if a woman in a lesbian relationship inserts the sperm of her husband into the vagina of a second woman who becomes pregnant, the baby will legally belong to the man as he is the owner of the sperm and the baby is not the result of direct unlawful sexual intercourse between a man and a woman. The pregnant woman will be the legal mother (response 785).

Regarding this last case, it is worthwhile mentioning the reasoning of the Lebanese *Shi'ite sheikh*, M.G. Mughniyyah,[50] who in a *fatwa* of the 1960s, uses a *hadith* by Hasan b. Ali (son of the fourth Caliph, Ali) to condemn all types of artificial fecundation (not better specified). The *hadith* spoke of a married woman who, during a homosexual relationship with a virgin immediately after intercourse with her husband, had fertilised the virgin with her husband's seed. Against this action Hasan had decreed as follows: the married woman must pay compensation for the lost virginity of the young woman; then the married woman is stoned as the relationship between the married woman and the virgin is equivalent to adultery; the younger woman then gives birth and the child is given to the father, the owner of the seed; lastly, the virgin is lashed for having committed fornication.[51]

In the light of the same *hadith* and using reasoning by analogy and the personal interpretation of the sources (which is permitted to *Shi'ites*), the *sheikh* examined the very new problem of assisted medical procreation prohibiting all types of artificial fecundation. In particular, the *sheikh* deems that artificial fecundation (not better specified) is prohibited on the basis of two observations:

1. The rules of the *Shari'a* indicate that the pudenda are lawful only with legal permission.
2. Koran 24.30 "Say to the believing men that they should lower their gaze and guard their modesty." Here God orders protecting the pudenda in general, without specifications (carnal union or other); therefore, the organ is to be

---

[50] Castro, art. cit., 222–225.

[51] Some Imamites (i.e. Duodeciman *Shi'ites*) did not agree on these rules. For example, for al-Tusi (d. 1068) and his followers, all four of these rules are valid; others refuse the second one (stoning to death) and accept the others such as al-Hilli (d. 1277) for whom the married woman should be flogged but not stoned. In fact, the majority of the Imamite *fuqaha* believe that the punishment (*hadd*) in the case of tribadism-lesbianism is of 100 whiplashes both for an unmarried and for a married woman, and for both the active and the passive parties. On the other hand, according to the *Hanbalite* Ibn Qudama (d. 1223) for the women there is no punishment *hadd* but discretionary punishment (*ta'zir*). Ibn Idris al-Shafi'i (767–820) also prefers flogging; he also refutes the seed owner's right to the baby as it was not procreated in a marital relationship or equivalent; lastly, Idris does not agree to the rule obliging the married woman to compensate the young woman for her lost virginity as the virgin gave herself spontaneously. See Castro, art. cit., 223.

protected from everything, including artificial fecundation. This passage is reinforced by 23.5–7: "5. Who abstain from sex, 6. except with those joined to them in the marriage bond, or (the captives) whom their right hands possess, for (in their case) they are free from blame, 7. but those whose desires exceed those limits are transgressors."

In the case of pregnancy following this prohibited fecundation, what would the legal status of the baby be? According to Mughniyyah the baby cannot be attributed to the owner of the seed because he did not personally have sexual intercourse in the form of marriage or a union that can be assimilated with marriage, but it is attributed to the woman who gave birth to it. Indeed, "the child, born from fornication, inherits from its mother and from her relatives and they inherit from him. If the child born from fornication is attributed to its mother, all the more reason the child born from artificial fecundation".[52]

Another *Shi'ite*, the Iraqi Grand *ayatollah* Sayed Mohsen at-Tabataba'i al-Hakim (1875–1970), maintains that the child born from fecundation (not better specified) is given to the mother; the baby is not attributed to the owner of the seed because he did not personally perform sexual intercourse both when the seed reached the woman's sexual organ and when the man's seed comes into contact with another woman in a lesbian relationship.[53]

In the positions listed so far in this section, the fact that the baby is not the result of a direct carnal contact between a man and a woman (as in the case of the techniques of artificial procreation or in a lesbian relationship, etc.) gives rise to various opinions on the status of the child. In the opinion of some, this child will legally belong to the owner of the seed and the pregnant woman will be the legal mother (response 785); for others it belongs only to the woman from whom it inherits (Mughniyyah and Tabataba'i).

The aforementioned *fatwa* (1999) by the Iranian *ayatollah* Ali Khamenei[54] deemed that the donation of gametes was not strictly prohibited. Amongst Iranian religious figures in favour of the donation of gametes, one school of thought limits heterologous procreation to the donation of ovules by a second woman when a man's first wife is unable to produce useful ones, or is unable to procreate.[55] The new couple should first of all obtain the approval of a religious

---

[52] In Castro, art. cit., 224.

[53] Ibid., 224.

[54] Inhorn Marcia C., Religion and Reproductive Technologies. IVF and Gamete Donation in the Muslim World, *Anthropology News*, 2005, 46 (2), 14 and 18. Inhorn Marcia C., Middle Eastern Masculinities in the Age of New Reproductive Technologies: Male Infertility and Stigma in Egypt and Lebanon, *Medical Anthropology Quarterly*, 2004, 18 (2), 162–182.

[55] On the debate in Iran, see also the report by Fatemi S.M.G.S., Egg Donation: A Comparative Study of Shi'i Fiqh and the Iranian Legal System, First International Egg Donation Conference, *Diversity Surrounding Egg Donation Issues*, London, UK, 7–10 February 2003; Jafarzadeh M., Ghaffari M., et al., The Attitude of Iranian Law to Egg Donation: Present Situation and Future Prospects, First International Egg Donation Conference, *Diversity Surrounding Egg Donation Issues*, London, UK, 7–10 February 2003.

court. Subsequently, the man must form a relationship with the "donor" through a temporary marriage (*mut'a*), sometimes defined a marriage "of pleasure", deemed lawful in *Shi'ite* Islam and by the Iranian Civil Code (art. 1075–1077). In this case, recourse to temporary marriage would be required exclusively to give the man a child within a marital relationship when the first wife is unable to do so. This is a consensual marriage subject to the same constraints as an ordinary marriage: the woman must not be married already; she cannot marry a non-Muslim man; the union is not valid between certain degrees of kinship.[56] The duration of the temporary marriage is established in the marriage contract between the partners and may be of one hour, one day, one month, one year or other periods, but it must not exceed the lifetime of those contracting the agreement, in which case it becomes a permanent marriage.

In 2003, the Iranian parliament approved a specific bill legalising the donation of embryos of married couples (with their consent) to other couples who are married but sterile, after permission is granted by a family court that assesses the capacity of the recipient couple. The law however, does not clearly state the position of the baby with respect to the donor and recipient couples. These ambiguities make approval by the Guardian Council problematic.[57]

The position of another prestigious *Shi'ite* authority who is well known due to the current political-military events in Iraq, namely al-Sistani, the Grand *ayatollah* of Iraq, is more rigid in his work al-Fatawa al-Muyassarah (Beirut, 2002). In his opinion, homologous artificial insemination and homologous in vitro fecundation are lawful. It is lawful to fecundate in vitro the seed of a husband with the ovules of a woman other than his wife in order then to implant the embryo in the uterus of his legitimate wife. The contrary, however, is unlawful, that is fecundation of the eggs of a woman with the sperm of a man other than her husband and then grafting the embryo into the uterus of the same woman. Lastly, it is lawful to inseminate a wife with the seed of her husband after his death; the child is attributed to its father but does not inherit anything from him.

PROBLEMS RELATIVE TO THE EMBRYO

In 1987, the conclusive Recommendations of the Symposium organised by the IOMS on "The Islamic Vision of Some Medical Practices" state that the majority of experts and scholars did not deem the surplus fertilised ovules inviolable but they could be destroyed. On the contrary, a minority maintained that the

---

[56] Abu-Sahlieh A.A., L'éthique sexuelle en droit musulman et arabe. Cas de l'Egypte, passé, présent et avenir, *Revue de Science Criminelle*, 1999, 1, January–March, 49–67.

[57] Fatemi S.M.G.S., Legalising Embryo Donation in Iran: The First but not the Last Move, in *Proceedings of the 2nd International Egg Donation Conference*, Valencia, Spain, February 2004; Azimaraghi O. and Stones-Abbasi A., Diversity of Attitudes Towards Egg Donation within the Iranian Communities, in *Proceedings of the 2nd International Egg Donation Conference*, Valencia, Spain, February 2004.

fertilised ovule represents the first phase of the new human being honoured by God; therefore, in the choice between destroying fertilised ova, using them for scientific research and leaving them to die naturally, the minority preferred to leave them to die without aggressive interventions.[58]

In 1990 (14–20 March) in Jeddah, the 6th session of the Fiqh Academy (Organisation of the Islamic Conference) decided that: it is allowed to inseminate the strictly indispensable number of eggs with the husband's semen. The surplus pre-embryos should be left to die spontaneously; the donation of fertilised eggs was judged unlawful.[59]

In the early 1990s, the Code of Ethics of the Egyptian Society of Obstetrics and Gynaecology, referring to IVFET, considered recourse to a third person unlawful as was experimentation on embryos. However, the same Society positively evaluated the cryopreservation of surplus embryos to be implanted into the mother from whom the gamete comes when the previous pregnancy has failed. This is justifiable only if, when the embryo is unfrozen and inserted into the mother's uterus, the husband is alive and the couple are married.[60] Lastly, an interesting element is that the Egyptian IVFET Centre stated that human life begins only when the embryo is implanted in the uterus and begins to take nourishment and does not begin on fertilisation.

There are other criteria on which experts agree. Free and informed consent by the couple is one of the indispensable conditions for recourse to artificial procreation. At the same time, doctors must refuse clinical cases that entail risks for any future offspring and they must be sure that there will be no negative effects, either on the parents or the children, and at the physical as well as psychological and mental levels.

More in harmony with the classic position of Muslim law, Isam Ghanem, a specialist in Muslim medical jurisprudence, recalls a well known "saying" of the Prophet, taken from Bukhari, according to which God breathes the soul into the human body only 120 days after fertilisation, with the foetus not yet being a real human being before this limit. As a consequence, in this period of time, the foetus belongs exclusively to the parents who may authorise research on it, just as they may freely dispose of its organs.[61]

The divergences that may be found between doctors and jurists, on understanding when human life begins, do not seem irreconcilable. Doctors, often more Westernised, consider that life begins with fertilisation or when the

[58] Mazkur (Al) K., Saif (Al) A., et al. (eds.), *The Islamic Vision of Some Medical Practices, Full Text of the Symposium*, 18–21 April 1987, Kuwait, Islamic Organisation for Medical Sciences (IOMS), 1989, Vol. III, 681.

[59] Decision no. 6/5/57 in IOMS (Organisation Islamique pour les Sciences Médicales), *Transplantation de certains organes humains du point de vue de la Charia, Actes du Sixième colloque organisé par l'IOMS*, Koweit, October 1989, Rabat, ISESCO, 1999, 474.

[60] Aboulghar M.A., Serour G.I., et al., Some Ethical and Legal Aspects of Medically Assisted Reproduction in Egypt, *International Journal of Bioethics*, 1990, 1 (4), 265–268.

[61] Ghanem I., Embryo Research: An Islamic Response, *Medical Science Law*, 1991, 1, 14.

embryo is implanted in the mother's uterus. The jurists, however, accept that biological life takes place with fertilisation but have traditionally directed their interest to the moment when God infuses the soul into the body, as only from that moment is human life believed to be formed. It is in fact clear that the "value" of a foetus, before ensoulment (independently of the time when this comes about) is minor compared to after ensoulment with the inevitable consequences on the rules established to regulate abortion and protect the embryo.

Regarding the problems of the techniques of assisted conception, an important point of reference for the Muslim world was established by the recommendations made at the end of the first international conference on the "Bioethics on Human Reproduction in the Muslim world" held in 1991 at the University of Al-Azhar in Cairo (Egypt), in the presence of 200 doctors of the law, theologians, doctors, sociologists, etc. The document[62] underlines the following points:

- Research on the stimulation of ovulation, IVF and artificial insemination are allowed only if the gametes belong to a married couple and on condition that the fertilised egg is transferred into the uterus of the wife to whom the egg belongs.
- The donation and commerce of sperm and eggs are forbidden.
- The transfer of the fertilised egg to a "surrogate mother" is prohibited even in a context of polygamy.
- The number of embryos transferred into the uterus must not exceed three to four.
- Any surplus fertilised eggs may be cryopreserved; they belong to the married couple and may be transferred to the wife in a subsequent cycle when the previous treatment has failed only if the marriage contract is still valid. These pre-embryos may be used for research if the couple give their consent.
- Research on pre-embryos should be limited to therapeutic research with the consent of the partners. Non-therapeutic research requires the free and informed consent of the couple and these pre-embryos may not be transferred to any uterus.
- Research with the aim of amending the hereditary characteristics of the foetus is prohibited, including to choose its sex.

The Islamic Medical Association of North America (IMANA) also considers that the surplus embryos produced by IVF between husband and wife "can be discarded or given for genetic research".[63]

Regarding foetal sex selection, the positions reported in the conclusive Recommendation of the Seminar on "Human Reproduction in Islam" (IOMS, Kuwait, 1983) were more articulate: "Foetal sex selection is unlawful when it is practised

---

[62] Serour G.I. (ed.), *Ethical Guidelines for Human Reproduction Research in the Muslim World*, Cairo, IICPSR, 1992, 29–31.

[63] IMANA Ethics Committee, *Islamic Medical Ethics: The IMANA Perspective* (PDF format), in www.imana.org, 1–12.

at a national level, while on an individual basis, some of the scholars participating in the seminar, believe there is nothing legally wrong with the attempt to fulfil the wish of a married couple to have a boy or a girl through available medical means, while others scholars believe it is unlawful for fear that one sex might outnumber the other."[64]

There was also discussion at the Workshop organised at the University of Al-Azhar in November 2000 on "Ethical Implications of ART" on the non-medical use of preimplantation genetic diagnosis (PGD) to select children's sex. The participants condemned recourse to PGD with a discriminatory purpose against females, but specified that total refusal of the practice entails serious risks for mothers in Middle Eastern societies, especially when the birth of a male child remains fundamental for the psycho-physical and social well-being of the mother and family. The use of PGD for family balancing can therefore be deemed lawful in individual cases.[65]

In 1997, the Committee of Bioethics of Lebanon recalled that IVFET is also lawful when the husband is Muslim whilst the wife is Muslim, Christian or Jewish, in order to "guarantee" Muslim offspring according to the rules of the *Shari'a*. The document[66] recommends fertilising only the necessary number of eggs; the surplus embryos remain the property of the couple who can decide to let them die or dispose of them for therapeutic research. The reduction of a multi-foetal pregnancy is also lawful if it is the only way to guarantee a normal pregnancy or to save the foetus or if the mother's life is at risk.

In the case of absolute sterility of the husband, amongst the new techniques to overcome this obstacle, discussion has begun on reproductive human cloning performed during the period of validity of the marriage contract, with the transfer of the nucleus from the husband's somatic cell to the enucleated ovum of his wife. Gamal Serour is in favour of this solution in extreme cases.[67]

The same technique was discussed at the aforementioned Workshop on "Ethical Implications of ART for the Treatment of Infertility" (Cairo, 2000) but no common ethical positions emerged on its use.[68]

SOCIETY AND LEGISLATION

The costs of new reproductive techniques (NRT) in "developing countries" seem much higher than in developed nations. In addition, local politicians do not consider infertility as an illness and this leads to the reluctance by the nations to cover (even partially) the expenses for the "treatment" of sterile individuals.

---

[64] *The Seminar on Reproduction in Islam. Minutes of the Recommendation Committee*, in IOMS, *Human Reproduction in Islam (1983)*, (Ed. al-Gindi A.R.), op. cit., 274–277.

[65] Serour G.I. (ed.), *Proceedings of the Workshop on Ethical Implications of Assisted Reproductive Technology for the Treatment of Infertility*, 2000, op. cit., 96–97 and 106.

[66] Comité de Bioéthique – Ordre des Médecins du Liban, op. cit., 12–13.

[67] Serour G.I., *Ethical Implications of Human Embryo Research*, Rabat, ISESCO, 2000, 23.

[68] Serour G.I. (ed.), *Proceeding of the Workshop on Ethical Implications of Assisted Reproductive Technology for the Treatment of Infertility*, op. cit., 107 and 102–103.

The first successful experiments of in vitro fertilisation (IVFET) in the Arab world were performed in Jordan, Egypt and Saudi Arabia from the mid-1980s. In Jordan, this was possible following a *fatwa* issued by the *mufti* of the Hashemite kingdom, *sheikh* Ezzedine Al-Khatib in 1985, who expressed himself in favour of the practice on condition that the parents were married; consequently, two male twins were born to two mothers. Since then, many other countries have opened similar centres, mainly private, because of their high running costs which, inevitably, limit recourse to them to more affluent parents. Despite the fact that the modern medical centres are often capable of using the most advanced conception techniques, artificial insemination remains little used due to the mistrust aroused in the masses on factors of a religious, cultural, social and, last but not least, economic order. Furthermore, these countries have to deal with more important priorities in the social and health field, such as demographic growth.

Egypt appears the leading Muslim country in the Middle East in the application of reproductive technologies but in a social context in which the opinion and effective willingness of citizens with regard to these techniques is contradictory. In other words, the local socio-economic, religious and political elements influence the perception of these Western reproductive practices. Due to the importance of the role played by Egypt in the area, we will dwell on these aspects.

The first IVFET centre in Egypt was opened in 1986 and in 1999 there were already 35 similar centres. In a book, Inhorn has described[69] several particular obstacles that inhibit or limit recourse to NRT in Egypt; this case-study – with the appropriate differences and due caution – can be applied to other Arab countries. We will dwell here on four of these obstacles: (1) cultural beliefs, (2) social class, (3) sex, (4) religion.

1. A significant cultural motivation for the refusal of NRT in Egypt seems to be entrenched in specific beliefs on the nature of human reproduction and on reproductive physiology. The contemporary Western procreative version involves a "duogenetic" model of procreation characterised by an equal contribution of spermatozoa and ovules, man and woman. This cultural model does not appear to be widely accepted by the majority of Egyptians from a poorer and illiterate background. They often agree with a "monogenetic" vision of procreation where women do not make a significant contribution.[70]

2. NRT are inaccessible to poor and middle-class sterile women also because today all the Egyptian IVFET centres, except one, are private therefore patients even have to pay for the drugs. The only exception is the IVFET Centre of the Shatby Hospital of the University of Alexandria where, however, a very limited number of poor women are treated.[71]

---

[69] Inhorn M.C., *Local Babies, Global Science: Gender, Religion and In Vitro Fertilization in Egypt*, New York, Routledge, 2003.

[70] Ibid, Global Infertility and the Globalization of New Reproductive Technologies: Illustrations from Egypt, art. cit., 1845.

[71] Ibid., 1855–1856.

    In other words, the patients who have recourse to NRT generally belong to the upper echelons of society and are of a high cultural level. The doctors also mainly belong to an elite who specialised in the West and have a more modern mentality than that found on average in the population.

3. The relations between the sexes and marital dynamics also influence access to NRT in the higher classes.[72] The sterile woman of any class fears in particular Muslim family law which easily grants the man unilateral divorce if the woman is sterile[73]; the wife of a sterile husband could also divorce (but having to appeal to the Court) but risks social stigmatisation which she tends to avoid except in unbearable situations (e.g. violence). In any case, many men prefer to divorce their sterile wives rather than undergo tests, make the necessary attempts and bear the costs required from the couple for IVFET.

    The introduction into Egypt in the 1990s of the technique of the "intracytoplasmic sperm injection" (ICSI)[74] has increased the number of clinics practising IVF but also, indirectly, the number of divorces. As many of the wives who remained for years with a sterile husband are too old to produce usable eggs for a microinjection, the ICSI puts the sterile Egyptian woman and the more mature sterile woman in precarious conditions as husbands may prefer to divorce and choose a younger woman to attempt paternity with the new technique.[75]

4. Muslim religious sources exalt maternity and filiation. However, the sterile woman who has recourse to NRT to become a mother can opt only for homologous practices whilst legal adoption is prohibited by Muslim law and by the positive law of modern nations.

    The apologetic approach of the Egyptian mass media (and in many other Arab-Muslim countries), starting with the presentation of films and articles aiming to highlight the moral decline of America and Europe characterised by a sample of immoral and ridiculous NRT, deserves separate consideration.[76] The aim of this is always oriented towards exalting Muslim morals which prevent similar degeneration as mothers can place full confidence in excellent and attentive Muslim doctors.

    The obsessive reporting of any laboratory mistakes also has the effect of inhibiting Egyptian couples from seeking a technological solution to the lack of children in the West, but also in Egypt or in other Muslim countries.

    The majority of Muslim countries with centres for medically assisted reproduction have not yet passed governmental decrees or legislation regulating their practice; exceptions include Tunisia, Turkey, Saudi Arabia and Egypt. In Jordan,[77]

---

[72] Ibid., 1847.

[73] Farahat M.N., The Legal System of Muslim's Marriage in Egypt: Some Keynote Remarks, *Mediterranean Journal of Human Rights*, 1997, 1 (2), 203–209.

[74] This is the microinjection of a sperm cell into an ovocyte.

[75] Inhorn, Global Infertility and the Globalization of New Reproductive Technologies: Illustrations from Egypt, art. cit., 1846.

[76] Ibid., 1847.

[77] Botiveau, art. cit., 90.

the positive law does not regulate these techniques, encouraging their development on condition that they do not oppose the limits established by medical deontology. The Farah Hospital of Amman performs some 3,000 IVFET each year, and some 1,000 children have been born since 1984; due to the fame created by its activity, many parents come from other Arab countries, especially from the Persian Gulf (70%). The hospital has given itself rules of self-control through an Ethical Committee made up of eight Jordanian doctors from outside the hospital who have the task of examining the problems that arise. In addition, foreign experts give their advice on more sophisticated operations and, lastly, the *fatwa* of the *mufti* of the Kingdom dated 2 January 1985 is used as the document approving the hospital's activity. Five other hospitals practise IVF in the country.

In the Saudi kingdom, in Indonesia and in Malaysia, authorisation for the creation of new centres comes from government bodies. In Lebanon and other countries, the directives of the professional bodies are followed.

The path taken by the Arab Republic of Libya is emblematic. Libya was the first Arab country to have legislation on this issue, law 175 of 7 December 1972 art. 403 a–b of the Penal Code which prohibited recourse to any technique of assisted procreation. The law reconfirmed its roots in the principles of the *Shari'a* beginning with the following words: "In the respect of the rules of the noble *Shari'a*". According to the law, anyone who obliges a woman to undergo a technique of artificial insemination risks up to 10 years imprisonment; if the woman is consenting, the penalty for the offender is halved. The penalty does not exceed 10 years imprisonment if the "crime" has been committed by a doctor. The woman who consents or who practises the technique on herself risks up to five years imprisonment. Her husband is punished with the same penalty if he is consenting, independently of the fact that the practice was performed by his wife or by an outsider.[78] Nevertheless, in 1986, law 17 was passed concerning medical responsibility through which homologous artificial insemination and homologous IVFET in the case of necessity are accepted, with the consent of the couple and when the marriage contract is in force.

In all Muslim countries where there is recourse to these practices, it is exclusively with married heterosexual couples.

Some legislations specify the number of embryos that can lawfully be transferred in the case of IVFET. For example, in Saudi Arabia no more than 3–4 embryos are allowed. It should be noted that Saudi Arabia is one of the countries in the world with the highest rate of babies born through IVF (4% against 1.1% in the USA).[79]

Cryopreservation of embryos is practised[80] in Iran, Jordan, Lebanon, Pakistan, Egypt, etc. In Saudi Arabia and Turkey, the technique is regulated by the law;

---

[78] Mayer A.E., Libyan Legislation in Defense of Arabo-Islamic Sexual Mores, *The American Journal of Comparative Law*, 1980, 2, 287–313.

[79] Friend T., Saudis take Lead on Stem-Cell Cloning, *USA Today*, 7 August 2002.

[80] Schenker J.G. and Shushan A., Ethical and Legal Aspects of Assisted Reproduction Practice in Asia, *Human Reproduction*, 1996, 11 (4), 908–911.

however in Saudi Arabia the maximum duration of cryopreservation is not fixed by law. In the Saudi kingdom, the cryopreservation of ovocytes, which is also tolerated in Egypt, is regulated. In the latter country, the ovocytes are preserved until the death of the woman, similarly to the preservation of the male sperm which is kept until the man's death.

Recourse to micromanipulation, especially the ICSI technique is very important in Muslim countries in the cases of severe male infertility as it is not lawful to have recourse to a sperm donor. ICSI is regulated by the law and/or practised in Egypt, Saudi Arabia, Turkey, Iran[81], Tunisia, Jordan, etc.

In 1997, the National Committee of Medical Ethics of Tunisia, in the absence of legislation, recalled that the embryo is a potential person; the frozen embryo should not be kept after the period of fertility of the couple or the death of one of the partners after a limit fixed by the law. Sperm banks are accepted, for example, to preserve the gametes of young people who are to undergo operations or therapy entailing sterility.

One example of the role kept directly or indirectly by the principles of family law based on the *Shari'a* in contemporary legislation is evident in the Law no. 93 of 7 August 2001 in Tunisia regulating reproductive medicine, with the law applying the most important positions formulated by the Muslim jurists, adding however some original positions in order to cope with particular situations. Article 4 limits recourse to reproductive medicine exclusively to a living married couple, of procreative age and with gametes belonging to them. According to art. 5, the fertilisation of gametes and the implantation of the embryo must be performed only in the presence of the members of the couple and after having obtained their written consent. Recourse to a third donor of gametes and the donation of embryos are prohibited. It is forbidden to insert an embryo in the uterus of another woman (art. 14–15). No. 6 allows an unmarried individual undergoing treatment that is harmful for the procreative capacity to freeze gametes to be used after marriage. Recourse to cloning techniques is forbidden (art. 8). Article 11 specifies that freezing the gametes or embryos is licit for therapeutic purposes. Freezing requires the written consent of the couple. The cryopreserved gametes and embryos are kept for five years, and may be extended for a further five years on written request of the person concerned (in the case of gametes) or of the couple (in the case of an embryo). At the end of this period (without extension) or in the event of the death of one member of the couple, the gametes must be destroyed and the freezing of the embryo suspended. After the decree of divorce, the two partners or one of the two may request the court to suspend the freezing of the embryos.

On 27 August 2002, the Iranian Parliament authorised IVF for couples unable to have children due to sterility. However, the bill had to be approved by the Council of Guardians. In 2003 the Parliament made sperm donation illegal.

---

[81] See for example Khalili M.A., Manouchehri M.A., et al., Treatment Outcome Following Intracytoplasmic Injection of Sperm Retrieved from Ejaculate, Epididymis or Testis of Infertile Men, *Archives of Iranian Medicine*, 2004, 7 (3), 232–236.

CONCLUSION

Sterility is very often experienced as a serious handicap, especially by women. The existence of children is fundamental for the psycho-physical well-being of Muslim parents and for the dignity of the family. The forms of "adoption" contemplated by Muslim law and by the law of Muslim countries are not very common and are not well accepted socially.

In the Muslim world, all heterologous procreation techniques are considered equivalent to the practice of *zina*, therefore prohibited; any children produced have the status of illegitimate children, i.e. without a legal relationship and inheritance rights with the father.

The discussion is more lively on a number of issues: the possibility of making a woman pregnant after menopause; the use of cryopreserved embryos by partners legitimately married; recourse to IVF or to cryopreserved embryos on a widow or divorced woman in the period of *idda*.

For some *Shi'ite* scholars, the absence of a direct carnal contact between unmarried partners in heterologous fecundation is fundamental; for them, when there is no carnal contact, the child remains related to the owner of the seed whilst the pregnant woman is the legal mother. For others (the majority) the child belongs only to the pregnant woman from whom it inherits.

For some *Shi'ites*, homologous IVF guarantees a legal father but not a mother for the child, probably because the fecundation has taken place in vitro.

Again in *Shi'ite* Islam, the positions expressed by some Iranian clerics and by the Great *ayatollah* of Iraq, al-Sistani, appear original.

In general, there is no unanimous opinion on the use, experimentation and destruction of surplus fertilised ovules. The opinion of leaving them to die naturally as a sign of respect for nascent life would appear to be that of the majority at present.

As far as the cryopreservation of surplus embryos is concerned, the opinion generally seems to be in favour of this. The embryos belong to the married couple they originate from and can be reimplanted only in the wife if the marriage contract is still valid.

Therapeutic research on embryos in the early phases is generally accepted with the parents' consent.

The criterion of the infusion of the soul can be decisive to allow experimentation or the use of pre-embryos before animation takes place.

THE DEVELOPMENT OF ORGAN TRANSPLANTS

ETHICAL-JURIDICAL PRINCIPLES   161
SOME FEATURES OF THE DEBATE   163
TRANSPLANTS FROM LIVING DONORS   168
TRANSPLANTS FROM CORPSES   170
THE DEBATE ON THE CRITERIA OF DEATH   174
THE ORGAN TRADE   178
UTERINE TRANSPLANTATION   180
XENOTRANSPLANTATION   181
SOME NATIONAL LEGISLATIONS   183
CONCLUSION   196

ETHICAL-JURIDICAL PRINCIPLES

Organ transplants in Muslim countries first started in the early 1970s and caused considerable popular concern as they raise at least two juridical-theological post-mortem problems:

1. Until the Resurrection of bodies on the Day of Judgement (one of the fundamental beliefs in Islam) Muslim law prescribes burial of the deceased as soon as possible and prohibits cremation and any mutilation of the corpse.
2. The Creator is the sole owner of everything, including the human body. Man merely exercises a sort of trusteeship on the latter, in other words, a conditioned ownership for which he is responsible before God.

After the initial doubts raised by jurists and in the public opinion by the introduction of transplants, there has been a progressive approval of these operations, even if some opposition remains to the practice in general or limited to some circumstances (especially transplants from corpses).

Both the principles deemed favourable and contrary to transplants are taken from Muslim law; thanks to reasoning by analogy (*qiyas*), it is easy to adapt these principles to concrete situations. Amongst the most widespread criteria and principles in favour of transplants, the following can be observed:

1. Koran 5.32 "If any one saved a life, it would be as if he saved the life of the whole people." The Word of God relates perfectly – for the purpose of justifying transplants – with two "sayings" of Muhammad quoted in the Islamic Code of Medical Ethics (Kuwait, 1981). The first states: "The faithful in their mutual love and compassion are like the body; if one member complains of an ailment all other members will rally in response"; the second recalls that the faithful are like the bricks of a house that support one another.

Dariusch Atighetchi, Islamic Bioethics: Problems and Perspectives.
© Springer Science+Business Media B.V. 2009

2. A *hadith* of the Prophet collected by Bukhari states: "There is no disease that Allah has created, except that He also has created its treatment."[1] This "saying" is interpreted as an explicit invitation to doctors to discover the treatment necessary to eradicate all pathologies.

3. The juridical principle that states: "Necessity is an exception to the rule and makes what would otherwise be prohibited licit"[2] allows overcoming the juridical obstacles which, in certain cases, interfere with saving lives; for this purpose, this principle may be privileged with respect to other *ahadith* such as that which states: "Allah has sent down both the disease and the cure, and He has appointed a cure for every disease, so treat yourselves medically, but use nothing unlawful."[3]

4. The principle of the "lesser evil" according to which harming (Ar. *darar akhaff*) a corpse (violating the deceased to remove an organ) is tolerated to prevent greater harm for a living individual who would die without that organ.

5. Caring for the ill comes under the responsibility of society and the donation of organs may be considered a social obligation (Ar. *fard kifaya*); if someone dies because an organ cannot be found, society is responsible. This criterion is based on the analogy with the behaviour of Omar, the second Caliph (d. 644), who ruled that if a man dies of hunger because society does not help him, then the latter must pay monetary compensation (*diyya*) as the penalty for its responsibility.

If everything belongs to God, can man donate an organ of his own? On this topic, a unanimous position was approved by the Indonesian Council of Ulama, together with the Indonesian Forum for Islamic Medical Studies and the Federation of the Islamic Medical Association (FIMA), when they met in Djakarta on 30 July 1996, in the final resolution of the International Seminar on Organ Transplantation. The Resolution specifies that organs belong to God (Koran 2.195 and 4.29) but, as divine assets were created for the benefit of men (Koran 2.39), man can use his organs for the benefit of the community. The interest of the community should be protected by the government which has the duty of guaranteeing stipulations of donation (both from living individuals and from corpses) in the respect of the indications of doctors.[4]

---

[1] Bukhari, *Sahih*, *Medicine*, Vol. 7, Book 71, no. 582, www.usc.edu/dept/MSA/fundamentals/hadithsunnah/bukhari/sbtintro.html

[2] Santillana D., *Istituzioni di diritto musulmano malichita con riguardo anche al sistema sciafiita*, Roma, Istituto per l'Oriente, 1926–1938, Vol. I, 68.

[3] Abu Dawud, *Sunan*, *Medicine*, Book 28, no. 3865, in www.usc.edu/dept/MSA/fundamentals/hadithsunnah/abudawud/satintro.html

[4] The Indonesian Ulema refer to three positions in Muslim patrimonial law. (1) The advocates of the school of thought defined *milk(u)-al- raqabah* (right over the substance or body, nude property) consider men to be the owners of their body and organs; men therefore are entitled to sell and lend their organs. (2) According to the advocates of *milk(u)-al-manfa'ah* (right to enjoyment of the thing, dominium utilitatis) man is entitled to use his organs and lend them, but not to sell them.

The opponents of transplants mainly base their arguments on three principles: the sanctity of the body and of human life; our bodies are given to us in trusteeship; the body is reduced to an object-material end. The most explicit support for this orientation is that expressed by the popular Egyptian preacher, the *mufti* al-Sha'rawi (in 1987), against every type of transplant as: (1) explantation from a corpse is the equivalent of mutilation therefore it is prohibited by the *Shari'a*; (2) explantation from a living individual still harms the donor; (3) the body is the property of God.[5]

The three juridical principles quoted refer to specific words of God and the Prophet and are interpreted unfavourably for transplants as is the case, for example, with a *hadith* of the Prophet (collected by Abu Dawud and Ibn Maja) that maintains that "Breaking a dead man's bone is like breaking it when he is alive"[6]. In actual fact, in the light of the principle of necessity, this *hadith* would have the intention of underlining that no unjustified harm must be inflicted on the deceased whilst certain "necessary" treatments would not come under those that are prohibited.[7] Islam nevertheless teaches that the corpse must always be the object of respect.

Another passage in the Koran used by the opponents of the practice is 2.195 in which God invites man not to destroy himself with his own hands. The interpretation in question includes the self-destruction of one's own organs by means of transplants. The criticism of this interpretation[8] underlines that donation is self-destructive only when it involves single organs of a living person making it the equivalent of suicide. On the other hand, if the organs are in pairs, it is the refusal of a possible donation of one of them that destroys the life of the potential needy recipient.

SOME FEATURES OF THE DEBATE

The type of objections raised by jurists and theologians against a new diagnostic-therapeutic practice can transcend the competences of doctors.

---

(3) According to the school supporting *milk(u) al-intifa* man is entitled to use his organs for his own benefit but does not have the right to lend them or sell them. The position of the Resolution is based on the third line of thought. See *Resolution of International Seminar on Organ Transplantation and Health Care Management from Islamic Perspective*, www.members.tripod. com/PPIM/ resolusi.htm

[5] On this specific aspect, Sha'rawi specified: "as far as the permissibility of donating human organs in case of death is concerned, we may say that if donation is prohibited for a man in his life, the very case, with greater reason, also applies in his death, because since man does not own his body in life, his inheritors do not, with greater reason, own his body after death". See Tantawi M.S., Judgement on Sale or Donation of Human Organs, in Mazkur (Al) K., Saif (Al) A., et al. (eds.), *The Islamic Vision of Some Medical Practices*, Kuwait, Islamic Organisation for Medical Sciences (IOMS), 1989, 287–296

[6] Abu Dawud, *Sunan, Funerals*, Book 20, no. 3201, in www.usc.edu/dept/MSA/fundamentals/ hadithsunnah/abudawud/satintro.html

[7] Qattan (Al) M.B.K., Islamic Jurisprudential Judgement on Human Organ Transplantation, *Saudi Medical Journal*, 1992, 13 (6), 483–487.

[8] Babu Sahib M.M.H., The Islamic Point of View on Transplantation of Organ, *Islamic and Comparative Law Quarterly*, 1987, VII (1), 128–131.

The *Shari'a* (the Law of God) divides humanity into two categories, the faithful
and the infidels. The former include Muslim believers whose specificity is sanctioned by the Koran 3.110 when it says: "Ye are the best of Peoples, evolved for
mankind." The infidels, on the other hand, are divided into the People of the Book
and idol-worshippers. The former are those who believe in a revealed book (Jews
and Christians) and who are recognised, in Muslim territory, as having the juridical status of *dhimmi* i.e. "protected" although being partially discriminated against
with respect to Muslims; the latter are the polytheists and idol-worshippers against
whom there should only be a situation of war with the aim of their conversion or
death.[9] Similarly, the territory is also divided into Muslim territory (*dar al-Islam*)
and the territory of war (*dar al-harb*).

On the subject of transplants, this classic juridical division of humanity into
the faithful and infidels arouses considerable interest: some "scholars" have
wondered whether donors and recipients of the organ have to belong to the
same religion or if between Muslims themselves justified differences of treatment can be made on some juridical-religious categories (e.g. in the case of
those guilty of very serious offences). In fact, whilst the criterion of choice of
the organs is essentially clinical for a doctor, other types of evaluation may be
followed by Muslim jurists. For example, a *fatwa* pronounced by the University
of Al-Azhar in the early 1970s maintained that the transplant of cornea from
a Muslim or from an infidel was allowed if the organ (although the cornea is
actually tissue) was given to a Muslim but prohibited if taken from a Muslim
to give to an infidel.[10] According to the Iranian *ayatollah* S. Makarem, it is lawful to donate organs to non-Muslims except when the recipient is at war with
Muslims.[11] In a *fatwa* of 24 June 2002, the authoritative *Sunni sheikh* Yusuf
al-Qaradawi declared it was unlawful for the Muslim to donate organs to a
non-Muslim that attacks Islam; similarly, it is forbidden to donate organs to an
apostate as he is a traitor to his religion and people,[12] which is the reason why
he deserves death.[13] In addition, if both a Muslim and a non-Muslim require
an organ or a blood donation, the Muslim has precedence according to the
Koranic verse that states; "The Believers, men and women, are protectors one
of another."[14]

---

[9] D'Emilia A., *Scritti di Diritto Islamico*, Roma, Istituto per l'Oriente, 1976, 27.

[10] Antes P., Medicine and the Living Tradition of Islam, in Sullivan L.E. (ed.), *Healing and Restoring*, New York, Macmillan, 1989, 173–202.

[11] Raza M. and Hedayat K.M., Some Sociocultural Aspects of Cadaver Organ Donation. Recent
Rulings from Iran, *Transplantation Proceedings*, 2004, 36, 2888–2890.

[12] Islam accepts religious freedom but does not tolerate abandoning the last and perfect revelation
to embrace other faiths. As apostasy is an offence to God and damaging for the entire community, the *Shari'a* can even prescribe death for apostates (*murtadd*). When these extremes are not
reached, Muslim law can prescribe the loss of juridical capacity (i.e. loss of property, marriage
is invalidated, etc.) for the *murtadd*.

[13] Qaradawi (Al) Y., *Donating Organs to non-Muslims*, in www.islamonline.net, Fatwa Bank.

[14] Ibid.

The Syrian *sheikh* Yaqubi deems that donors should be sought, in the first place, from amongst the *harbis*, namely the members of the enemy party as their body can be violated or from amongst those who deserve death for serious crimes such as murder and apostasy.[15] This interpretation is also close to that of S. Tantawi, Grand *mufti* of the Arab Republic of Egypt (1990), according to whom it is licit to take organs from the corpses of those sentenced to death after execution.[16] A serious offence punishable by death seems to reduce the rights of the offender to the integrity of his body; both Yaqubi and Tantawi insinuated that explantation is dangerous for the deceased and so it should preferably be carried out on those who are dangerous for Islam (e.g. non-Muslim enemies) or on those who are condemned to death by the Law (Muslims guilty of serious offences) as it is less problematic to handle their bodies to the advantage of the community. A view similar to that shared by Yaqubi can be identified in the law of 1972 (and law no. 43 of 20 December 1996) on organ transplants in the Arab Republic of Syria where, amongst the conditions allowing explantation from a corpse, we find that the operation is licit if death is the result of capital punishment (art. 3.3).[17] In July 1992, the press agencies reported the news of the execution in Egypt of a convict from whom, immediately after his death, the kidneys and heart were explanted to save Egyptian citizens[18]; in this case, explantation from a man sentenced to death (i.e. for very serious offences) can be configured as a post-mortem expiation-atonement by the offender, with respect to the community.

References to the distinction made by the *Shari'a* – Muslims, People of the Book (*dhimmis*) and infidels (*kafirs*) – can also be found in the Decision on transplants no. 99 of the Senior *Ulama* Commission (Saudi Arabia) of 25 August 1982 in which they unanimously maintain the lawfulness of self-transplantation for a Muslim or for a *dhimmi*; the majority accepted explantation from a corpse or from a living person but for the benefit of a Muslim.[19] Regarding the trips to the USA by Saudi patients to receive organs from corpses, Otaibi and Khader stated that this implies a great economic cost as well as entailing considerable "social disruption".

A subsequent Saudi document introduces an ethical pre-selection of the beneficiary of a transplant. This is the case of resolution no. 26 (1/4) of the Council of the Academy of Islamic Law (Jeddah, IV Session, 6–11 February 1988) where point 2 of the "Definitions and classifications" requires the beneficiary of a transplant to be leading an honest existence in accordance with the *Shari'a*. According to the Algerian A. Ossoukine as well, the jurists demand that the

[15] Rispler-Chaim V., *Islamic Medical Ethics in the 20th Century*, Leiden, Brill, 1993, 35.
[16] Ibid.
[17] Syrian Arab Republic, Law no. 31 of 23 August 1972 on the Removal and Transplantation of Organs from the Human Body, *International Digest of Health Legislation*, 1977, 28, 132–134.
[18] Ennaifer H., Éthique Médicale, Universalité et Culture, *IBLA*, 1993, 56 (172), 221–232.
[19] Otaibi (Al) K., Khader (Al) A., et al., The First Saudi Cadaveric Kidney Donation, *Saudi Medical Journal*, 1985, 6(3), 217–223.

people who can benefit from the explanted organ are those who have shown evidence of a worthy and exemplary life.[20] *Sheikh* al-Qaradawi specifies that, in the case of necessity of an organ, a good Muslim has precedence over a sinner who would use the divine gift (the organ) to disobey God and harm his brothers in religion.[21]

This "distinction" between the organs of Muslims with respect to those of infidels is the effect of a literal application of specific juridical principles of the *Shari'a* to the donation of organs (see the "political dimension of Islamic bioethics", chapter 2) and could lead to discriminatory attitudes towards some patients. In this regard, the account of the Saudi *sheikh* al-Qattan is interesting, in that he privileges the patient's belonging to the *umma* (the community of Muslims) over the consideration of the health of the patient, independently of his religious faith. More specifically, he states: "Muslim Law exhorts the members of the Muslim Nation to reinforce its bonds with love and charity so that they become similar to a single body. The donation of an organ of the body to save the life of one of your brothers without causing harm to yourself represents the apex of the unity of the nation and a good example of cooperation between its members."[22] If we consider that altruism is one of the cornerstones of Muslim medical ethics, al-Qattan applies this principle above all to the Muslim community.

The different behaviour to be followed with Muslims and non-Muslims can also be seen in the *Shi'ite* context as shown by a collection of contemporary religious decrees.[23] No. 784 states: It is licit to amputate the corpse of a non-Muslim or of someone of whom it is not known whether he is a Muslim or not, in order to transplant an organ or anything else into the body of a Muslim; this rule is similar to that which makes explantation from an impure animal (*najis*) to a person lawful. No. 781 forbids explantation from the corpse of a Muslim except to save the life of another Muslim but the person carrying out the operation, in order to remedy the violation of the corpse, must pay the *diyya*, the monetary compensation.

At this stage it is essential to note that similar potentially discriminatory attitudes (at least regarding non-Muslims) appear a minority as shown, still in Saudi Arabia, by M.A. Albar, according to whom the donation of organs is an act of charity, altruism and love towards humankind.[24] All the opinions of this kind return to the humanitarian inclination of medical ethics connected with specific passages in the Koran (e.g. 5.32) and the *Sunna* oriented in this direction.

---

[20] Ossoukine A., Les prélèvements d'organes vus par le fiqh islamique, *Journal International de Bioéthique*, 1998, 9 (1–2), 67–73.

[21] Qaradawi, *Donating Organs to non-Muslims*, op. cit.

[22] Qattan, art. cit., 486. Similar positions are also upheld by Shahat (El) Y.I.M., Islamic Viewpoint of Organ Transplantation, *Transplantation Proceedings*, 1999, 31, 3271–3274.

[23] Bostani (al) A. (ed.), *La guide du musulman*, Paris, Publication du Séminaire Islamique, 1991, 268–269.

[24] Albar M.A., Organ Transplantation – An Islamic Perspective, *Saudi Medical Journal*, 1991, 12(4), 280–284.

However, in principle, this latter view does not necessarily contrast with the (nevertheless widespread) orientation of those who prefer to have recourse in the first place to Muslim donors. For example, the Comité de Bioéthique and the Ordre des Médecins du Liban have spoken from this standpoint, specifying that any organ from a non-Muslim can be transplanted into a Muslim if organs of other Muslims are not available.[25] Lastly, some consensus is still raised by the condition (of arduous application) put forward by some jurists who accept the transplant from a Muslim to a non-Muslim but, in the event that the latter decides to be cremated after death, the Muslim's organ must first be re-explanted and buried separately to respect the prohibition of the *Shari'a* on cremation.

In general regarding blood transfusion, the criterion of necessity allows blood transfusions from non-Muslims according to the orientation of the *Hanafite*, *Shafi'ite* and *Hanbalite* schools, whilst the *Malikites* allow it if a Muslim donor is not available.[26] A *fatwa* by the Religious Rulings Committee of Al-Azhar (published in 1368 H., AD 1949) recalls that when the recovery or life of a patient depends on a transfusion, the practice is lawful "even if he [the donor] is non-Muslim".[27] Amongst the opinions of the individual doctors of the Law, *mufti* Muhammad Shafi judges transfusion a forbidden practice (*haram*) as blood is part and parcel of the human body and, furthermore, blood is impure (*najas*).[28] However, the practice becomes permissible (*ja'iz*) when life is endangered or recovery is impossible without a transfusion. According to *sheikh* A.F. Abu Sanah a blood transfusion is lawful as it is not prohibited by the Sacred Sources and it is also indispensable to save human lives.[29] Abu Sanah is also favourable to blood blanks. Both judge buying blood unlawful, except when it is the only solution available to save a life. This specific position is also agreed with by al-Sistani,[30] the Grand *ayatollah* of Iraq.

Regarding exchanges with the faithful of a different religion, M. Shafi deems it preferable to avoid receiving blood from a non-Muslim, by analogy with the fact that it is preferable for a woman with a bad character not to breastfeed a child as the blood can convey bad inclinations.[31] According to the jurist Abd Allah al-Basam, the body of a Muslim and of an infidel are pure both alive and dead. This makes the organ transplant from non-Muslims to Muslims lawful; by analogy, a blood transfusion from non-Muslims to a Muslim is allowed.[32]

---

[25] Comité de Bioéthique et Ordre des Médecins du Liban, *Bioéthique et Jurisprudence Islamique, Lettre du Comité de Bioéthique*, 1997 January, Beirut, 1–19.

[26] Ibn Ahmad B., The Role of Islamic Shari'a on Medical Innovations, *Contemporary Jurisprudence Research Journal*, 1993, 3, 24–34.

[27] See the text in: Abdullah A.M., The End of Human Life, in Mazkur (Al) K., Saif (Al) A., et al. (eds.), *Human Life: Its Inception and End as viewed by Islam*, Kuwait, IOMS, 1989, 367–373.

[28] Ebrahim A.F.M., *Islamic Jurisprudence and Blood Transfusion*, Mobeni, Islamic Medical Association of South Africa, 1990, 13–16.

[29] Ibid., 16–17.

[30] Sistani (Al) U.S.A.H., *Al-Fatwa al-Muyassarah*, Beirut, 2002, 412.

[31] Ebrahim, *Islamic Jurisprudence and Blood Transfusion*, op. cit., 23.

[32] Ibid., 24.

Doctors tend to neglect the juridical distinction between Muslims and non-Muslims on the subject of blood transfusions[33] according to the principle of the necessity to save human lives.

All these different attitudes regarding Muslims as donors and recipients of transplants may have an influence on the behaviour of Muslim immigrants to the West when the explantation of organs from a deceased Muslim is requested for transplant into non-Muslims. Amongst these immigrants there are those who consider donation as religiously licit or a duty. On the contrary, there are those who may refuse donation in order to protect the integrity of the corpse of a Muslim[34]; or, out of mistrust of organ donation to infidels, on the real intentions of doctors or for unspecified reasons but, at least for a Muslim, which are almost always inseparable from the religious factor. However, the priority concern always appears to be that of being the victim of some form of "physical" exploitation by non-Muslim medical staff.

What is to be done when several patients with the same serious condition of the pathology urgently require a kidney transplant and there is only one organ available? The rule could be that of drawing lots for the patient who will benefit from the organ by analogy with the rule followed by Prophet Muhammad when he drew lots for the bride that was to accompany him on a journey.[35]

TRANSPLANTS FROM LIVING DONORS

The main conditions set out by the jurists to legitimise a transplant from a living donor take on special significance as the majority of transplants carried out in the Middle East and North Africa come from living related donors (LRD), that is people genetically related to the recipients (such as parents, brothers and sisters, grandparents), or from living close relatives (emotionally related donor or ERD such as spouse, in-law, adopted sibling) both because these are operations that are less complex than transplants from a corpse and due to the great difficulty in finding the organs of the latter type. Organs from living non-related donors (LNRD) or living unrelated donors (LURD) are often sold; due to the serious ethical problems related to the trade in organs, transplants from LNRD are generally hindered in Muslim countries.

According to the Islamic Code of Medical Ethics (Kuwait 1981), the donation of an organ has to be the effect of a free and voluntary act that can be performed when the donor does not run any risk for his life whilst the harm suffered is minimal.

---

[33] Hathout H., Blood Transfusion and Religion, *Journal of the Kuwait Medical Association*, 1969, 3, 1–2.

[34] The objections put forward by an imam in Milan (Italy) were of this tone on the explantation of organs from the corpse of a young Moroccan immigrant (Asnaghi L., Trapiantati gli organi dell'immigrato. Il corpo è sacro, l'hanno violato, *La Repubblica* of 8 November 1991, 21).

[35] Comité de Bioéthique du Liban, op. cit., 8.

The majority of jurists seem to be in favour of transferring one of the double organs if the survival of the donor is guaranteed, the damage is not serious and the purpose is humanitarian.[36] Donation must not lead "intentionally" to death or to the disablement of the donor even if it is carried out to save another person. The Grand *mufti* of Egypt Tantawi, during a Symposium organized by the "Egyptian Organization for the Treatment of Patients suffering from Kidney Disorders", stated that if the donor suffers 5% damage whilst the recipient has a 95% benefit, this action will be particularly appreciated by God.[37]

In the late 1980s, the *sheikh* of Al-Azhar Jad al-Haq stated: "if an experienced Muslim doctor . . . asserts that cutting the body of a living man with his permission to take an organ . . . for transference to another body of a living man for treatment purposes will not originally entail serious harm to the donor, as harm should not be eliminated by another harm, and will be of benefit to the recipient, then such transfer is permissible in the *Shari'a*, on condition that transfer is not effected by way of sale or against material recompense".[38]

Present-day Arab-Muslim governments must be acknowledged as having considerable skill in formally introducing very modern health regulations modelled on Western ones. A significant case is that of Syria where Law no. 31 of 23 August 1972 (updated by Law no. 43 of 20 December 1996), regarding the conditions to be respected in a transplant from a living donor, anticipates many of the provisions imitated in the following years by other Muslim countries. The provisions include the condition that the explanted organ must not be vital, even if the donor himself requests its explantation; three specialists are consulted to assess the risks of the removal of an organ from the living individual; the donor has to provide an explicit written authorisation, stating his free choice, and he has to be in full possession of his mental faculties; the organ can be explanted from a minor when both the recipient and the donor are twins with the consent of the parents or guardians; traffic in organs is prohibited and, the final condition, the donor is entitled to free care in a public hospital. The features of the Syrian legislation include art. 8 (Law of 1972) which indicates that a transplant is lawful only with the explicit written consent of the recipient of the organ (or of the family if the recipient is incapable of expressing his wishes); art. 7 lays down that the hospitalisation, the removal, the transplant and the subsequent care are free of charge for Syrian citizens.

As is physiological in a context characterised by the absence of a religious hierarchy (*Sunni* Islam), there are divergent opinions on transplants from living donors. For example, Mokhtar Sellami, Grand *mufti* of the Arab Republic of Tunisia recalled that explantation of a double organ (e.g. a cornea, a kidney) from a living donor nevertheless implies a certain degree of invalidity for the

---

[36] Furqan A., Organ Transplantation in Islamic Law, *Islamic and Comparative Law Quarterly*, 1987, VII (1), 132–136.

[37] Cairo Press Review, 29 March 1988.

[38] In Tantawi, Judgement on Sale or Donation of Human Organs, art. cit., 293.

donor. Therefore it is preferable not to transfer parts of the body as the human being is not its exclusive owner.[39] In any case, continued the Tunisian *mufti*, if explantation takes place, the organs have to be the object of donation (as laid down by Tunisian law) and not of trade. The Egyptian *mufti* al-Sha'rawi also said he was against transplants from living donors because the intactness of a Muslim is sacred. For the Egyptian Abd al-Salam al-Sukkari the double organs meet a design by the Creator therefore their number must not be modified even for the benefit of one's neighbour. The Islamic Fiqh Academy of New Delhi at its session on 8–11 December 1989 limited the lawfulness of the kidney donation from a living donor exclusively to ill relatives. In the latter case this was an opinion expressed in the light of the great concern over the traffic in organs which was highly developed in that region.

Even more limiting is the opinion of a *Shi'ite* jurist opposed to the explantation of "major parts" of the body (e.g. eye, hand, foot) from a living donor but favourable in the case of portions of skin, flesh, etc. In this circumstance, it would even be licit to pay compensation to the donor.[40]

The role played by the religious figures who are critical of transplants does not appear capable of stopping the development of transplants in many countries, but succeeds in hindering it with considerable concern by the national health authorities. At the same time, these criticisms, as they express concerns which are widespread in Muslim communities, are important as they are assimilated by the favourable jurists and the State legislations in order to defend the intactness of the donor, whether living or deceased.

TRANSPLANTS FROM CORPSES

In the Arabian peninsula, until the time of Muhammad (d. 632), there was the widespread custom of dismembering (as a sign of contempt) the bodies of the enemies killed during struggles between the different tribes. Islam prohibited all acts of violence or irreverence on corpses, laying down a correct procedure for the burial of the corpse.

As a consequence of these conceptions (and the objections listed in the paragraph "Ethical-juridical principles"), in the past decades the idea has spread that transplants from corpses are prohibited by the *Shari'a*. To eliminate this erroneous conviction, the contribution of the religious authorities has been constructive, even if this has only limited, but not eliminated, popular resistance and prejudice.[41]

---

[39] Sellami M.M., Islamic Position on Organ Donation and Transplantation, *Transplantation Proceedings*, 1993, 25 (3), 2307–2309.

[40] Bostani, op. cit., 269.

[41] Faqih (Al) S.R., The Influence of Islamic Views on Public Attitudes Towards Kidney Transplant Donation in a Saudi Arabian Community, *Public Health*, 1991, 105, 161–165.

The fundamental principles on which the advocates of transplants from corpses base their arguments are the Koran 5.32 and the principle of the need to save human life, understood as an operative criterion that is above any other (including the principle of the protection of the corpse), save the duty to respect the body of a dead man as if it were that of a living person. In 1989, the Egyptian *mufti* al-Sha'rawi, generally opposed to any type of transplant, nevertheless stated he was in favour of transplants from a corpse as an extreme solution, similar to the premise of the *Shari'a* to feed on a corpse in order to survive. His fellow Egyptian Muhammad Abu Shadi maintains that, as the Muslims should sacrifice their life in the "holy war" (*jihad*) for the collective good, similarly the donation of organs of the deceased for the good of the community represents a noble act.[42] According to the Syrian *sheikh* Mustafa Zarqa in the work "The removal of organs from the deceased for the salvation of the living" the use of organs from corpses to save the living is always licit.

Opponents of transplants from corpses also appeal to the Sources of the *Shari'a*. Amongst the main arguments, reference is made to the juridical principle whereby everything belongs to God, including the human body which is not at man's complete disposal. This argument was answered by the Grand *mufti* of Egypt, Tantawi, in a *fatwa* in the newspaper *Al-Wafd* in early 1989, declaring that the entire universe belongs to God who, however, allows man to dispose of his body for the purposes of good. The opponents insist with a *hadith* in Abu Dawud and Ibn Maja in which Muhammad is said to have considered breaking the bones of a dead man as serious an act as that of breaking the bones of a living individual. According to the counter-reply, this prohibition does not apply when the violation of a corpse has the aim of saving a life whilst the rest of the body is buried correctly.

A further motive for refusing explantation from a corpse was recalled by the Islamic Fiqh Academy of New Delhi in 1989. The Academy considered non-obliging a testamentary provision in which the person allows the post-mortem donation of organs. This opinion is based on the consideration that we cannot give away what does not completely belong to us (the only real owner is the Creator); in addition, a human organ has an inestimable value and its inclusion in a will makes it invalid.[43] In this case too, this opinion is not shared by other religious bodies.

The greatest resistance has been noted in the Muslim countries of the Indian subcontinent (Pakistan, Bangladesh and the Muslim community in India) compared to other Muslim countries, regarding transplants. This tendency comes from two main reasons: (a) from the fear of the facility with which the transplants are accompanied by trafficking in organs, (b) from the deep-rooted attention to the prohibition of any lesion on the human body and the corpse.

---

[42] Rispler-Chaim, op. cit., 33.

[43] Ebrahim A.F.M., Organ Transplantation: Contemporary Sunni Muslim Legal and Ethical Perspectives, *Bioethics*, 1995, 9(3/4), 291–302; www.ummah.org.uk/bicnews/Articles/organ2.htm

Two different examples testify to the difficulties put forward by jurists to the new practice: the situation in Pakistan and in Singapore. In 1967 Fazlur Rahman, at that time Director of the Islamic Research Institute in Pakistan, when asked about the lawfulness of the transplant of eyes from corpses, expressed himself favourably on the basis of the priority of the benefit of the living compared with the protection of the corpse. This position was contested by the main religious authorities[44] but without succeeding in preventing the doctors from performing the first transplants in the country.

In 1972 the Committee of the *Fatwa* (the authority that decides on the religious questions of the Muslim community) of Singapore (a country with a non-Muslim majority), chaired by the *mufti* Syed Isa Semait, had deemed the transplant of kidneys contrary to the principles of Islam. Fifteen years later (in 1987), the Committee of *Fatwa*, chaired by the same *mufti*, influenced by the positive therapeutic results obtained by the practice (also thanks to the introduction of cyclosporin to fight rejection), recognised the transplant of kidneys as licit in cases of emergency if the donor had accepted in his lifetime signing a special donor's card (*opting-in*), similar to the case in Saudi Arabia in similar circumstances.[45] The authorities in Singapore recognised a "specific" health regulation for Muslims (there is a twofold set of rules on transplants) compared to the other nationals; for the latter there is presumed consent (*opting out*) which allows taking the organ from a deceased individual except where the subject had expressly refused post-mortem donation.

In fact, the first favourable opinions to transplants from corpses in the Muslim world were recorded in 1959 when *sheikh* Hassan Maamoon, Grand *mufti* of Egypt, approved the transplant of cornea from an unidentified corpse or from a person who had given his consent during his lifetime.

In the present context, the majority of "scholars" seem favourable to explantation from a corpse (the real problem concerns the definition of the criteria of death) on condition that the person authorised it during his lifetime or, alternatively, his relatives give their consent[46]; in the event that it is not possible to identify the corpse or if there are no heirs, authorisation to remove the organ is required from the head of the Muslim community (*wali al-muslimin*)

---

[44] Rahman F., *Health and Medicine in the Islamic Tradition*, New York, Crossroad, 1989, 106–107.

[45] Rasheed H.Z.A., Organ Donation and Transplantation – A Muslim Viewpoint, *Transplantation Proceedings*, 1992, 24(5), 2116–2117.

[46] Regarding relatives, the decision-making role follows, in some legislations, different decreasing orders: sons, father, mother, spouse, siblings, legal guardian (Tunisian law of 25 March 1991); father, mother, spouse, sons, sibling, legal guardian (Algerian law of 31 July 1990). The Libyan law of 10 March 1982 mentioned the permission for explantation granted by relatives up to the fourth degree. Jordanian law (no. 23 of 1977 amended by law no. 17 of 1980) requires the consent of the father or the mother, which implies the silence of one of them; the opposition of only one entails prohibition of the removal. Syrian law (1972) requires the consent of the family without further specifications; in addition the removal is authorised without consent if nobody comes to claim the body even when the identity of the deceased is known.

(Academy of Islamic Law, Saudi Arabia, Jeddah, IVth session, Resolution no. 6, 6–11 February 1988).[47]

The presumed consent to explant organs from a corpse in the absence of the explicit consent of the deceased expressed during his lifetime meets strong opposition in the Muslim world. This represents a serious obstacle to the diffusion of transplants in this cultural context. In fact, a post-mortem explantation that is the "automatic" result of silence in life would diminish the respect due to the deceased and the gravity of the act of explantation. In addition, strong family bonds make it very difficult to overlook the explicit consent to explantation by the relatives of the deceased in the absence of previous consent during the deceased's lifetime. This is why in the Muslim world people are generally asked to express their desire to donate their organs when they are alive (e.g., by signing a donor's card as in Saudi Arabia)[48] whilst, failing that, the authorisation of the relatives remains binding; amongst the exceptions there are the laws of Turkey of 21 January 1982 (art. 1) and Algeria, but limitedly to the cornea and the kidneys in cases of emergency (31 July 1990).

On the same subject the Indonesian Ulama follow a particular reasoning.[49] God is the only owner of the organs, whilst man is entitled to use the organs for his own benefit but does not have the right to lend or sell them. As God owns them, it follows that permission for donation from a corpse by the relatives of the donor is not binding. However, for the purpose of avoiding incomprehension or legal recourse against doctors, the stipulations must include the consent of the family concerned. Furthermore, the government can establish compensation for the donors. Lastly, written consent from the living donor is compulsory; those who want to donate an organ post-mortem are strongly recommended to state this willingness in writing.

A presupposition of a cultural nature influences the development of the practice of transplants throughout the Muslim world. The greater inclination shown by Muslims to donate organs during life to blood relatives (i.e. LRD) or relatives and, to a lesser extent, to non-blood relatives (i.e. LNRD), compared to the scarce inclination shown towards post-mortem donation can also be explained by the strong family bonds characteristic of Arab-Muslim culture.

On the practical level, the position that contemplates the obligation of the consent of the relatives, in the absence of that expressed by the living, reflects the best ethical position but for social, cultural and bureaucratic reasons (in practice, due to scarce general interest) this is translated into too few organs being available.

---

[47] IOMS, *Transplantation de certains organes humains du point de vue de la Charia, Actes du Sixième colloque organisé par l'IOMS, Koweit, Octobre 1989*, Rabat, ISESCO, 1999, 468.

[48] Shammari (Al) S., Public Attitude and Accessibility Concerning Kidney Donation Card and Living Related Donation in Riyadh, Saudi Arabia, *Annals of Saudi Medicine*, 1991, 11 (3), 336–340.

[49] *Resolution of International Seminar on Organ Transplantation and Health Care Management from Islamic Perspective*, op. cit.

As we will show in Chapter 10 "The End of Life", on telling the truth to the seriously and/ or terminally ill patient, in many areas of Muslim tradition, the family still acts as a protective shield from the unhappiness and psychological pain of the patient. This protection often requires not revealing to the patient his real condition in order to leave some hope. Similar caution was taken in Saudi Arabia on the first kidney transplants from a corpse when the doctors recommended not broaching the subject of explantation from the corpse with a close relative, as it was preferable to begin contacting a more distant relative of the deceased (e.g. the grandfather rather than the father) who in turn would have informed the close relative. Furthermore, doctors were advised not to speak of donation from a corpse on the first approach with the family of the deceased. The law regulating explantation from a corpse following cerebral death in Oman dates back to 1994; between 1995 and 1997 the relatives of 13 deceased prohibited explantation from corpses. The reason comes from disagreement between the family members: as the family is a large group, the widest possible consent is sought but has the result of increasing contrasting opinions[50] until explantation is abandoned as the technical time available for explantation is very brief.

THE DEBATE ON THE CRITERIA OF DEATH

Historically, the criterion for ascertaining the death of an individual coincided with the arrest of the cardio-respiratory functions. According to a classic version (dating back to Shafi'i) death could be identified by means of the progressive weakening of the sight, heaviness in the legs, pinching of the nose, the pallor of the temples, the loss of suppleness of the skin on the face and when breathing stopped, which indicates that the soul has left the body. Similarly in 1971, the Rector of the Institut Musulman de la Mosquée de Paris, Hamza Boubakeur, in refusing heart transplants, as the therapeutic results were still to be reached whilst the violation of the completeness of the body was apparent, made a partial return to the classic criteria: "Death for Muslim doctors, jurists and theologians results – without any divergence – from the recognition, not of the precocious or immediate signs (circulatory, respiratory arrest, absence of subcervical reflexes, etc.), but from the later and more edifying signs of death (pallor, hypostasis, drop in temperature compared to the ambient temperature, rigor mortis)."[51]

Today we know that the brain acts as the coordinating and unifying centre of the human body; the total and irreversible destruction of the brain (i.e. of the cortex and of the encephalic trunk)[52] marks the death of the person as a "whole"

---

[50] Kehinde E.O., Attitude to Cadaveric Organ Donation in Oman: Preliminary Report, *Transplantation Proceedings*, 1998, 30 (7), 3624–3625.
[51] Boubakeur H., La greffe du coeur. Point de vue de l'Islam, *Revue d'Histoire de la Médecine Hebraïque*, 1971, 95, 153–154.
[52] The cortex controls the spontaneous movements and is the seat of consciousness and the cognitive capacities. The cerebral trunk is in the lower part of the brain from which it activates the cortex and activates fundamental functions such as breathing, temperature regulation, the sleeping-waking cycle and the blood pressure.

organism, even when some of his organs, if maintained artificially in resuscitation, can continue to function (e.g. beating heart, kidneys and liver) for a brief period during which it is possible to explant the organ to be transplanted (statistically speaking, the definition of cerebral death is met by only about 1% of all deaths).

One of the fundamental ethical commitments of doctors in transplants consists precisely of explanting an organ in a good condition from individuals whose cerebral death is certain whilst, at the same time, oxygenation of the blood of the organs must be maintained so that they can be used by the recipient.

Against the brain death criteria, which are indispensable in order to perform a transplant from a corpse, Muslim jurists initially remained faithful to the classic criteria deeming it shocking to consider deceased an individual who showed signs of life, even if induced by resuscitation, such as heartbeat, nutrition, excretion, the growth of hair and nails. The acceptance of the criteria of brain death has not been easy. For example, according to the Council of the Islamic Jurisprudence Academy of the Muslim World League (Mecca, October 1987) life-support equipment can be switched off when three doctors ascertain the irreversible absence of any cerebral activity; it is added that the death of the person can be pronounced only after breathing and the heartbeat have stopped. In fact, this implies the impossibility of explanting organs as the absence of oxygen deteriorates them irremediably.[53]

Similar uncertainties are reconfirmed by a study on the common positions on the subject by a sample of 50 doctors of the Law from Kuwait, Saudi Arabia, Iran, Egypt, Lebanon and Oman.[54]

The situation in Egypt and Pakistan remains controversial. These countries do not yet have national laws regulating explantation from corpses, especially due to the resistance by religious circles in accepting the criteria of cerebral death.

The jurists in favour of the brain death criteria can overcome the obstacle having recourse to the analogy with the "movements of the slain or slaughtered".[55] This old rule, often agreed with by *Shafi'ite* scholars, states that if an assailant injures an individual causing fatal wounds, the movements of the dying victim are the "movements of the slain". If a second assailant finishes off the victim, the accusation of murder concerns only the first assailant whilst the second may be prosecuted exclusively for aggression of the corpse.[56] By analogy, the heartbeat,

---

[53] Ebrahim A.F.M., Islamic Jurisprudence and the End of Human Life, *Medicine and Law*, 1998, 17, 189–196.

[54] Mousawi (Al) M., Hamed T., et al., Views of Muslim Scholars on Organ Donation and Brain Death, *Transplantation Proceedings*, 1997, 29 (8), 3217. Only 32 doctors of the Law answered the questions with the following results: 90.6% (29) accept the donation of organs in life only out of serious need; 87.5% (28) accept port-mortem donation under the previously mentioned conditions; 90.6% (29) initially refused the brain death criteria as death is mainly conceived as the cessation of all the physical (not cellular) activities including breathing, movement, heartbeat, hearing, etc. Many maintain that the cerebrally dead person is not yet really dead.

[55] Abdullah, art. cit., 370–372.

[56] Some scholars believe that not only the first aggressor but also the second should undergo retaliation; conversely, others maintain that only the second should be killed by retaliation as he killed the victim who still showed signs of life.

the movement of the lungs, etc. that are supported mechanically accompanied by the condition of brain death may be considered equivalent to the "movements of the slain".

Today, in clinical situations suitable for transplants, the jurists seem more willing to accept the brain death criteria, especially against medical assurance. This has been a progressive acceptance, accompanied by a lively debate, although limited to the experts.

The first important juridical-religious recognition of the lawfulness of the new criteria is Resolution no. 17 (5–3) by the Council of the Islamic Fiqh Academy at the Third International Conference of Muslim Jurists (part of the Organisation of the Islamic Conference – OIC) in Amman (Jordan) in 1986. According to the Resolution a person is legally dead when there is one of the following signs: (1) complete cardiac arrest and breathing has stopped, and the doctors certify that this state is irreversible; (2) total cessation of all the functions of the brain, the doctors decide that this state is irreversible whilst the brain has begun to degenerate. Under these conditions it is lawful to disconnect the life-support treatment even if some organs continue to function automatically (e.g. the heart) under their effect.[57]

A further problem concerns the definition of brain death. At present – when accepted – it seems mainly defined with the same rigour as Western countries, i.e. ascertaining the death of the neocortex and brain stem through five signs: total absence of consciousness; fixed bilateral mydriasis, absence of reflexes and immobility of the eye bulbs; absence of spontaneous respiration, absence of reflexes depending on the trunk and a flat electroencephalogram for at least six hours.

However, there are a substantial number of bodies that opt for the "British Criteria", i.e. identifying brain death with brain stem death. The first transplant of a kidney from a Saudi corpse (in 1985) was performed adopting the criteria of the Royal British Medical Colleges. A similar criterion was preferred by countries which had been under the British protectorate: Kuwait, Oman, United Arab Emirates and Bahrain. The IOMS meeting in Kuwait in 1985 pronounced itself in favour of the brain stem death criterion,[58] stating that death corresponds to the absence of life by the part of the brain responsible for the vital functions, i.e. the brain stem. Only after brain-stem death is it lawful to remove resuscitation equipment. According to the Indonesian Council of Ulama, the Indonesian Forum for Islamic Medical Studies and the Federation of Islamic Medical Association (Jakarta, 30 July 1996), the most modern technique to establish death

---

[57] See www.islamibankbd.com/ page/ oicres.htm#17 (5–3); Hassaballah A.M., Definition of Death, Organ Donation and Interruption of Treatment in Islam, *Nephrology Dialysis Transplantation*, 1996, 11 (6), 964–965; Daar A.S., Organ Donation – World Experience: The Middle East, *Transplantation Proceedings*, 1991, 23 (5), 2505–2507.

[58] Conclusive Recommendation of the Seminar on *Human Life: Its Inception and End as Viewed by Islam*, Kuwait, 1989, Vol. II, 627–629.

lies in determining the absence of the functions of the brain stem by means of EEG associated with the irreversible breakdown of the functions of the heart and lungs.[59]

Explicitly influenced by the British criteria is the position assumed in 1995 by the UK's Muslim Law Council after two years of discussions by both *Sunni* and *Shi'ite* experts.[60] The resolution also states: (1) doctors represent the suitable authority to determine the valid criteria of death; (2) modern science considers the death of the brain stem as the most suitable definition of death in view of explanting organs; (3) the Council is in favour of transplants; (4) Muslims may make use of a donor card (DC); (5) in the absence of a DC or the specific wish of the deceased to donate his organs, the next of kin may grant their permission; (6) trade in organs is prohibited. The fact that presumed consent is not accepted should be noted.

Like the rules that have been developed in the West, death must be recognised by specialists, including a neurologist and excluding the surgeon who will perform the transplant. The latter measure reappears (with variants) in several national legislations such as the Syrian law of 1972, the Algerian law of 31 July 1990, the Tunisian ruling of 25 March 1991 and Turkish law no. 2238 (art. 12) of 1979.

There does not yet exist a prevailing decision on the explantation of organs and tissues from a dead foetus. According to some scholars, this operation (e.g. bone marrow) is licit only if it has positive therapeutic effects on the recipient. Mokhtar Sellami, Grand *mufti* of Tunisia, is much more critical as recourse to foetuses would be an incentive to a market of pregnancies and abortions to the detriment of the protection of the foetuses.[61]

At an international conference held in Kuwait from 23 to 26 October 1989, and organized by the IOMS in collaboration with the Islamic Fiqh Academy of Jeddah, some recommendations on three particular situations were approved[62]: (a) transplant of nerve tissues and brain cells, (b) explantation from an anencephalic foetus and (c) transplant of sexual glands. (a) In the attempt to treat some pathologies, the first type of operation is licit if the nerve cells come from the marrow of the suprarenal glands of the subject to be treated.[63] (b) A live

---

[59] *Resolution of International Seminar on Organ Transplantation and Health Care Management from Islamic Perspective*, op. cit.

[60] A Juristic Ruling regarding Organ Transplant, *Islamic Voice*, August 1998, Vol. 12, no. 140.

[61] Sellami, art. cit., 2307–2308.

[62] See the text in IOMS , *Transplantation de certains organes humains du point de vue de la Charia, Actes du Sixième colloque organisé par l'IOMS, Koweit, Octobre 1989*, Rabat, ISESCO, 1999, op. cit., 466–468.

[63] According to the conference, the brain cells for a transplant can come from the brain of a precocious living embryo, between the tenth and eleventh week of pregnancy. These tissues and cells may be obtained in three ways: (1) from animal embryos – according to Muslim experts, this technique is lawful when its efficacy is guaranteed; (2) from embryos in the uterus obtained by hysterectomy – as this causes their death, the experts consider it unlawful except in the case of a therapeutic abortion; (3) the culture of brain cells – this practice is lawful if the cultures are obtained legally.

anencephalic child may be used for transplant only after the brain stem death. (c) It is illegal to transplant testicles or ovaries in another person as the offspring would not be the result of gametes coming from legally wedded parents.

In the following year the Islamic Fiqh Academy of Jeddah (14–20/3/1990) with Resolutions no. 6/5/56 and 6/8/59 approved the content of the previous Recommendations.[64]

The refusal of transplants which give rise to "genealogical hybrids" was reconfirmed by the Academy of Islamic Research of Al-Azhar in 1997.

THE ORGAN TRADE

The growing reliability of transplants associated with the increasing demand for organs comes into conflict with the scarce availability of these (for the technical and cultural reasons mentioned). This shortage has caused the development of traffic in organs (in particular of kidneys) with high profits for facilitators, private hospitals and unscrupulous doctors. India, Egypt and the Philippines currently appear to be the countries most involved in this traffic. The greatest problems that accompany the sale of kidneys concern the therapeutic results and the high mortality rate although transplants from LNRD , if performed with optimal procedures, produce valid results.

There is also a difficulty in distinguishing the sale of an organ from compensation received for donation.

Generally, the trade in organs is condemned by the doctors of Muslim law who take as their starting point the concept of the human body as a divine gift to be respected. However, several jurists have wondered how to behave when the only alternative to save a life remains precisely that of purchasing the organ. A response was formulated in 1987 by the majority of the experts of the IOMS in Kuwait. They distinguished between trafficking (prohibited) and trade in the absence of alternatives; in this case there may be a licit purchase as it has the purpose of saving a life. Nevertheless, in order to avoid injustice against the neediest, the mediation and control by State bodies is requested in order to pay the figure asked and carry out all the indispensable screenings on the donor.

At the same conference in 1987, the Grand *mufti* of Egypt, M.S. Tantawi, recalled the different positions on the subject of the purchase of organs.[65] In his opinion, there are jurists who make no difference between donation and sale of organs, prohibiting both because it is not lawful to sell or donate what we do not possess (only God is the owner of the human body), e.g. Sha'rawi. Other jurists accept donation (not sale) only if there is no serious damage for the donor and the recipient's being saved depends only on the donation (e.g. the *sheikh* of Al-Azhar, Jad al-Haq). Others accept buying and selling an organ solely when it

---

[64] See the text in IOMS, *Transplantation de certains organes humains du point de vue de la Charia*, op. cit., 472–473 and 476.

[65] Tantawi, art. cit., 292–295.

is the only solution to save the patient[66]; in this case, strict necessity makes the unlawful lawful, but it is the doctors' decision. Tantawi seems to accept this third position.

Resolution no. 26 (1/ 4) on organ transplants by the Council of the Academy of Islamic Law (Jeddah, 6–11 February 1988) under point 7 reasserted the prohibition of the trade in organs but took care to consider as a subject for further reflection the possibility for the recipient to bear the expenses in order to obtain the organ that is indispensable for his survival or to pay compensation as a sign of gratitude to the donor.[67]

In a document by the Committee of Bioethics of Lebanon of January 1997, recourse is made to the principle of "extreme necessity" to validate the purchase of an organ made in order to save a patient in the absence of alternatives.[68] The document pointed out how in current practice in Lebanon the living donor who is not a close relative frequently receives compensation for the altruistic act.

Out of 32 jurists questioned (the study mentioned above)[69] on the ways of carrying out transplants, all agree on condemning trade in organs, however 22 (68.7%) allow the purchase of the organ if the donor insists and there are no alternatives to save the recipient; 21 jurists allowed the donor to ask the recipient for compensation.

Condemnation of the trade in human organs is stronger in the State legislations on the model of art. 7 of the Unified Arab Draft Law on Human Organ Transplants adopted by the Council of the Ministers of Arab countries (Khartoum, 1987), according to which the purchase or the donation of organs against remuneration is always prohibited and no doctor must transplant such an organ if informed of such negotiations. An example of the motivations justifying this severity is given by the situations that arise for many of the patients who have purchased an organ. In the period 1986–1990, 72 Kuwaiti patients bought kidneys abroad and subsequently, on their return home, had to go to the Kuwait University Transplant Centre due to the serious complications that were a result of the transplants. 15% of the patients operated on died within three months, whilst more than 50% of the remaining patients showed consequences including rejection, infections, contagious diseases, tuberculosis and various forms of hepatitis; four of the patients were HIV-positive and two of these died of AIDS.

The organ trade also has consequences on the health organisation as the easiness of procuring organs abroad (e.g. India) ends up by raising an obstacle to the

---

[66] Mokhtar al-Mahdi, Chairman of Neurosurgery at Ibn Sina Hospital in Kuwait in 1987, indicated the figure to pay for a kidney according to *Shari'a* criteria: 5000 Kuwaiti dinars, by coincidence the same sum paid by the Health Ministry of Kuwait to procure a kidney from abroad. Mahdi (Al) M., Donation, Sale and Unbequeathed Possession of Human Organs, in Mazkur (Al), Saif (Al), et al. (eds.), *The Islamic Vision of Some Medical Practices*, op. cit., 280–286.

[67] IOMS, *Transplantation de certains organes humains du point de vue de la Charia*, op. cit., 467–468.

[68] Comité de Bioéthique du Liban, op. cit., 7.

[69] Mousawi (Al) and Hamed, art. cit., 3217.

implementation of programmes of transplants from corpses or from the living in the countries of origin of the patients, reducing the experience and skill of the medical staff in the sector. In addition, the possibility of finding organs through commercial negotiations reduces the inclination of people and the relatives of the patient from giving organs out of altruism. In the light of similar difficulties, it comes as no surprise that the government programmes for transplants from corpses of many Arab-Muslim states are finding it difficult to develop, despite the commitment of men of religion and doctors to convince the faithful through the mass media of the legitimacy for Muslim ethics of transplants from corpses.

UTERINE TRANSPLANTATION

The constant modernisation of the Saudi health system exposes it to new challenges in the observance of the ethical values. An example is a uterus transplant performed in Saudi Arabia (in Jeddah) in 2000. The operation was successful, "the recipient had two menstrual cycles after the operation" but "after 99 days, a vascular occlusion led to the removal of the transplanted uterus."[70] The motive of this surgical intervention is presented by the same medical team:[71] uterine transplantation can be "useful in the treatment of infertility, especially in communities where the surrogate mother concept is unacceptable from a religious or ethical point of view".

Commenting upon the operation, M. Aboulghar, Professor of Gynecology at the Cairo University said: "I am astonished to find the Islamic theologists are supporting all types of assisted reproduction including possible uterine donation, and they do not move forces to support a law for organ donation in Egypt. Every time the issue is raised in the parliament, the majority of Moslem theologists do not support the law and are very critical to organ donation."[72]

The problem of the uterine transplant had already been dealt with by the IOMS and by the Academy of Muslim Law of Jeddah in the conclusive Recommendation[73] at an international conference on organ transplants in Kuwait from 23 to 26 October 1989. This Recommendation was wholly taken up the year afterwards (1990) in Resolution no. 6/ 8/ 59 of the Academy of Muslim Law of Jeddah.[74] The two documents were divided into two parts: (1) the transplant

---

[70] Apology to Dr. Wafa M. Khalil Fageeh, gynecologist and assistant professor at King Abdilaziz University, *Lancet*, 29 September 2001, 358, 1076.

[71] Fageeh W., Raffa H., et al., Transplantation of the Human Uterus, *International Journal of Gynaecology and Obstetrics*, 2002, 76, 245–251.

[72] Aboulghar M., Uterine Transplantation, in Serour G.I. (ed.), *Proceeding of the Workshop on Ethical Implications of Assisted Reproductive Technology for the Treatment of Infertility, November 22–25, 2000*, Cairo, Al-Ahram Press, 2002, 91–94.

[73] See the text in IOMS , *Transplantation de certains organes humains du point de vue de la Charia, Actes du Sixième colloque organisé par l'IOMS, Koweit, Octobre 1989*, Rabat, ISESCO, 1999, op. cit., 470–471.

[74] Decision no. 6/8/59 in IOMS , *Transplantation de certains organes humains du point de vue de la Charia*, op. cit., 476.

of genital glands (ovaries and testicles) is juridically prohibited as the glands continue to produce gametes which transmit the genetic heritage of the donor even after their transfer to another person. (2) The transplant of the external genital parts or most shameful parts (*awrât mughallaza*) is also prohibited. The expression *awrât mughallaza* does not appear to include the uterus. This means that the Recommendation (1989) and the Resolution (1990) do not appear to expressly prohibit uterine transplant. Indeed, for both the documents it is lawful to transplant other genital organs in case of need, except genital glands (ovaries and testicles) and the external genital parts (*awrât mughallaza*).

In November 2000 the University of Al-Azhar in Cairo, the IICPSR (International Islamic Centre for Population Studies and Research of Al-Azhar) and ISESCO (Islamic Educational, Scientific and Cultural Organisation) organised an International Workshop on "Ethical Implications of Assisted Reproductive Technology (ART)." The conclusive Recommendations advised further research on uterine transplantations in animal models only. However, in the event that the practice were to prove to be useful for man, the Islamic Research Council should be summoned to express itself on the juridical-religious lawfulness due to the many problems raised by this practice.[75]

In May 2004, the Egyptian *sheikh* Abdel Rahman al-Adawy, head of the Council's Jurisprudence Research Committee, pronounced himself as against uterus transplants (see the transplant performed in Saudi Arabia) on the basis of a previous *fatwa* of the Islamic Research Council forbidding the donation and transplant of sexual organs as well as the donation of individual organs (e.g. heart, liver, pancreas). In the *sheikh's* opinion, the uterus is not only a shelter for the child but has its own genealogical characteristics and, as a consequence "as the infant takes its characteristics from the mother's womb, if it grows in a womb donated by a stranger, he/ she will carry some of the stranger's genealogical characteristics. Then the question has to be raised as to who the child's mother is".[76] The *sheikh* also declared that he was personally opposed to the donation of any organ from a living donor as neither the donor nor the recipient, more often than not, can lead a normal life.

XENOTRANSPLANTATION

The chronic shortage of available organs in Muslim countries has aroused the interest in xenotransplantation on which several authorities and juridical-religious bodies have pronounced themselves. Against the objection represented by verse

---

[75] Serour G.I. (ed.), *Proceedings of the Workshop on Ethical Implications of Assisted Reproductive Technology for the Treatment of Infertility*, op. cit., 106.

[76] Yomna Kamel Middle East Times Staff, *Azhar Scholars say One Womb per Woman and No Transplants*, www.yomnakamel.blogspot.com/ 2004/ 05/ azhar-scholars- say- one- womb-per-woman.html

4.119 in the Koran,[77] some reply that xenotransplantation actually respects the Koran 5.32 by saving a human life and, in addition, this is justified by the principle of need. According to the latter principle, the option that brings the greater benefits must be chosen between two conflicting interests.

The Indonesian Resolution on "Organ Transplantation from the Islamic Perspective" (Jakarta, 1996) mentioned above, distinguished[78] the categories of animals the organs of which may lawfully or unlawfully be transplanted to man according to the categories of animals the flesh of which can juridically be consumed or are prohibited in Muslim law. According to the Resolution: (1) the use of organs of animals the consumption of which is lawful (*halal*) – for instance, sheep, goat and cow – is allowed if these animals have been slaughtered according to the rules of the *Shari'a*. (2) Regarding animals that may not be consumed (*haram* animals, e.g. pigs, dogs, mice, snakes), the Resolution notes that the animal involved in the explantation is not eaten but used. Consequently, the Ulama and doctors unanimously deemed lawful the use of forbidden (*haram*) and ritually impure (*najis*) animals if they represent the only cure available, precisely because the animal is not eaten but used.

Sheikh Qaradawi deems explantations from animals positively (including from a pig) but "only in the case of necessity". The conclusive document of the Islamic Fiqh Academy of New Delhi dedicated to transplants (8–11 December 1989)[79] obtained the unanimous approval of the participants who recognised the lawfulness of recourse to organs of *halal* animals (i.e. the consumption of which is allowed, e.g. sheep, goat, cow) slaughtered according to the rules of the *Shari'a*. Conversely, when life is at risk and there are no alternatives, there may be recourse to the organs of animals the flesh of which is *haram* (prohibited) or animals the flesh of which is *halal* (allowed) but which have not been ritually slaughtered. Pig organs and tissue can be used only if there exists a serious danger for life. Questioned by the Islamic Medical Association of South Africa, the *Majlis al-Ulama* (Council of Ulama) of Port Elizabeth confirmed[80] the lawfulness of a transplant from animal to man to save the life of the latter or improve his quality of life but only with recourse to *halal* animals slaughtered in the Muslim fashion (*dhabh*). However, unlike the Indian *ulama*, the South Africans refused the possibility of drawing any benefit, even medical, from the pig as its parts are completely impure.

The key point of the disagreement between the jurists on recourse to the pig lies in the difficulty of reconciling the *hadith* in Bukhari according to which God

---

[77] Satan says: "I will mislead them, and I will create in them false desires; I will order them to slit the ears of cattle, and to deface the (fair) nature created by God."

[78] *Resolution of International Seminar on Organ Transplantation and Health Care Management from Islamic Perspective*, op. cit.

[79] www.ummah.org.uk/ bicnews/ Articles/ organ2.htm

[80] Ebrahim A.F.M. and Vawda A., *Islamic Guidelines on Animal Experimentation*, Mobeni, IMASA, 1992, 29–30.

does not provide man with remedies with what He has forbidden (e.g. the pig), with the approach shared by several jurists who equate the medical need with the dietary one and, as Muslim law allows consuming pork in order not to starve to death, similarly, medical use of the pig to save a life would become licit.

SOME NATIONAL LEGISLATIONS

We will present some data on legislation and the development of transplants in certain Muslim countries showing that the majority of organs come from living donors (the majority of whom are LRD rather than LNRD) followed by those from corpses. Some countries (e.g. Syria) prohibit transplants from LNRD due to the associated ethical problems, others hinder it or practise it in limited cases. Other countries (e.g. Iran) openly have recourse to LNRD but try to control organ trafficking.

At present, of the Middle Eastern and North African countries only Yemen has not launched a programme on transplants. Explantations from a living related donor still represent the majority of operations but, since the 1990s, there has been a significant increase of transplants from corpses.

The development of transplants in Muslim countries has gone through similar phases. At first, patients would be sent to the USA and to Europe, then some centres began to perform transplants locally from LRD with the help of European and American doctors. Subsequently, a number of countries implanted organs from corpses coming from abroad (generally North European and American). These organs could often not be used in the West according to Western standards.[81] Other countries started to perform transplants from local corpses. Lastly, multi-organ transplants were begun in some countries.

The most frequent clinical obstacles in Muslim countries include[82]: (a) the high rate of HCV (hepatitis C) in the population in dialysis; (b) tuberculosis is endemic in many of these countries and is 5–10 times more common in subjects having had a transplant than in the West; (c) due to the high number of marriages between blood relatives there is a greater frequency of hereditary renal pathologies and this is important in identifying potential donors amongst relatives.

Saudi Arabia

In Saudi Arabia, transplants were approved by the Committee of the *Senior Ulama* in Resolution no. 99 of 25 August 1982 in which it was unanimously accepted that an organ, or a part thereof, could be removed from a Muslim or *dhimmi* living person and grafted on to him (self-transplantation). In addition,

---

[81] Daar A.S., Organ Donation – World Experience: The Middle East, *Transplantation Proceedings*, 1991, 23 (5), 2505–2507.

[82] Khader (Al) A.A., The Iranian Transplant Programme: Comment from an Islamic Perspective, *Nephrology Dialysis Transplantation*, 2002, 17, 213–215.

the majority approved explantation from a corpse and from a living donor for the benefit of Muslims.

Resolution no. 26 (1/4) on organ transplants by the Council of the Academy of Islamic Law (Jeddah, 6–11 February 1988)[83] deemed licit the transplant of parts of the body that are renewed spontaneously (e.g. blood, skin) from one person to another (1). Furthermore, the transplant from a corpse is licit but the pre-mortem authorisation of the donor is necessary or, after his death, the consent of the close relatives (6). Trade in organs is condemned; however, the possibility for the recipient to pay costs in order to obtain the vital organ or to pay compensation to the donor is the object of reflection (7).

The first two cases of kidney transplant from a living related donor (in the first case a daughter of the patient, in the second, the brother) took place in 1979 following agreements of technical-scientific assistance with the Renal Unit of St. Thomas's Hospital, London. In 1983 about 60 kidney explants from European corpses were performed whilst since 1984 Saudi corpses have been used. In 1985 the first kidney transplant from a Saudi corpse was performed, taking the brain stem criterion of death accepted by the Royal College of the UK (therefore there is no recourse to the electroencephalogram-EEG). At present the diagnosis of brain death contemplates the clinical confirmation of the absence of the functions of the brain stem by means of an EEG and/or cerebral angiography.[84] As in other countries of the Arabian Peninsula, recourse to brain stem criteria derives from the assistance provided by British hospitals and universities in starting transplant programmes. The National Kidney Foundation (NKF) was also created in 1985 to coordinate kidney transplants in the Kingdom; in 1993 it was renamed the Saudi Centre for Organ Transplantation (SCOT) for the purpose of coordinating all types of transplants.

The religious authorities have conducted a strong awareness-raising campaign to inform Saudi citizens of the positive attitude of Islam towards transplants, both from corpses and from living donors. Despite the good intentions, the availability of organs from corpses remains low due to the scarce inclination of relatives to donate the organs of their deceased relatives to strangers. At the same time there is greater willingness to be LRD instead of signing a Donor Card (DC) authorising explantation from the corpse and this has the principal aim of protecting the relatives, in the name of the great value that the family maintains in the Saudi context.[85]

In the period between 1996 and 1999, 669 kidney transplants were performed from a living donor and 331 from corpses. In 1998–1999 all the 348 kidney transplants from living donors were between blood relatives (LRD). In 1997 out

---

[83] See the text in IOMS, *Transplantation de certains organes humains du point de vue de la Charia*, op. cit., 467–468; www.islamibankbd.com/page/oicres.htm#26 (1/4).

[84] Souqiyyeh M.Z., Shaheen F.A.M., et al., Trends for Successfully Documented Cases of Brain Death in Intensive Care Units in Saudi Arabia, *Transplantation Proceedings*, 1996, 28 (1), 380.

[85] Shammari, art. cit., 339.

of 171 kidney transplants from living donors, 161 were from LRD while the remaining 10 were from the spouse (LNRD).

There is a tendency in Saudi Arabia to refuse transplants from LNRD[86] and presumed consent is not accepted for explantation from the corpse. Explicit authorisation for explantation is required, with the signature of a DC; in its absence, the consent of relatives is necessary. The importance of the family in Saudi society means its consent is required to remove an organ from the deceased. Here is the wording of the consent form signed by a father in Saudi Arabia for the first kidney donation from a corpse[87]: "I the undersigned . . . hereby agree to donate my son's kidneys for transplantation in favour of those in need, hereto I seek only God's reward and blessing." The kidneys in question were given to two Saudi women. Note the clear reference to God and His blessing for this clinical act.

The present version (1420 H, AD 1999) of the Saudi Uniform Donor Card shows an explicit reference to the family in place of God. Here is the text[88]: "I . . . have spoken with my Family about organ and tissue donation. Upon my death, I wish to donate:[] Any needed organs and tissues [] Only the following organs and tissues: kidneys, liver, hearth, lungs, corneas."

The attention paid to the family is shown by the rules of NKF-SCOT to convince the kin of a deceased person to donate the organs. The formation of a committee in hospitals headed by the Director of the Hospital and including social workers "and, if possible, a religious minded person and/or a person well known in the city" is recommended. The member of the committee considered most suitable speaks to the family. An immediate reply is not expected. It is useful to show the relatives Decision no. 99 of the Senior Ulama Commission of 1982 and other religious documents in favour of post-mortem donation. Written consent of the kin must be signed by the closest relative (the order is: father, mother, wife, brother, cousin of the brain-dead patient) and by two witnesses (one must be another relative).

As there are numerous foreign workers in the Kingdom, the indications of the SCOT also deal with the procedures for post-mortem donation by them. The name, address and telephone number of the sponsor or of the company should be obtained; in addition those of relatives or friends in the Kingdom and, if possible, the telephone number of a close relative in the home country. After the confirmation of brain death the family must be contacted but "the matter of organ donation should not be immediately discussed"; the family must be left the time to recover from the shock. Later, the NKF coordinator will contact the family to obtain consent for the explantation. This must be sent by telex or fax and must be "attested" by the embassy of the country of origin. Lastly, as a sign

---

[86] Shaheen F.A.M., Souqiyyeh M.Z., et al., The Saudi Centre for Organ Transplantation: An Ideal Model for Arabic Countries to Improve Treatment of End-Stage Organ Failure, *Transplantation Proceedings*, 1996, 28 (1), 247–249.

[87] See Otaibi and Khader, art. cit., 219.

[88] See the website of the Saudi Centre for Organ Transplantation (SCOT), www.scot.org.sa/

of respect and gratitude to the deceased donor, the Ministry for Health will pay the cost of transport of the corpse to the home country with the accompaniment of a chaperone.

Donation from a living individual requires free and informed consent by the donor.

Iran

The Iranian situation is anomalous. The first kidney transplant in the Middle East (1968) took place in Iran, but a real programme has only been in force since the mid-1980s. Before this date, the kidneys arrived mainly from Eurotransplant (as was the case in Saudi Arabia, the United Arab Emirates, Turkey, etc.). Between 1985 and 1993, 2627 kidney transplants were performed, with 28.2% of the organs coming from LRD, 68.27 from LNRD and 3.6% from ERD such as the spouse, friends, etc. The number of patients in dialysis is constantly increasing and their treatment is paid for by the State. Explantation from a corpse is hindered by resistances of a cultural, social and religious nature; nevertheless, several transplants from a corpse have been performed. The first heart transplant was performed in 1993 as was the first liver transplant. In the period between 1992 and 1993, 9 multi-organ transplants[89] were performed from corpses with the consent of the relatives; brain death was established by means of clinical criteria, apnoea and electroencephalogram.

Apart from the many exceptions, resistance to the donation from the deceased explains the higher number of transplants from LNRD as they are indispensable in order to obtain life-saving organs as the organs that can be found with LRD are inevitably few in number. To limit commercial exchanges, there is an attempt to respect some rules. For example, the donor and the recipient could be asked to declare in writing that no commercial transaction has taken place and, in the case that the contrary is discovered, the surgeons can refuse to perform the operation.[90] In fact, the situation appears contradictory. Out of 32 donors interviewed in a study, only one admitted having donated a kidney out of altruism, for the others there was always a commercial dimension.[91] Poverty is often considered the main cause of LNRD donation and organ trafficking in the country. To increase the number of organs available, a law approved on 2 February 1997 by the Rafsanjani Cabinet and ratified by the Iranian Parliament established that

---

[89] Malek Hosseini S.A., Salahi H., et al., First Report of Nine Consecutive Multiorgan Cadaver Donor Transplants in Iran, *Transplantation Proceedings*, 1995, 27 (5), 2770.

[90] Simforoosh N., Bassiri A., et al., Living-Unrelated Renal Transplantation, *Transplantation Proceedings*, 1992, 24 (6), 2421–2422.

[91] Mention is to be made of the case of a husband who, as soon as he was discharged from the hospital where he had donated a kidney, took his wife to donate one of her kidneys as the payment received by the man was not sufficient to settle their debts. See Broumand B., Living Donors: The Iran Experience, *Nephrology, Dialysis, Transplantation*, 1997 (12), 1830–1831.

kidney donors should receive, after the operation, 10 million rials from the Charity Foundation for Special Diseases (CFSD), a non-governmental organization (NGO) financed by the government.[92] However, alongside this official compensation (defined a "gift for altruism"), there is often a personal and uncontrollable economic negotiation between the donor and the recipient. In addition, the costs for pre-operation tests on the donor are the result of negotiations: sometimes the donor undertakes to pay the costs of these tests himself if, at the pre-operation phase, the explantation is refused; in other cases, these costs are paid by the recipient. Elsewhere the donor, although he receives 10 million rials from the government for the donation, immediately pays for the pre-operation tests but then requests reimbursement from the recipient.[93] Even donation between relatives is often accompanied by a payment of money. Nevertheless, recourse to organs from LNRD seems to be a compulsory choice for Iran due to the high rate of end-stage renal disease (ESRD) patients.[94]

Precisely due to the ethical problems raised by the LNRD (too often associated with commercialisation and prohibited by other Muslim countries) the advocates of transplants from corpses are on the rise.

On 21 May 1989 the *ayatollah* Khomeini made a pronouncement on the removal of organs from a cerebrally dead person declaring "if it (removal of organs from a brain-dead body) allows the life of someone else to be saved, it is not forbidden, but it needs the permission of the 'owner' of these organs".[95]

In 1992 a religious decree by the *ayatollah* Ali Khamenei authorised explantation from corpses.

In 1993 the Iranian Parliament (*Majlis*) had voted against a bill aimed at legalising explantation from a cerebrally dead donor. On 23 October 1995 the Transplantation Act was rejected by the fourth Parliament. On 9 November 1999 the Parliament gave its preliminary approval to a bill ("Deceased or Brain Dead Patients Organ Transplantation Act") similar to the previously defeated one. On 5 April 2000 the 270 members of Parliament definitively recognized brain death and cadaveric organ transplantation. The bill required final approval from the Iranian constitutional overseeing body, the Guardian Council, a body controlled by *Shi'ite* religious authorities and, therefore, more conservative. The following December the Council approved cadaveric organ transplants. Article 3 of the law of April 2000 specified that transplant operators do not have to pay the *diyya*

---

[92] Zargooshi J., Iranian Kidney Donors: Motivations and Relations with Recipients, *The Journal of Urology*, 2001, 165 (2), 386–392, 388. Akrami S.M., Osati Z., et al., Brain Death: Recent Ethical and Religious Considerations in Iran, *Transplantation Proceedings*, 2004, 36.

[93] Zargooshi, art. cit., 387–388.

[94] Larijani B., Zahedi F., et al., Rewarded Gift for Living Renal Donor, *Transplantation Proceedings*, 2004, 36, 2539–2542.

[95] See similar opinion of other religious figures in Salahi H., Ghahramani N., et al., Religious Sanctions Regarding Cadaveric Organ Transplantation in Iran, *Transplantation Proceedings*, 1998, 30, 769–770.

(the blood price) as monetary compensation for lesions to the body and corpse due to the removal of organs.

On 15 May 2002 the Cabinet Council approved the Executive Bylaw on "Deceased or Brain Dead Patients Organ Transplantation Act." The most important articles of the Executive Bylaw state[96]:

1. Cerebral death consists of the irreversible interruption of cortical, subcortical and brain stem functions to be determined by means of a subsequent Protocol of the Ministry of Health and Medical Education (see below).
2. Four clinicians must perform the diagnosis: a neurologist, a neurosurgeon, an internist, an anaesthetist and it must be confirmed by a pathologist.
3. The doctors that perform the diagnosis must not take part in the transplant.
4. The wishes of the patient (when alive) on explantation may be oral or written and may be approved by a written communication by an heir. If the wishes of the deceased are not available, a form can be signed by informed heirs. The deceased's guardians are heirs apparent; the consent of the heir is mandatory and should be in writing.

The Protocol of the Ministry of Health (MHME) containing the criteria of cerebral death specified that to determine the irreversible cessation of all the cerebral functions, clinical parameters and complementary paraclinical tests must be used. The latter include the positive apnoea test and two electroencephalograms (EEG) of 20 minutes each at an interval of at least six hours; if both EEGs are iso-electric, they confirm brain death.

Despite the constraints on donation from a corpse in the national law, there are religious authorities willing to infringe these limits to obtain organs and save a life.[97]

Kuwait

To try and remedy the shortage of organs from corpses, in 1980 the Kuwait University Transplant Centre launched a programme of organ donation from LNRD limited to ERD (e.g. spouses and close friends). A scientific committee had to evaluate beforehand the sincere disinterestedness underlying the donor's altruistic intention. The first operations were successful and induced surgeons to extend the programme to non-related donors without emotional links. When however it was discovered that some donors had been paid by the relatives of the recipients, transplants from LNRD were definitively stopped.

---

[96] Akrami, Osati, et al., art. cit., 2885–2886.
[97] For example, if someone has made a post-mortem pro-donation will but the family is opposed, it is lawful to explant to save the life of a Muslim (*ayatollah* Lankarani and Safi-Gulpayegani); it is not indispensable to obtain permission for the explantation from a corpse if the relatives of the deceased cannot be found and it is urgent to explant to save somebody's life (*ayatollah* Khamenei S.A.); if a person gives instructions during his lifetime or his family so desires, it is lawful to transplant the organs of the deceased to settle debts contracted in his lifetime or as an act of charity (*ayatollah* Lankarani); a patient can give instructions that after his death his organs are sold so that the money can be spent on works of charity (*ayatollah* Sanei), etc. See in Raza and Hedayat, art. cit., 2888–2890.

Precisely in order to avoid similar risks, Law Decree no. 55 (20 December 1987) of Kuwait on transplants[98] forbids under art. 7 any type of trade in organs and prohibits doctors from transplanting as soon as they are informed of the commercial context of the operation. A similar position had been taken a few months earlier in the Draft Law adopted by the Council of the Ministers of Health of Arab countries (1987) and, earlier still, by Decree no. 698 of 1986 of Iraq.

Other significant points of Kuwait's Law Decree no. 55 are art. 2 according to which each individual may donate organs or dispose of them according to his wishes expressed during his lifetime in the presence of two valid witnesses. According to art. 5 in the absence of these indications, the removal of organs from the corpse requires the written consent of the relatives and is valid if the deceased did not express during his lifetime any written opposition to their removal in the presence of two valid witnesses; in the case of differing opinions between relatives of the same degree, the consent of the majority is required.

In fact, the need for organs reintroduced transplants between totally unrelated individuals. However, a commission of the Ministry of Health must first interview the donor to dispel any doubt on his disinterested act.[99]

The Gulf War temporarily stopped the programme on transplants in Kuwait which was immediately resumed as shown by the 49 kidney transplants (6 of which from corpses) performed between November 1993 and April 1996.

At present, out of almost 2 million inhabitants, some 200–250 new cases of renal insufficiency a year are reported; about 100 are suitable for transplant, but only about forty find blood relative donors available.

Between March 1996 and October 1998, 110 suspected cases of brain death were reported, confirmed in 69 cases. In the remaining 41, failure was caused mainly by delays in carrying out the tests. Consent for donation was obtained in 22 cases (32%) whilst the organs were collected in 19 cases. Regarding the 50 unsuccessful cases, in 40 the families refused whilst in 7 cases the problems were of an administrative nature. The refusal of the families is generally caused by emotional and, sometimes, tribal and religious reasons.[100]

Egypt

In Egypt the first kidney transplant was performed in 1976 (LRD, from a mother to a daughter), the first bone marrow transplant in 1989 and the first liver transplant in 1992. Chronic renal failure causes over 200 deaths a year per million inhabitants.[101] Only 80 patients out of a million currently receive support therapy (dialysis). The most serious problem concerns the absence of a programme of

---

[98] See the Decree in *Recueil International de Législation Sanitaire*, 1988, 39 (4), 907–909.

[99] Samhan M., Mousawi (Al) M., et al., Renal Transplantation from Living Unrelated Donors, *Transplantation Proceedings*, 2001, 33, 2642–2643.

[100] Mousawi (Al) M., Samhan M., et al., Cadaver Organ Procurement in Kuwait, *Transplantation Proceedings*, 1999, 31, 3375–3376.

[101] Barsoum R.S., The Egyptian Transplant Experience, *Transplantation Proceedings*, 1992, 24, 6, 2417–2420.

donation from corpses and a law on the subject. In this context, the first organs were from LRD but, due to the need for organs, since the 1980s transplants from LNRD have begun and in the 1990s represented some 80% of all the operations performed.[102]

At the same time, the progressive liberalisation of the health system has multiplied the number of private clinics and laboratories, encouraging the performance of transplants outside any ethical context, with considerable risks for the people involved. To limit the most dangerous situations, the National Medical Council has set up internal rules (1988) so that the Medical Association is the only body qualified to provide doctors with the authorisation to take part in transplant operations.[103] In addition, in order to avoid the sale of organs by poor Egyptians to rich patients from abroad, the Association has prohibited the transplant of organs from an Egyptian to a non-Egyptian; in any case, the donor should be a blood relative of the patient. This last measure intends to limit the arrival of the poor from neighbouring countries (e.g. Sudan, Somalia) willing to sell one of their organs in Egypt.

To stem the expansion of the trade of organs linked to LNRD, the Egyptian Society of Nephrology (ESN) also prohibited (1 July 1992) its members from taking part in transplants from non-related donors. Here the term "related" indicates "close family members'. The penalty for transgressors is expulsion from the Society and the withdrawal of the authorisation to practise as a nephrologist. This is a policy shared by the Egyptian Medical Association which extends the penalty to all doctors, surgeons, etc.

In spring 1997, the *sheikh* of Al-Azhar, M.S. Tantawi and the Academy of Islamic Research of Al-Azhar made separate pronouncements on the subject. The *fatwa* by the Academy (announced by the then *mufti* of the Republic N.F. Wassel) stated[104] that the transplant of organs was licit (*halal*) exclusively through donation whilst trade in organs was categorically prohibited. Tantawi said he was ready to donate his own organs after death but on condition that he was "completely dead", i.e. that all his vital functions had ceased (biological death), including complete cardiac arrest. This implies the impossibility of proceeding with explantations from a corpse (as the organs would be compromised due to the absence of oxygen) with the exception of the corneas. In other words, the *sheikh* refuses "clinical death" (i.e. the arrest of the brain's functions) in favour of "biological death". The Academy also pronounced that it was in favour of biological death; furthermore, explantation from a corpse is allowed if the deceased expressed his assent in a will or two heirs give evidence of this

---

[102] Barsoum, art. cit., 2417.

[103] Chiffoleau S., Le débat égyptien sur le don et la transplantation d'organes, *Journal International de Bioéthique*, 1998, 9 (1–2), 111–116. Tadros M., Wanted Dead or Alive, *Al-Ahram Weekly*, 12–18 December 1996.

[104] Shaheb S., Dispute over Defining Death, *Al-Ahram Weekly*, 8–14 May 1997. Youssef Y., Les transplantations auront bientôt leur loi, *Al-Ahram Hebdo*, 26 March–1 April 1997.

desire. In the absence of these conditions, explantation is possible with the consent of the "authorities concerned".

Muslim scholars, even when they are favourable to transplants, show resistance to accepting "clinical death" whilst the Egyptian Medical Association opposes this position deeply rooted in theologians as it prevents explantation from a corpse. In spring 1997, at the end of a study Seminar[105] in the presence of 30 of the most authoritative Egyptian doctors and surgeons, it was declared that the majority of the participants agreed on the fact that with the arrest of the cerebral functions, the person is dead. Opponents included the anesthetist Safwat Hasan Lutfi and the neurosurgeon Mamdouh Salama. The anaesthetist maintained that the detachment of the soul appears with the cooling of the body accompanied by the cessation of the vital functions. Brain death does not eliminate the heat of life, which indicates the presence of the soul: the heart beats and the kidneys and liver still function. In addition, international literature indicates the possibility of a return of cerebral activity after its interruption.

A draft bill of 1997 put forward by the Minister of Health Ismail Sallam does not establish when death takes place, merely specifying "the fact that death has taken place should be established by three doctors – including a neurologist – who do not take part in the transplant operation".[106]

In spring 1998 the Health Commission of the Assembly of the Egyptian People discussed the umpteenth bill by the Minister of Health aiming to authorise the explantation from a patient who is "clinically dead." The family of the patient must sign their own consent form. The bill does not allow the explantation as long as there is a minimal hope of the patient's recovery. A certain number of Members of Parliament refused the Bill as it was against Islam. The group, led by Mohamed Qoéta refused the brain death criteria, preferring biological death with the arrest of all the functions. The neurologist Nabil Gobrane added that the explantation of organs from a clinically dead patient is equivalent to premeditated murder: the patient is still alive because his limbs are still working.[107]

The difficult relationship existing on transplants between the Medical Association, the political world and religious authorities is illustrated by the direct words of Hamdy al-Sayed, Chairman of the Egyptian Medical Association in 1996 with reference to the continuous defeats of the bills in favour of transplants in the Egyptian parliament: "Every time we try, our efforts are thwarted by those religious elements who consider transplanting an organ from a dead person to a living one to be against Islam. They are powerful voices in the People's Assembly... Sometimes we wonder who is financing this, because it is obviously a very costly campaign."[108]

---

[105] Shaheb, art. cit.
[106] Shaheb, art. cit.
[107] Youssef M., La greffe d'organes soulève un tollé parlementaire, *Al-Ahram Hebdo*, 15–21 April 1998.
[108] Tadros, art. cit.

Turkey

The first kidney transplant was performed on 3 November 1975 (from a mother to her son). The first kidney transplant from a corpse coming from Europe was performed on 10 October 1978. In the first few years, in the absence of specific rules, organs from corpses from Eurotransplant and from the SEOPF (South Eastern Organ Procurement Foundation) were used, whilst relatives of the first degree with their consent were used for transplants from living donors.

Both the political and the religious Turkish authorities soon proved to be sensitive to the need to develop the practice of transplants with the appropriate technical structures and the training of specialised personnel; nevertheless, there remains a glaring difference between this awareness and the strong popular scepticism that limits the collection of organs. Through a resolution of 6 March 1980, the High Council for the Religious Affairs of Turkey spoke explicitly in favour of transplants.[109] The previous year (3 June 1979) State law no. 2238 had been approved, in which art. 6 requires both verbal and written consent expressed before at least two witnesses by the adult who wishes to donate his organs. Article 7 requires the donor to receive detailed information on the risks of the operation, forbids removing organs from the mentally handicapped and any commercial or opportunistic transaction and forbids the disclosure of the identity of the donor (with the exception of a transplant between spouses and close relatives). The same law states that death must be established unanimously by four doctors (a cardiologist, a neurologist, a neurosurgeon and an anaesthesiologist). The doctor performing the transplant may not be one of those who certified death. The next law no. 2594 (21 January 1982) specifies that in an emergency, if relatives do not exist or cannot be found,[110] in the respect of the brain death criteria, the organs can be explanted without the permission of the next of kin.

In the period between 3 November 1975 and 1 January 1998, 1167 kidney transplants were carried out in Turkey.[111] Regarding the last 846 of these, performed at the Baskent University Hospital, 713 were from living donors whilst the remaining 133 were from corpses. Of the 713 transplants from a living donor, 579 (81%) were from relatives of the first degree; 27 from relatives of the second degree (e.g. cousins) and 50 were children of the relatives of the second degree considered genetically linked to the recipients. The remaining 57 (8%) genetically "unrelated" donors were linked to the recipients only by an emotional-affective

[109] Haberal M., Historical Evolution of Kidney and Liver Transplantation, *Transplantation Proceedings*, 1995, 27 (5), 2771–2774.

[110] Haberal M., Development of Transplantation in Turkey, *Transplantation Proceedings*, 2001, 33, 3027–3029. Haberal, Historical Evolution of Kidney and Liver Transplantation, art. cit., 2772.

[111] Haberal M., Karakayali H., et al., Results of Living-Unrelated Donor Kidney Transplantation at our Centre, *Transplantation Proceedings*, 1999, 31, 3124–3125. Haberal M., Demirag A., et al., Cadaver Kidney Transplantation in Turkey, *Transplantation Proceedings*, 1995, 27 (5), 2768–2769. Haberal, Historical Evolution of Kidney and Liver Transplantation, art. cit., 2772–2774.

bond; this donation in 41 cases was from a wife to a husband, in 7 cases from a husband to a wife, in 8 cases between friends and in one case from an aunt to a nephew. In order to avoid the underground trade in organs, Baskent University Hospital does not accept "living unrelated donors" except between spouses.

Updated figures[112] reveal a significant growth in transplants from corpses in the country. From November 1975 to July 2001 there were 4,709 kidney transplants of which 3,622 from a living donor and 1,091 from cadaveric donor (almost one-quarter of the total); in addition, there were also 311 liver transplants of which 62 from a living donor and 249 from a cadaveric Donor.

Morocco

Law no. 16–98 of 25 August 1999 on donation, removal and transplant of human organs and tissues regulates as a whole the multiple aspects related with the practice.

According to this law, the transplant of human organs (defined as both regenerating elements of the body and not, including human tissue but excluding those associated with reproduction) may be performed only for therapeutic or scientific purposes.

Organs can be removed only on the previous consent of the donor who remains free to withdraw it at any time. The donation or legacy of an organ must be free and must not be remunerated or the object of a transaction in any circumstance or form (except for the reimbursement of expenses).

The donor and his relatives cannot be informed of the identity of the recipient. The donor must be fully informed of the risks inherent to the removal of the organ and any possible consequences.

Explantation from a living donor may be performed only for therapeutic purposes of the recipient, whose relationship has been ascertained (or in favour of the spouse of the donor on condition that the partners have been married for at least one year). Explantation from a living minor or from a living adult under legal protection is illegal.

Any adult individual in full possession of his capacities may, during his lifetime and in the lawful ways, express the desire to accept or forbid the removal of his organs after his death. The authorised public hospitals can remove organs for scientific or therapeutic purposes from a deceased person who, when he was alive, did not refuse this removal, unless there is opposition by the spouse or, when competent, by the members of the family in an ascendant or descendant line.

The removal of organs may not be performed until the brain death of the donor has been established and it has been ascertained that the causes of death do not include any suspicious circumstances.

Before the transplant, a doctor must ensure that the recipient agrees.

---

[112] Haberal, Development of Transplantation in Turkey, art. cit., 3027–3029.

Pakistan

Each year in Pakistan[113] at least 12,000 new patients with ESRD are identified who are added to the previous patients with the same pathology but who are incurable due to the low number of transplants performed each year. Only one-quarter of patients can have recourse to dialysis due to the shortage of equipment and money, as the public health service does not generally offer it free of charge. The difficulty of their situation is accentuated by the lack of a programme for donation from corpses, moreover under discussion at the level of legislation and which is a type of transplant which has little social approval. The first consequence is represented by the search for LRD amongst family members up to the spouses. Between November 1985 and October 1994, 300 kidney explantations were performed from LRDs at the Dow Medical College of Karachi (the major centre for transplants in the country), whilst transplants from LNRDs barely seem practised as they are deemed unethical on the basis of the worrying example of commercial transplants in India. In the period between 1993 and 1997, 665 kidney transplants were carried out from a living donor and only two from a corpse.

The Pakistani surgeons underline the urgency to increase transplants from LRD and from corpses as this is much less expensive for the State than chronic dialysis, not to mention the unquestionable therapeutic improvement and a better quality of life for the patient. The family-based social system encourages collecting organs in large families in which the older members often have the final word, being able to persuade a younger member to donate a kidney to a relative in need. The literate and employed member tends to be luckier as he is not called on as a possible donor. At the same time it is easier for a literate and employed person to obtain an organ in the case of need, in order to return to productive activity.

Updated figures[114] indicate 725 kidney transplants in the country since 1999: 401 were fromLRD; 43 from spouses and 281 from LURD. 14 transplants from corpses were performed at the Sindh Institute of Urology and Transplantation (SIUT) with kidneys supplied by the Eurotransplant Centre. Despite the absence of a law on brain death, the only two transplants from local corpses (according

---

[113] Naqvi S.A., Hussain M., et al., Economics of Renal Rehabilitation in Pakistan: a Case for Increasing Transplantation Activity, *Transplantation Proceedings*, 1992, 24 (5), 2125–2126; Naqvi S.A. and Rizvi S.A.H., Renal Transplantation in Pakistan, *Transplantation Proceedings*, 1995, 27 (5), 2778; Naqvi S.A., Mazhar F., et al., Limitation in Selection of Donors in a Living-Related Renal Transplant Programme, *Transplantation Proceedings*, 1998, 30, 2286–2288; Akhtar F., Mazhar F., et al., Donor Selection in Living Donors: Prospects and Problems, *Transplantation Proceedings*, 1999, 31, 3385.

[114] Rizvi A. and Naqui A., Current Issues and Future Problems in Transplantation in East Asia, *Transplantation Proceedings*, 2001, 33, 2623–2625. Organs Donated by Brain-Dead Girl Give New Lease of Life to Four, *Karachi News*, 26 January 2005, in www.jang.com.pk/thenews/ jan2005-daily/ 26-01–2005/metro/karachi.htm

to the criteria of brain death) were performed at the SIUT respecting the explicit wishes of the relatives and the deceased individuals. The first case, in 1998, concerned a young kidney donor, the second case (January 2005) a multi-organ explantation (two cornea and two kidneys) from a young woman of 22. The SIUT is the only Pakistani public sector institute where treatments for urology, nephrology and transplantation patients are provided free of charge (including lifetime immunosuppressive medications).[115]

As for brain death criteria, paragraph 21.2 of the Code of Ethics (2001–2002) of the Pakistan Medical and Dental Council specifies: "Prior to considering transplant from the dead donor, brain death should be diagnosed, using currently accepted criteria, by at least two independent and appropriately qualified clinicians, who are also independent of the transplant team. . . ."[116]

United Arab Emirates

Federal Law no. 15 of 1993 regulates the donation from a living donor and from a corpse.[117] For a transplant from a LRD, the blood relationship between the donor and the recipient must not be less than the second degree. The donation of vital organs is prohibited. Donation must be exempt of any social or financial pressure and requires the written consent of the donor who may change his or her mind before the operation. The donor's medical record must show that he or she was sufficiently informed of all the possible consequences of the donation.

Regarding explantation from a corpse, brain death is attested according to the parameters laid down by the law. No doctor involved in the transplant may take part in the diagnosis of brain death. Consent must be given by the relatives. Removal is not lawful if the deceased had expressed during his or her lifetime, by means of a will (with the attestation of two witnesses), the refusal to donate his or her organs post-mortem. Trade of organs is prohibited.

Tunisia

The first cornea transplant was carried out in 1948 at the Charles-Nicolle Hospital. In 1996 the first national eye bank was set up in Tunisia run by the CNPTO (Centre National pour la promotion de la transplantation d'organes). Since 1986 there has been a kidney transplant programme in Tunisia which has reduced journeys to Europe by patients in search of an organ. Between June 1986 and June 1992, 100 kidney transplants were performed in the country, 91 of which were from living related donors and the remaining 9 from corpses.[118]

---

[115] SIUT Performs Kidney Transplant on Under Ten Children, *Karachi News*, 30 September 2005 in www.jang.com.pk/thenews/sep2005-daily/ 30–09–2005/metro/karachi.htm

[116] www.pmdc.org.pk/ethics.htm

[117] Shahat, *Islamic Viewpoint of Organ Transplantation*, art. cit., 3274.

[118] Matri (El) A., et al., Organ Transplantation in Tunisia, *Transplantation Proceedings*, 1993, 25 (3), 2350.

The law of 25 March 1991 specifies that explantation from a corpse is allowed only if the subject is cerebrally dead (art. 15) and states that (art. 3): "Organs may be removed . . . on the condition that, during his lifetime, he has not indicated his objection to such removal and that, after his death, there has been no objection to such removal on the part of the following persons, enjoying full legal capacity, in the following order: the children, father, mother, spouse, brothers and sisters, legal guardian."[119] It should be noted that here the wording of consent consists of "not expressing opposition" to explantation either by the person when alive or by his or her relatives, rather than expressing "explicit consent". The wording guarantees a greater possibility of finding organs (this is because hardly any people express themselves during their lifetime), although there remains the insurmountable obstacle of the possible opposition of family members.

In 1999 a new law allowed citizens to include the word "donor" on their identity cards.

Jordan

Law 23 of 24 April 1977,[120] on transplants from a living donor, prohibits the explantation of vital organs even if there is the consent of the donor (art. 4-A1) and prohibits all commercial trade (art. 4B). The transplant from a corpse is lawful if the deceased had made a will in this sense before dying (art. 5-A). Conversely, if the deceased had not expressed himself, the explantation is lawful when there is the consent of "one of the two parents of the deceased if they are alive" or the legal representative if the parents are not alive (art. 5-B). The article mentioned here seems to mean that the silence of the spouse is without influence, whilst the opposition of one of the two appears to prevent the operation. Lastly, if the deceased is not identified and nobody claims the body within 24 hours of the death, the transplant is lawful if there is the consent of the appropriate authorities (art. 5-C). Article 8 specifies that explantation is lawful only when death is certified by a doctor but not by the same one who will perform the transplant.

Amongst the amendments of the subsequent law no. 17 of 20 July 1980, art. 3 specifies that it is prohibited to transplant an organ if the operation mutilates the corpse to the extent of impairing its dignity.[121]

CONCLUSION

Amongst the numerous problems discussed which influence or hinder the development of transplants, the following can be recalled: the inadequate general awareness of the importance of transplants; the activity of forces and authorities (including religious ones) that negatively influence public opinion on transplants;

---

[119] See the Law in *International Digest of Health Legislation*, 1991, 42(3), 449–452.

[120] In *Al-Jarida al-Rasmiyya*, 1 June 1977, no. 2703, 1320–1321 (in Arabic).

[121] In *Al-Jarida al-Rasmiyya*, 1 September 1980, no. 2955, 1278–1279 (in Arabic).

organ trafficking; the strong pressure of foreign doctors (especially in the Gulf countries) which damages the experience and development of local transplantology. It is worth adding some more general obstacles[122] such as the poor effort in the prevention of pathologies that cause the demand for life-saving organs; health costs do not represent a priority in national budgets; the absence of group spirit amongst doctors and cooperation between the various centres due to conflicts and personal interests; the still limited presence of coordinating centres for organ donation; the deficient national health insurance.

---

[122] Shaheen F.A.M., Souqiyyeh M.Z., et al., Current Issues and Problems of Transplantation in the Middle East: The Arabian Gulf, *Transplantation Proceedings*, 2001, 33, 2621–2622.

AIDS

INTRODUCTION   199
PARTICULAR ASPECTS   207
THE COUNTRIES   210

INTRODUCTION

The theological principle of divine omnipotence was for a long time a compulsory reference in attempts to explain the causes of disease, pestilence and epidemics in Muslim culture. The dominant attitudes on this principle have undergone variations throughout the centuries in parallel with developments in medical science and epidemiological knowledge. The traditions (*ahadith*) dating back to the Prophet appear partially contrasting. In the light of divine omnipotence, the fact that a pestilence did not affect all men or animals gave rise historically to two trends of thought: one refusing the causal nexus between contagion and effects (in which case the disease should have affected everyone) and every relation was caused by God (reference to the theological voluntarism of the *Ash'arite* school); the other accepting the natural nexus of cause and effect.

In past centuries, for many Muslims, Islam represented a sort of sanctuary where refuge could be taken to fight illness and disease; for others, however, Islam was capable of adapting to Western technical–scientific discoveries. When a serious epidemic of the plague broke out in Tunisia in 1784–1785, the majority of the population did not accept the sanitary measures advised by the Europeans (isolation, fire, destruction of objects, etc.), preferring to take refuge in acts of worship. However, the aforementioned measures were adopted by the European residents in Tunisia with the result that there were twice as many deaths amongst the Tunisians as amongst the Europeans.[1] A century later, when there was an epidemic of typhoid and cholera, the religious authorities had lost a great deal of their authority in the health domain whilst there was greater acceptance of European practices (with the appended values), including by local doctors who were, at the same time, less traditionalist.[2]

Today, coinciding with the progress of medicine, the problem of the will of God is felt less strongly or is put into parentheses whilst the Prophetic traditions

---

[1] Nanji A.A., Medical Ethics and the Islamic Tradition, *Journal of Medicine and Philosophy*, 1988, 13 (3), 257–275.

[2] Ibid., 272–273.

Dariusch Atighetchi, Islamic Bioethics: Problems and Perspectives.
© Springer Science+Business Media B.V. 2009

that refused the natural dimension of contagion are not considered in the thera-peutic-scientific context.[3]

As is intuitable, there does not exist a contemporary disease that lends itself, like AIDS, to an apologetic approach aimed at exalting the values of Islam. Initially, the infection was presented to Muslim (and non-Muslim) public opinion as the effect of homosexual or extramarital relations and drug addiction. The low percentage of HIV infection in many Muslim areas was considered the practical consequence of the customs of Islam showing a salvific efficacy. The Word of God offers clear indications against two of the main causes responsible for AIDS, i.e. fornication and homosexuality. On fornication, Koran 17.32 states: "Nor come nigh to adultery: for it is a shameful (deed) and an evil opening the road (to other evils)." Regarding homosexuality, the passage 26.165–166 can be quoted: "Of all the creatures in the world, will ye approach males, and leave those whom God has created for you to be your mates? Nay, ye are a people transgressing (all limits)!"[4]

The Muslim solution to the risk of AIDS consists of premarital abstinence and precocious marriage. Although aware of the slow (but constant) growth of HIV infection in the majority of Muslim countries, opportunities are never lost to recall that these countries have been lucky as their religious, cultural and social values teach self-respect and morality.[5]

In 1985, the magazine *Al-Itisam* wrote that Europeans and Americans today understand that AIDS is a punishment from Heaven.[6] A *hadith* (in the collection of Ibn Maja) recalls that when a monstrous sin is diffused in a community, it is struck by the plague and other unknown diseases; AIDS today becomes a punishment for homosexuals. The Koran 7.33 states: "Say: The things that my Lord hath indeed forbidden are: shameful deeds, whether open or secret; . . .", a passage which is often taken as a reminder that those who follow the Koran avoid infection. The majority of the *fatwas* since the mid-1980s appear to ignore the problems connected with the care of the patients to the advantage of the apologetic approach where it is shown that the infection would not exist if a Muslim lifestyle were adopted. The Egyptian weekly *Al-Liwa al-Islami* (The flag of Islam) under the headline "To follow the Path of Islam is the best way not to get infected" indicated AIDS as God's punishment for those who sully the country with their sins.[7] The South African *mufti* Zubair Bayat states that anyone who destroys chastity with indecent behaviour is affected by various

---

[3] Rahman F., Islam and Medicine: A General Overview, *Perspectives in Biology and Medicine*, 1984, 27 (4), 585–597.

[4] See also the Koran 27. 54–58 and 7.80–84.

[5] Sayeed S.A., AIDS in the Light of Islam: Malady & the Panacea, *The Muslim World League Journal*, 1996, 23 (11), 45–48.

[6] See Rispler-Chaim V., Islamic Medical Ethics in the 20th Century, *Journal of Medical Ethics*, 1989, 15 (4), 203–208.

[7] Gawhary (El) K., Breaking a Social Taboo – AIDS Hotline in Cairo, *Middle East Report*, 1998, Spring, 18–19.

illnesses, including AIDS.[8] El-Marrouri was extremely critical of contraception and AIDS in the paper *Al-Sahwa* of the radical Moroccan Islamic association *Al-Adl wa-l-Ihsan* (Justice and Reform). AIDS is described as divine punishment for fornicators, which is why it is essential to summon young people back to Islam instead of fighting the pathology with condoms, as advised by the Ministry of Health because "the condom is the vector of the pandemic of fornication and lewdness."[9]

Resolution no. 82/13/8 of the Academy of Muslim Law of Jeddah (8th Session in Brunei, 21–27 June 1993), in the preamble, agreed with the opinion according to which adultery and sodomy are the main causes of the epidemic.[10] To fight it, it is necessary to combat perversion, the sinful tendencies of the mass media, as well as the problems caused by tourism.

One of the most radical interpretations is offered by the Algerian M. Aniba[11] who includes the phenomenon of AIDS in a total rejection of everything that comes from Western culture, with AIDS alleged to be a product of its behavioural models. The introduction of condoms is refused as this would lead to a relaxing of customs, whilst the only solution for the incurable pathology is a return to the S*hari'a* and the purity of Islamic morals, both of which have been betrayed since the independence of Algeria. The pathology is alleged to have started in the West (USA, then England, France, Germany, etc.) with the consequence of currently representing the most serious danger in the history of Western civilisation which "has not been capable of improving the intellectual, moral and spiritual qualities of its societies". The free circulation of the sexes, sexual freedom, adultery and prostitution are the immediate causes, whilst in the name of the respect of "private life", freedom and democracy, the body can be used without limits, in the same way as animals. For these reasons, Western civilisation is contaminating the whole world with the damage caused, inter alia, by the emancipation of women, whose inclusion in Algerian public life, for example, has increased debauchery, corruption, rape, the number of illegal abortions and illegitimate children.[12]

The dependence between AIDS and sin is underlined in several medical documents. For example, in 1993, the Recommendations of the Seventh Islamic Medical Seminar[13] on "AIDS – related social problems – an Islamic perspective" (Kuwait, 6–8 December 1993), jointly organised by the Supreme Council for Islamic Fiqh (Jeddah) and by the WHO, recalled that to fight AIDS: "Promotion of virtue,

[8] Bayat Z., The Road Towards Good Health, *Al-Jamiat*, 2000, 4 (4), in www.jamiat.org.za/al-jamiat/aljam44.html

[9] Dialmy A., Moroccan Youth, Sex and Islam, *Middle East Report*, 1998, Spring, 16–17.

[10] See www.islamibankbd.com/page/oicres.htm#82/13/8; Ossoukine A., La prise de parole en bioéthique ou l'affirmation d'une identité culturelle, *Journal International de Bioéthique*, 2000, 11 (3/4/5), 191–199.

[11] Aniba M., *Le mal du siècle vu par l'Islam*, Kafana, 1992.

[12] Ibid.

[13] www.islamset.com/bioethics/ aids1/index.html

public decency and lawful sexual behaviour and morality are the natural preventive measures" (point VIII). The medical–religious project was then stated: "The Seminar calls for the implementation of Islamic Law as an effective safeguard against the spread of such diseases which threaten individuals and societies" (VIII.5). The Deputy Minister of Health for Preventive Medicine of Saudi Arabia, Yacob al-Masruwah, at the XXVI Special Session of the General Assembly of the United Nations dedicated to HIV/AIDS (New York, 27 June 2001) recalled the reasons for the low number of infected persons in the Kingdom: "The reason for this is the adherence to the religion of Islam which prohibits sexual relations outside the confines of marriage."[14] The Egyptian Minister of Health and Population, Ismail Sallam, expressed himself similarly on the same occasion (New York, 26 June 2001): "Moral and religious values have protected many countries and we should not omit these resources when it is now desperately needed."[15]

Amongst Muslims in Europe, the opinion of Larbi Kéchat, Rector of the Addawa Mosque (rue Tanger) in Paris,[16] is more moderate (tolerating contraceptives), although oriented in the same direction. In his opinion, God does not like suffering and wants health for man who, as soon as he transgresses the laws that regulate his body and the environment in which he lives, is exposed to imbalances. AIDS is an epidemic caused by the cultural climate dominating in the West; a climate that excites frustrated instincts, leading to a search for pleasure at all costs. AIDS is alleged to be the result of disorder in sexual relations, "it is the responsibility of society regarding the invention of freedom" and it is the result of an absence of values. The condom is not the solution but it may be useful to limit damage; however, this must not be translated into advertising campaigns oriented at exalting sexual freedom that "perturb" young Muslims. Intensifying the provocative stimuli inevitably causes imbalances and the latter cause the disease. Marriage, concludes Kéchat, is the only solution that also allows health to be protected.

The preconceptions on HIV-positive individuals and AIDS patients seem strong and widespread in much of the Muslim world. The attitude tends to become discriminatory in the health regulations of certain countries, with regard to the huge foreign labour force.

Whilst in the West there is discussion on how to help AIDS patients clinically and psychologically, where to care for them, for how long, who is to pay for the care, how to protect their rights at work, in society, etc., Muslim medical ethics have concentrated preferably on the causes of the pathology and the lesson to be drawn to underline the reconfirmed moral superiority of Islam. Even the Islamic Code of Medical Ethics (Kuwait, 1981) does not avoid the temptation of apologetics. The text, in fact, at a certain point presents once again the classic

---

[14] www.un.org/ga/aids/ statements/docs/saE.html

[15] www.un.org/ga/aids/ statements/docs/egyptE.html

[16] Bertini B., *La Représentation du SIDA chez les Femmes d'Origine Maghrébine résidantes en France*, Paris, Mémoire de DEA, EHESS, 1996, 58–59.

principles of Muslim medical ethics according to which being a doctor is in the first place an act of charity; medicine is a gift that all must enjoy: rich and poor, sinners and virtuous alike. The doctor is an instrument that God uses to alleviate the suffering of men: the patient is the master and the doctor is at his service. The Code subsequently recalls that the Muslim doctor must not only heal but also prevent illness according to the Koran "Let not your own hands push you into destruction", i.e. he must intervene on any behaviour of the patient that is harmful: smoking, alcohol, dirtiness, environmental pollution, etc. Lastly, the document (Chapter IX) comes to venereal diseases with a clear apologetic reflection. It says that the prophylaxis against venereal diseases intends to re-evaluate values such as chastity, purity and self-control and adds, with a note of controversy: in certain "developed" countries, the spread of gonorrhoea and syphilis has reached epidemic proportions, to the extent of inducing the health authorities to declare a state of national emergency. Nevertheless, the Code continues, Western doctors continue to underestimate the problem whilst the information media do not dare present chastity as the only solution to wipe out these diseases. Whilst Western doctors may publicly oppose pollution, smoking, obesity, etc., moral–sexual permissiveness is the only field precluded to them, where they cannot be moralistic, having to restrict themselves to healing.[17]

As underlined in the chapter on Medical Ethics, in a context in which religious law ought to be totalising, as the doctor is an instrument of God his task cannot be limited to healing disease, but he must deal with the whole person including the values orienting daily behaviour, even penetrating (to a certain extent) the personal sphere. The "totalising" value of Islam is reconfirmed by the document published by the EMRO (Eastern Mediterranean Regional Office of the WHO) dedicated to the influence of religions (in the first place Islam) in wiping out the scourge of AIDS and sexually transmitted diseases (STDs).[18] The text recuperates some general considerations on the relationship between Islam and the West in which it is recalled that whilst in the West the separation between public and private morals is advocated, Islam judges this separation "artificial" as the legal authorities of a society must always maintain control over the ethical–moral values in it and their application. Whilst in the West social values may change under the pressure of autonomy and freedom of the individual, a modification that must be taken into account by the legislators, Muslim jurists, deem that the law must protect the ethical values of a religious society. The same document refers to art. 24 of the *Declaration of Human Rights in Islam*, approved in Cairo by the Organisation of the Islamic Conference on 5 August 1990: "All rights and freedoms guaranteed by this Declaration are governed by the rules of Islamic Law." Furthermore, art. 17 responds to the concept of "privacy" dominant in the West: "everyone has the right to live in an

---

[17] *Islamic Code of Medical Ethics*, Kuwait, 1981, in www.islamset.com/ethics/code/
[18] WHO, *The Role of Religion and Ethics in the Prevention and Control of AIDS*, Regional Office for the Eastern Mediterranean, 1992, 22.

environment that is free from moral corruption and disease, which enables him or her to develop his or her personality morally, the society and the state being bound by law to guarantee that right.". In other words, the reference to human rights does not allow models of behaviour that threaten the integrity and the principles of a Muslim society to prevail.[19]

The rules of lifestyle in Arab-Muslim countries may sometimes be in antithesis with some of the fundamental principles of the Charter of Human Rights.[20] For example, it is very difficult to defend an individual's right to freely live his or her sexual choices. Talking about condoms, multiple relations or extramarital relations is considered an incitement to debauchery. Muslims do not accept sex education and a policy of AIDS prevention based on free and safe (from the health point of view) sexual activity.

In 1991 M.H. Wahdan, Director of the Disease Prevention and Control Department (WHO, EMRO), with reference to the spread of AIDS in the Middle East said[21] that there were "specific indications of the fast spread of the disease locally in almost all countries of the region, especially among certain groups such as drug addicts, prostitutes and homosexuals where rates of infection have soared in recent years"; the greatest problem lies in identifying these individuals in societies where official data on the subject appears unreliable.

The number of infections reported at the end of 1996 in the Middle East and North Africa corresponds to less than 1% of the world total, and the majority of this percentage appears to have been originally caused by infected transfusions and haemoderivatives. In these two areas, 2,700 people died from AIDS and cor-related diseases, compared to 170,000 in Europe and 4.6 million in sub-Saharan Africa.[22]

The data from the WHO and UNAIDS concerning the total estimates of HIV/AIDS-infected individuals in 2003 in northern Africa and the Middle East[23] shows a figure between 470,000–730,000; the vast majority of these live in Sudan with about 500,000 cases, followed by Pakistan with 80,000 estimated cases, Somalia with 43,000 cases, Iran with 30,000, Morocco with 15,000 and Yemen with 11,227 cases. In south-eastern Asia, Indonesia must be mentioned with about 200,000 people infected by HIV and Malaysia (which has a Muslim majority) with 52,000 people infected. Compared to its population (68 million), Egypt reports a very low number of infected cases, namely 3,584. Today heterosexual relations are the most common means of transmission in the region, with about 55% of cases; the spread of the infection is also growing strongly

---

[19] WHO, , op. cit., 22–23.
[20] Himmich H. and Imane L., Ethique et Sida en terre d'Islam: un combat difficile, *Sociologie Santé*, 1993, 9, 36–41.
[21] WHO, op. cit., 5.
[22] Lenton C., Will Egypt Escape the AIDS Epidemic? *The Lancet*, 5 April 1997, 349 (9057), 1005.
[23] UNAIDS/WHO, *AIDS Epidemic Update, December 2003*, Geneva, 2003, 5. See some comments on the data in Kim J.Y., HIV/AIDS in the Eastern Mediterranean: A False Immunity? *Eastern Mediterranean Health Journal*, 2002, 8 (6).

due to injected drugs, the number of which quintupled between 1999 and 2002. The divergence on the figures presented by different international and national organisations expresses the internal weakness of the surveillance systems.

An element that is common to almost all Muslim countries, or those with a Muslim majority, is the underestimation of the phenomenon and its numerical dimension (but nevertheless low as a whole) as shown by the fact that the estimates of the WHO and international organisations are always much higher than the figures provided by the Muslim governments whose information – especially in the past – was in the best hypothesis incomplete whilst some governments even refused to provide any.

The reasons for the modesty-fear shown in reporting the spread of AIDS derive from certain characteristics of the Muslim world. Scientific research on the concrete sexual behaviour of samples of the population is difficult, as the traditional mentality refuses to communicate to outsiders about sexual conduct highlighting "licit" and "deviant" aspects. The political and religious powers are opposed to translating into figures "illicit" behaviour (premarital relations, multipartners, homosexuality, masturbation, zoophilia, etc.) because this statistical information implies the acknowledgement of the extent of this immoral conduct. Such a situation is difficult to accept in countries where governments are exposed to the accusation of betraying the principles of Islam by radical or fundamentalist religious movements (the close connection between politics and religion cannot be avoided). When a country tolerates similar surveys under international pressure or to legitimise its aspirations to modernity, great discretion in the disclosure of the results will be required, although this information is essential precisely to launch specific preventive measures. In the Arab context, AIDS (and STDs) lacks visibility due to the cultural climate that judges taboo everything that relates to the private and sexual sphere, in particular the related diseases.

In these contexts, social pressure is very strong, and in particular the discredit associated with pathologies of sexual origin is feared. Revealing the infection (as AIDS is linked to drugs and sexual deviance) means admitting deviant behaviour with respect to the social norm, with the risk of losing employment and being shamed in front of the family who may react by marginalising the patient both due to the shameful characteristic linked with the disease and out of fear of contagion; this rejection generally takes place precisely when the person who has contracted the disease is in greatest need of psychological and moral support.[24] The disease

---

[24] Regarding the situation of HIV-positive Muslim immigrants in France, Larbi Kéchat recalls that they are very lonely individuals. His opinion is very critical Towards the dominant attitude Towards sexuality, both in the West and in the Muslim world, but for opposing reasons. In the West, sexuality is presented with "nauseous exhibitionism", whilst Muslims express themselves with "mortal silence", synonymous with hypocrisy. It is such a taboo that no Muslim seems to want sexual intercourse! In Kéchat's opinion, this concealing of human nature inevitably produces deviant behaviour. See Bertini, op. cit., 77–79.

is almost always accompanied by a strong sense of guilt with regard to the family and spouse, such as to impose silence with obvious risks for the healthy spouse. On the contrary, some maintain that the wife of an infected husband should be warned of the danger precisely because the husband-patient feels responsible to her and, if he does not reveal the infection, he tends, at least, to abstain from sexual intercourse. When he discovers his infection, he may have two choices: (a) to abandon his family without informing them, to avoid shame and sorrow; (b) to hide his AIDS as long as possible, possibly declaring another pathology. When it is not possible to remain silent, only a close group of relatives is informed to avoid the dishonour falling on the whole family, both in the country of emigration and in the home country. Sometimes when the patient is looked after by his family, the origin of the disease, which is never named, is ignored. If the family knows the truth, nobody else must know. Despite these basic trends, there are indications of acceptance and solidarity by the family, even if this is the case when there is nothing more to be done. Solidarity is easier when the infection is not caused by behaviour "at risk" (but, for example, by a transfusion), in which case the infected person reveals his or her illness more easily to the family.

Despite this problematic context which is very hostile to AIDS patients, in December 2004 the HIV/AIDS Regional Programme in the Arab States (HARPAS) issued the Cairo Declaration signed by 80 authoritative religious leaders (Muslims and Christians) from 19 Arab countries. This declaration calls for fighting all prejudices. The most demanding point stated: "People living with HIV/AIDS and their families are worthy of care, support, treatment and education whether they are responsible for their illness or not. We call for our religious institutions in cooperation with other institutions to provide spiritual, psychological and economic support to these people."[25]

AIDS is discussed amongst young immigrants more easily and openly whilst the subject is taboo with the elder members of the family. It is not even the same thing to be a young HIV-positive male or female; the latter case appears particularly immoral and women can be sent back to Africa without telling the truth to the people they will stay with, or "disappear" due to their "contacts" with people who are "not normal" (sex, drugs, etc.). In national contexts where the woman is juridically and socially inferior to man, the woman with HIV/AIDS is at an even greater disadvantage.

Faith in absolute divine omnipotence also makes itself felt in the perception of the pathology, as it is unlikely that the Muslim patient loses all hope in help from God. If the Muslim does not hope in God, it means that he doubts His omnipotence and questions the faith itself.

A further fact, perhaps, contributes to the limitation of the spread of the infection in Muslim contexts: following the example of Prophet Muhammad, Muslim males are circumcised and this custom seems to help reduce the spread of HIV.

---

[25] Compassion and AIDS in the Arab World, *Al-Ahram Weekly Online*, www.weekly.ahram.org. eg/2005/765/sc9.htm

A survey of 27 international clinical studies on the relationship between male circumcision and AIDS was carried out by specialists at the London School of Tropical Medicine and Hygiene. The result published in the October 2000 issue of the journal *AIDS* concludes that circumcision is associated with a significant reduction of the risk of AIDS amongst men in sub-Saharan Africa, in particular amongst those most exposed to the risk of infection.[26] Furthermore, the suggestion is made to include male circumcision amongst the additional measures for the prevention of infection in the areas where circumcision is not traditionally present, but only on condition that all the hygienic-sanitary guarantees are respected.[27]

PARTICULAR ASPECTS

According to the experts of the EMRO-WHO, doctors ought to inform patients of the nature of their pathology and take the adequate precautions to prevent contagion of the spouse or of other relatives. However, this principle, in contexts where the legal protection of the individual is often more problematic, risks leading to an underestimation of the wishes of the ill individual, although for the purpose of protecting the health of others.

In Muslim law, if the pathology is incurable and is harmful for the other partner, marriage should be prohibited (*haram*) for the patient.[28] A healthy person should not marry an HIV- or AIDS-infected partner even if he or she knows their conditions, both in order to protect his or her health and to defend any offspring from the risks of infection. If both partners are infected, there are no obstacles to their marriage; however, precautions should be taken in order not to infect any children.[29]

Concerning married couples, Resolution no. 82/13/8 of the Academy of Muslim Law (Organisaton of the Islamic Conference, 8th Session in Brunei, 21–27 June 1993) ordered that the infected spouse must inform the other and cooperate with him or her in all protective measures.[30] According to the Islamic Fiqh Academy of India, if an infected person gets married without informing his wife of the infection, the woman can divorce; the same applies if infection takes place after marriage.[31] The Islamic Code for Medical Ethics of the IOMS in 2004

---

[26] Weiss H.A., Quigley M.A., et al., Male Circumcision and Risk of HIV Infection in Sub-Saharan Africa: A Systematic Review and Meta-Analysis, *AIDS*, 2000, 14, 2361–2370.

[27] Moses S., Plummer F.A., et al., The Association between Lack of Male Circumcision and Risk for HIV Infection: A Review of Epidemiological Data, *Sexually Transmitted Diseases*, 1994, 21, 201–210. See in addition Szabo R. and Short R.V., How does Male Circumcision Protect against HIV Infection? *British Medical Journal*, 320, 1592–1594.

[28] Rahman Z., AIDS and Relevant Issues in a Muslim Marriage, *Malayan Law Journal*, in www.mlj.com.my/free/articles.asp

[29] Ibid.

[30] www.islamibankbd.com/page/ oicres.htm#82/13/8.

[31] See Islamic Fiqh Academy (India), *Important Fiqh Decisions*, in www.ifa-india.org/english/

(art. 29(d) and art. 60) states that a doctor can disclose to the healthy spouse the infection of the other spouse but the announcement must be made in the presence of both spouses.[32]

If this is the ideal approach, the concrete situation is influenced by others factors that make disclosure to the partner difficult. As AIDS is transmitted in particular through sexual intercourse, the healthy partner can refuse; in the case that he or she wants to have intercourse, the condom must be used to limit the risk of infecting himself or herself or their children.[33]

In Muslim law, sexual intercourse generally represents a duty for the couple according to Koran 2.222.[34] The refusal by either of the spouses can be a lawful reason for divorce. In particular, refusal by the wife, without a valid reason, precludes her right to maintenance by the husband.[35] Today, amongst the reasons generally acknowledged by Muslim countries to refuse sexual relations with a spouse or for a divorce is infection by HIV or AIDS of one spouse. For example, in Malaysia, section 52(1) (f) of the Islamic Family Law (Federal Territories) Act 1984 provides: "A woman married . . . shall be entitled to obtain an order for the dissolution of marriage . . . on any one or more of the following grounds, namely, that the husband has been insane for a period of two years or is suffering from leprosy or vitiligo or is suffering from a venereal disease in a communicable form."[36]

Breastfeeding a baby by an infected mother entails little risk except when the nipples bleed or are damaged. This explains why the Seventh Islamic Medical Seminar on "AIDS-related Social Problems" (1993) is in favour of maternal breastfeeding. However, in the case of available alternatives, infected mothers can abstain from nursing.

Is abortion lawful when the mother is infected? Before infusion of the soul, the positions of the medical and juridical-religious bodies are not unanimous. The Recommendations of the Seventh Islamic Medical Seminar reconfirmed the position expressed in the final document of the IOMS in its 1983 international conference on "Human Reproduction".[37] In 1983, the participants confirmed the prohibition of abortion especially after 120 days but also recalled that "some participants, however, disagreed, and believe that abortion before the fortieth day, particularly when there is justification, is lawful". The 1993 Recommendations reconfirm this decision and also applied it to AIDS-infected mothers.[38]

---

[32] IOMS, *The Islamic Code for Medical and Health Ethics*, 2004, in www.islamset.com/ioms/code2004/index.html

[33] Recommendations of the 7th Islamic Medical Seminar on "AIDS-related Social Problems" (Kuwait, 1993), in www.islamset.com/bioethics/aids1/index.html

[34] "But when they have purified themselves, ye may approach them in any manner, time or place ordained for you by God."

[35] See for example what Section 59 (2) of the Islamic Family Law (Federal Territories) Act 1984 of Malaysia says, in Rahman Z., art. cit.

[36] Rahman Z., art. cit.

[37] Recommendations in Gindi (Al) A.R. (ed.), *Human Reproduction in Islam*, Kuwait, Islamic Organization for Medical Sciences (IOMS), 1989, 276.

[38] See www.islamset.com/bioethics/aids1/index.html

Resolution no. 90/7/9 of the Academy of Muslim Law (Fiqh) of Jeddah during its 9th session (Abu Dhabi, 1–6 April 1995) appears stricter on the subject: considering that the transmission of AIDS takes place, in the overwhelming majority of cases, only at an advanced stage of pregnancy (after the foetus is invested with life) or during delivery, it is therefore not permissible to abort the foetus.[39]

The Islamic Fiqh (Muslim Law) Academy of India decided in favour of the abortion of foetuses of mothers with AIDS before the infusion of the soul (120 days).[40]

The Academy of Muslim Law (Fiqh) of Jeddah in the aforementioned Resolution no. 90/7/9 (Abu Dhabi, 1–6 April 1995) dealt with the case of the deliberate transmission of the disease distinguishing between: (1) voluntary transmission to a single individual and (2) the intention to spread the pathology in society. Both are prohibited acts (*haram*). (1) In the first case, i.e. the person with AIDS who voluntarily infects another person, if the person is infected but is still alive, the party responsible is subjected to *ta'zir* sanctions (dissuasive punishments decided by judges); if the infected person dies, the responsible party should be punished with death for premeditated murder. In the case in which the voluntary attempt to infect another person fails, the responsible party is liable to *ta'zir* penalty. (2) In the case that the infected person wants to spread the infection in society as a whole, it is a "*hiraba*" act (i.e. a crime against humanity) for which the penalties in the Koran 5.33[41] are applied.

Similarly, in a *fatwa* in November 2003, the Iranian *ayatollah* Ozma Yusef Sanei stated that AIDS patients who wilfully spread the HIV must be punished with death as it is wilful murder.

There is often the request by the faithful, health practitioners and religious figures to isolate an infected patient to avoid risks for society. The aforementioned Resolution no. 90/7/9 of the Islamic Fiqh Academy (1–6 April 1995) is contrary and specifies that if there are no risks of contagion, isolating victims is not a necessity in the eyes of *Shari'a*.[42] Isolation or quarantine of the infected person is easily accompanied by discriminatory attitudes of various kinds towards the patient. The Recommendations of the Seventh Islamic Medical Seminar on AIDS (Kuwait, 1993) were against these discriminatory attitudes: "Regardless of how a person contracts AIDS, everyone has the same right to the necessary medical treatment, psychological support and health care. . . . No AIDS sufferer should have to endure any injustice or discrimination or humiliation as a result of his or her predicament."[43]

---

[39] www.islamibankbd.com/page/ oicres.htm#90/7/9

[40] www.ifa-india.org/ english/

[41] Koran 5,33 (trans. Yusuf Ali): "The punishment of those who wage war against God and His Apostle, and strive with might and main for mischief through the land is: execution, or crucifixion, or the cutting off of hands and feet from opposite sides, or exile from the land: that is their disgrace in this world, and a heavy punishment is theirs in the Hereafter."

[42] www.islamibankbd.com/page/ oicres.htm#90/7/9

[43] See also the Cairo Declaration by the HIV/AIDS Regional Programme in the Arab States (HARPAS), December 2004, quoted earlier.

One last consideration: the item "economic availability" or "health budget" has a great importance in bioethical thought today, but it assumes a decisive value in poor countries, which have very limited health budgets. This forces doctors to choose the pathologies for the allocation of funds, to the detriment of other pathologies. The more AIDS spreads, the more it is a factor of poverty and an obstacle to development, due to its high costs in social terms and in terms of health and employment. The new "antiretroviral" treatments can greatly slow down the course of the infection, thanks to protease-inhibiting drugs: people with HIV get better, their weight increases, whilst AIDS patients are also beginning to have hope. However, the new cocktail of drugs is too expensive for many Muslim countries so that concretely, the practical solution for them remains increasing and advertising prevention. The very cost of the anti-HIV tests (ELISA and Western Blot) makes it impossible to screen vast groups of the population except at an unbearable cost.

THE COUNTRIES

Protecting the community prevails over the rights of the individual whenever there are potential risks for the community. This gives rise to great tension in Muslim countries between the two points of view, but where the balance, despite the declarations on principle, is always inclined to the advantage of the community. In this regard, it is sufficient to observe the rules for the protection of the healthy spouse or dismissal of the infected person from many productive activities (if not all). The "defensive" measures regarding foreigners entering the country come under this protective way of thinking. The vast majority of Muslim countries, especially Arab countries, have very strict laws or regulations to control and/or expel foreigners infected with HIV/AIDS coming into the country (for tourism, work or study) or who have been resident for more than a certain period of time in the country. Some countries require tests on its citizens when they return from periods spent abroad. Until 2005, the few Muslim countries without any particular regulations on HIV testing on incoming foreigners or residents were Indonesia, Morocco, Senegal, Turkey, Niger, Mauritania and Azerbaijan.[44]

In other words, against AIDS, Arab-Muslim countries seem to be divided between the autarchic approach that follows the logic of social control and repression in the respect of Islamic values, and the approach of respecting the rights of the individual.[45] Algeria, Morocco and Tunisia can be put into the second group, as they have launched policies of prevention based on respecting the anonymity and rights of infected people. Algeria and Morocco have refused the temptation of controls at their frontiers and systematic screening. The values of fidelity and abstinence are obviously underlined but without excluding

---

[44] Wiessner P. and Lemmen K., *Quick Reference. Travel and Residence Regulations for People with HIV and AIDS, 2005*, Berlin, Deutsche AIDS-Hilfe e. V., 2005.

[45] Himmich and Imane, art. cit., 41.

the instruments of dialogue and tolerance. In addition, the notions of anonymity, confidentiality and medical secrecy are promoted in health policies, despite many ethical infringements by doctors exposed by the press.

It is urgent to dispel the illusion common at social, cultural and political level that Islam is sufficient to offer protection from AIDS.

Lastly, almost all HIV/AIDS patients in many Arab-Muslim countries are invisible (obviously not only there). The possibility of effectively fighting AIDS requires certain conditions. These patients must become visible and participate openly in social life. The social, cultural and religious stigma must be challenged[46]; for this purpose, interpersonal contact between healthy and infected people represents a fundamental instrument. Unfortunately, all this requires greater courage by both the political and religious authorities.

Lastly, amongst Arab countries, Morocco, Tunisia and Lebanon have succeeded in obtaining antiretroviral drugs at much lower prices through negotiations with pharmaceutical companies.

Egypt

The attacks and prejudices in the mass media against AIDS patients are very frequent as the pathology is considered peculiar to foreigners, drug addicts, homosexuals and prostitutes. Similar attitudes also appear common in the medical field. In 1993, the Chairman of the Egyptian Medical Association, Hamdy al-Sayed, asked for infected doctors to be treated but, at the same time, isolated from society.[47] A specialist from the Abasa Fever Hospital expressed himself in favour of national screening in Egypt, proposing the segregation of all HIV-positive patients in a colony. In addition, isolation of people infected by contagious pathologies in their own homes or in another place is allowed by Egyptian legislation. Some journalists have asked for the creation of isolated prisons for AIDS patients (*Al-Wafd*, newspaper, 20 June 1991). A survey conducted on 330 Egyptian doctors in Alexandria and another 144 doctors operating in Saudi Arabia (Asir Region) found[48] that, on average, one out of four doctors prefers isolation for the infected patient (specifically 25.8% in the first group and 28.5% in the second); almost 60% suggested admittance to hospital; a minority (16.7% and 11.8%, respectively) requested admittance to a community.

---

[46] On this subject, Hussein Gezairy, the Regional Director of the WHO-Eastern Mediterranean Region, has said: "the fight against the spread of HIV remains limited in our region because of the fear and taboos associated with infection and the disease"; "stigma blocks the way of the most vulnerable to know their sero-status and hinders the timely access of people living with AIDS to care", in Shahine G., Facing up to AIDS, *Al-Ahram Weekly Online*, 4–10 March 2004, 679 (4).

[47] Kandela P., Arab Nations: Attitudes to AIDS, *The Lancet*, 1993, 341, 884–885. See also Sakr H., Airing Taboos, *Al-Ahram Weekly Online*, 646, 10–16 July 2003.

[48] Sallam S.A., Mahfouz A.A.R., et al., Continuing Medical Education Needs Regarding AIDS among Egyptian Physicians in Alexandria, Egypt and in the Asir Region, Saudi Arabia, *AIDS Care*, 1995, 7 (1), 49–54.

The Ministry for Health reports a total of 720 cases of HIV infection whilst the Director of the National AIDS Control Programme deems that the most credible number may be double or triple that figure.[49] The WHO-UNAIDS data for 2002–2003 estimates the presence of 3,584 cases of HIV/AIDS; other figures reach 8,000 or 12,000 infected patients. However, apart from the predictable variety of figures, these are very small numbers in consideration of a population of almost 70 million. There is no possibility of a systematic screening of a certain amplitude, whilst it is to be remembered that an autopsy is carried out only if there is the suspicion that a death is caused by a criminal act, which is why many deaths due to AIDS may not be included in the statistics. In 1994, there was no special legislation and therefore to deal with the problems raised by AIDS it was necessary to have recourse to ordinary legislation. In 1986, an Executive Committee for the control of AIDS was created (Decision no. 434 of the Ministry of Health) with the task of providing recommendations to protect citizens. In the same year (1986), the pathology was included amongst the contagious diseases for which declaration was compulsory (Decision no. 435).

If AIDS is greatly feared, it is known that other STDs are much more common in Egypt, to the extent that in 1991 M.H. Wahdan (WHO and EMRO) stated, on the spread of these in the Middle East, that "infection and spread rates continue to rise"; the factors that cause this include the increased mobility of the population, urbanisation and the increase in tourism, each of which has effects on the morals and behavioural models of the population.

The Egyptian National AIDS Programme has made public the number of blood units tested HIV-positive in recent years.[50] We learn that in 1990 out of 136,422 blood samples tested, only four were positive; in 1996 out of 250,000 units tested, only three were positive. As the number of infections identified in the "risk" categories remains low, this would confirm the low rate of infection existing in society.[51]

One of the most serious problems for Egyptian health concerns precisely the level of the safety of available blood samples, as at least half of the usable blood comes from paid donors, at the service of private blood banks: their screening of samples is not as selective as that carried out by state blood banks. The other major contribution to the blood supply, essential for operations and transfusions, comes generally from donations by the patient's relatives (independently of the type of illness). In April 1999, the Minister of Health, Ismail Sallam, even

---

[49] Khalil A., Viral Threats Unchecked, *Cairo Times*, 27 May–9 June 1999, 3 (7), 8–9.

[50] Lenton, art. cit., 1005.

[51] A study on HIV, carried out in 1986–1987, on 86 prostitutes in the women's prison of El-Kanater, near Cairo, showed only one case of HIV-positivity to the ELISA test and this positivity was disproved by the following Western Blot test. Of the 86 women, 12 were syphilitic, 13 suffered from vaginal discharges and 10 from hepatitis B. See Bassily S., Mikhael M.N., et al., Female Prostitutes: A Risk Group for Infection with Immunodeficiency Virus (HIV), *Journal of the Egyptian Medical Association*, 1987, 70 (9–12), 553–561.

spoke of a "Blood Mafia" with reference to the scandal that broke out over the infection contracted by a 70-year-old woman after a transfusion; the blood came from a donor who then turned out to be homosexual and HIV-positive and who had given blood for payment to a private blood bank no fewer than 144 times in the last year. The minister replied by closing several private blood banks and the prohibition of any type of retribution for the donor.[52]

The articulation between voluntary and compulsory screening is of interest.[53] Each individual can freely have a blood test to identify HIV in the laboratories accredited by the Ministry of Health. Amongst the reasons for doing the test, there is the certificate of negativity which is essential to work abroad as certain Arab countries (especially in the Gulf) require it. The result of the test is notified to the person concerned. When the voluntary test is positive, the result is notified to the Regional Deputy Director for Prevention (together with the personal data of the person concerned) as AIDS is amongst the contagious pathologies that must be reported. A second blood sample is taken at a later date to confirm the positive result with the Western Blot method (Ministerial Circular of 4 April 1990). Although the information on the identity of who takes the test (and on the results) is theoretically confidential, many people do not accept having the test because it is compulsory to provide an identity photo.

There are different types of compulsory tests (Circular of 15 February 1988 of the Deputy Ministry of Health)[54]:

(a) Compulsory test for freely accepted situations such as blood donors or prisoner donors (prisoners are generally considered subjects "at risk")

(b) Compulsory test unknown to the individual, e.g. drug addicts, patients in psychiatric clinics; the informed consent of these subjects is not a preliminary condition for the test

(c) Compulsory test for people "at risk", e.g. individuals with venereal diseases; prisoners convicted for drug offences or vice; foreigners who intend staying more than one month[55] and convicted foreigners in prison

---

[52] Khalil, art. cit., 9.

[53] Chazli (El) F., Le SIDA au regard du Droit Egyptien, in Foyer J. and Khaïat L. (eds.), *Droit et SIDA*, Paris, CNRS, 1994, 147–171.

[54] Ibid., 153–154 and 157.

[55] Regarding foreigners, all those who live in the country for more than one month must take the test. According to a communication of the US Agency for International Development (USAID) of 27 April 1989, the rule applies to heads of families working or studying in Egypt (but it is also possible for dependents). The test is performed only in Egypt. The blood samples are initially analysed at the Ministry of Health's Central Laboratory in Cairo. Positive samples are sent for tests of confirmation to the US Naval Medical Research Unit (NAMRU-3) in Abbasiya. People whose test is confirmed as positive must leave Egypt. See USAID, Egypt, *AIDS Testing*, 27 April 1989. Further details are given in USAID, Egypt, *HIV Free (AIDS) Test Certificate*, 25 August 1993, where an unofficial translation of Circular no. 1 of 1993 of the Egyptian Ministry of Manpower and Training is given. The US Department of State, Consular Information Sheet, Egypt, 23 February 2004 confirms: "Evidence of an AIDS test is required for everyone staying over 30 days."

(d) Compulsory test for people exposed to the virus: haemophiliacs, individuals who have had transfusions abroad, patients under dialysis who have the test repeated every month, individuals who come into contact with AIDS patients and HIV-positive people

Professional secrecy is obligatory for doctors under art. 310 of the Egyptian Penal Code except when doctors are obliged by the law to disclose specific information. Articles 12 and 13 of the law on contagious diseases lay down that the name of the infected person must be notified to the competent authorities.

As a premarital health certificate is not required in Egypt, there is no compulsory test to determine the possible HIV-positivity of a future couple. The test may be made on the request of those directly concerned. If the result of the test is positive for one of the future spouses, the other partner will be informed by the health authorities and will then decide whether to get married or not. The doctor must warn the latter according to the principle of the "state of necessity" laid down by art. 61 of the Egyptian Penal Code.[56]

Furthermore, the doctor advises the HIV-positive partner to inform the spouse of the situation. If the former fails to do so, the doctor informs the family in order to protect them from further risks according to art. 17 of the Code of Medical Ethics.

According to art. 19 of the law on contagious diseases, the authorities can check the individuals who live in contact with AIDS patients, e.g. their relatives, obliging them to have tests. Article 21 allows the health authorities to exclude AIDS patients or HIV-positive individuals from jobs where there is contact with the preparation, sale and transport of food and drink.[57] The anti-HIV test is not compulsory for pregnant women who are not deemed to be at risk; however, the test is not used very much due to the absence of information and its non-reimbursable cost. Voluntary abortion is allowed only in the case of necessity according to art. 61 of the Penal Code. It is up to the legislator to express himself on the possibility of abortion for an HIV-positive mother or with AIDS to avoid possible transmission to her children.

The wilful contamination of someone with the virus may be prosecuted under art. 233 of the Egyptian Penal Code which states: "Anyone who kills another person with substances that cause death . . . is a murderer . . . independently of the way in which such substances have been used and will be punished with the death sentence." The wording specifies that the substances causing death include poisons of animal, vegetable and mineral origin. The inoculation of the virus of a fatal pathology can be assimilated with the administration of a poison.[58]

---

[56] Damage for others – avoidance of which is desired infringing the processional secret, must be serious and imminent and may not be avoided any other way (Ruling of the Supreme Court of 24 March 1983). See Chazli, art. cit., 158.
[57] Chazli, art. cit., 155.
[58] Ibid., 166–167.

Two conclusive evaluations can be made: (1) the right to the respect of private life, guaranteed by art. 45 of the Egyptian Constitution, is in contrast with the generalisation of the compulsory tests, the limitation of the medical secret and the obligation to inform the partner; and (2) the equal rights of citizens, laid down by art. 40 of the Constitution, is in contrast with the discriminations in schools, prisons and hospitals as well as in the possibility of refusing employment to HIV-positive individuals or dismissing them.

One initiative makes Egypt a special case in the Arab-Muslim panorama: since September 1996 the first *AIDS Hotline* in the country has been active in Cairo, with the telephone number being advertised on public transport and in newspapers. In the first two years of activity, it received about 17,000 calls.[59] It imitates similar services active in the USA and Great Britain and these services are now also present in other Muslim countries. The calls come not only from all over Egypt but also from the Gulf countries and even from Europe. The phone call is not recorded and everything remains anonymous. Users can ask for the information they want, such as on the ways of infection, on homosexuality, on the use of the condom: all subjects which are still taboo in a conservative society such as Egypt. The majority of the callers are young, single and incapable of discussing the risks of premarital sex, homosexuality and drug consumption. As of 2002, it was reckoned that about 50,000 telephone calls had been made.

The seminars held at the University of Cairo to provide information on AIDS used technical language in order to avoid accusations of "immoral propaganda" aimed at encouraging premarital relations. It is pointless to deny that to adequately deal with the danger of AIDS and STDs, freedom of information and dialogue on the topics of sexuality is indispensable. On the contrary, the few studies in Egypt show a worrying closure and disinformation. A survey of students and workers carried out by the Faculty of Mass Communication of the University of Cairo reveals that almost one-third believed that AIDS can be transmitted by insects; one-fifth feared catching the infection in public toilets and kissing in greeting; the concept of "safe sex" is not common and the majority of students do not know what a condom is.[60]

A sample of 4,000 women[61] who attended four clinical service improvement centres in Alexandria over the three-month period from October to December 1994 appeared to have a better knowledge of AIDS. Among them 96% knew about AIDS in particular from television and radio. Only 2% of women had heard of it from doctors or nurses. The vast majority knew that it was an

---

[59] Gawhary, art. cit., 18.

[60] Ibid.

[61] Mageid A.A., sheikh (EL) S., et al., Knowledge and Attitudes about Reproductive Health and HIV/AIDS among Family Planning Clients, *Eastern Mediterranean Health Journal*, 1996, 2 (3), 459–469.

infection caused by a virus which is transmitted through sexual relations with an infected person. They knew that infection is avoidable. At the same time, 61% thought that infected persons should be segregated and avoided; 35% did not know that the condom could prevent infection.

Even if the Egyptian rate of infection is very low, a survey by the USAID states that Egypt is still at the initial stage of development of the epidemic. This is explained by some factors such as[62]: the delayed appearance of the illness which allowed the ministries concerned to launch the first countermeasures; the general refusal of homosexuality; the prohibition of every form of procuring (Law no. 10 of 1961) and the adhesion to traditional religious values, therefore sexual relations outside marriage, still appear limited. Nevertheless, other factors are worsening the situation: a sexually more active youth, 2–3 million Egyptians who work abroad without their families often in a state of sexual promiscuity, "S's": sun, sea and sex), a large number of girls and refugees from high-risk countries, and scarcity of instruments of treatment and prevention. The effective extent of the drug and homosexuality phenomena remains unknown whilst prostitution is not centrally organised.

The majority of HIV-positive people conceal their condition. Due to prejudice it becomes almost impossible to deal with the problem publicly or discuss it openly, as would be required by any programme of prevention and information that has to be even slightly effective to the public. Two examples[63] reveal these difficulties. An interview with the first HIV-positive woman in Egypt was on a very popular programme on TV: the woman's identity was kept secret, she was veiled and filmed from the back. Being HIV-positive for 14 years, she contracted the virus through a blood transfusion in a Gulf state. Second example: it was necessary to wait more than 10 years from the discovery of the first HIV case in Egypt for a carrier of the virus to be able to speak openly at the First National Conference on AIDS in Cairo (April 1997) in front of 300 doctors, nurses and religious authorities about how it happened.

The limited knowledge of the ways the infection is spread also involves the sample of 474 Egyptian doctors studied by Sallam,[64] 330 of whom operated in Alexandria and the others in Saudi Arabia. Kissing as a probable source of HIV infection was considered by about 42% of those present in both groups; mosquito bites by 20% and 21.5% of the doctors, respectively; daily contact with HIV-positive people represented a risk for 9% of both groups; eating with an HIV-positive person and using the same cutlery was a risk of infection for 5.2% and 5.6% of the doctors; using their clothes 5.2% and 4.2% of the doctors, touching an infected person 5.2% and 2.1%; and shaking hands 3.9% and 2.9%. Precisely to make up for this misinformation, more than 60% of both groups

[62] Gawhary, art. cit., 18–19.
[63] Ibid., 19.
[64] Sallam and Mahfouz, art. cit., 52.

called for the need for health education; at the same time, mass screening of the population was suggested by 40.6% and by 45.8%.

In the country, the availability of antiretroviral drugs is extremely scarce; therefore patients look for them on the black market or at large private pharmacies where, however, the prices are prohibitive. The paradox is that whilst in Europe there has been a drop in the number of deaths and life expectancy has increased since 1997, thanks to the introduction of antiretroviral drugs, patients in Egypt are dying due to the lack of drugs.

Pakistan

An evaluation of the situation in this country is hindered by the lack of information available. The mark of infamy that brands patients with AIDS and STDs induces infected people to conceal their condition. Everything that concerns sexual behaviour and sexual diseases is rarely discussed publicly, leading to scarce knowledge of AIDS, of the ways of contagion and, inevitably, strategies of prevention. The traditional religious contexts, local customs and the pressure of the religious parties influence this environment.

According to official sources, the first infections in the country are alleged to have been caused by transfusions of infected blood. The estimates by WHO-EMRO and UNAIDS speak of some 80,000 people with HIV/AIDS up to 2002-2003.

A study in 1996 revealed that in prostitution circles in Karachi 60% of the women had heard of AIDS but only 44% of them were aware of the possibility of sexual transmission.[65] Prostitution and homosexuality are present, even if this is not officially acknowledged and there are still few official figures. There is a strong increase in drug addiction. Blood is donated mainly in two ways: donation for payment (the commonest form but with few controls) and donations between relatives or within the circle of friends. The "safer" alternative to find blood that is not infected remains recourse to donor-relatives on the basis of the conviction that it is unlikely that a relative who is infected or frightened of being infected would give blood to a relative whose blood is healthy.[66] In fact, in daily practice, relatives often give blood out of the fear of causing the death of the ill relative or friend, worsening their clinical conditions or under the pressure of the family, friends and doctors; under this pressure, they end up by hiding or underestimating previous illnesses and "dangerous" experiences.[67] The public health services, having an

---

[65] Hyder A.A. and Khan O.A., HIV/AIDS in Pakistan: The Context and Magnitude of an Emerging Threat, *Journal of Epidemiology and Community Health*, 1998, 52, 579–585.

[66] Mujeeb S.A. and Mehmood K., Prevalence of HBV, HCV and HIV Infections among Family Blood Donors, *Annals of Saudi Medicine*, 1996, 16 (6), 702–703.

[67] From a study on 839 family blood donors at the Blood Transfusion Services, Jinnah Postgraduate Medical Centre, Karachi, from 1 August to 30 September 1995, 58 (6.9%) were reactive to the tests for blood infections. Of these 41 were HBV (hepatitis B) reactive (4.9%), 20 HCV (hepatitis C) reactive (2.4%) whilst no HIV-positive cases were found. Tests carried out on donors for payment showed 10% of HBV infections and a significant rate of HIV-positivity. See Mujeeb and Mehmood, art. cit., 702–703.

urgent need to find blood, do not create excessive problems over the health and experiences of the donor. The problem of blood donation by illiterate people is not to be overlooked as it is necessary to check whether they effectively understand the meaning of expressions such as behaviour "at risk", etc.

In 1987, the National AIDS Control Programme (NACP) was launched to control infection in the country with almost 30 centres for blood screening. The widest screenings[68] involved more than 1,350,000 blood samples analysed by 1995 (of which 869 were HIV-positive) and a previous survey in 1992 on 250,000 samples (of which 129 HIV-positive); the rate was 52 out of 100,000 inhabitants in 1992 and 64 out of 100,000 (i.e. 0.064%) in 1995.

According to official projections, at least 5,000 people have died of AIDS every year since 1995 and this figure is destined to increase due to the strong demographic increase in progress. The categories at greatest "risk" are men with extramarital relations, prisoners and patients who have undergone several transfusions. The high fertility rate accompanied by a low use of contraceptives (mainly the pill and condoms) represents a further danger for the spread of AIDS. The ratio between infected men and women is 5:1. NACP has started a campaign to increase knowledge of AIDS in society but the high illiteracy rate (>60%) excludes the majority of the population. A hotline for questions and advice has also been set up. Other serious risk factors are rapid urbanisation, the low status of women and their exploitation, the absence of effective treatment, hunger and, in the last place, the great development of drug addiction.

In a "red light" district in Karachi in the period between November 1993 and June 1994, voluntary tests were carried out on a limited sample of commercial sex workers. Due to the very high rate of illiteracy, only verbal consent was obtained. The sample of commercial sex workers (CSW) included prostitutes–singers–dancers (82 cases), full-time prostitutes (7 cases) and 4 transvestites. There were 81 participants in the test (70, 7 and 4, respectively), all negative to HIV-1 antibodies; 4 (5%) were reactive to a Venereal Disease Research Laboratory (VDRL) test and to fluorescent treponemal antibody absorbed (FTA-ABS), a test for the diagnosis of syphilis) tests.[69]

According to government figures, in 1993 there were more than 3 million drug addicts, the majority of whom were heroin addicts. Out of a sample of 316 drug addicts (of whom 120 intravenous) all were negative to HIV-1 antibodies, whilst

---

[68] Hyder and Khan, art. cit., 581.
[69] Here are some figures: of the 93 participants, 92 were Muslims (the vast majority *Shi'ite*), 78% illiterate. The average age of sexual intercourse for the first time was 15 for the singers–dancers; 14 for the full-time CSWs and 11 for the transvestite males. Two-thirds (i.e. 65%) of the female CSWs did not use contraceptives. Seven out of the 82 CSWs inserted sponges soaked in phenolic antiseptic (Dettol) into the vagina to avoid pregnancy and infections. The majority of the CSWs never used the condom and the clients of the 3 groups rarely asked for it. Of 86 CSWs, 19% had received blood transfusions and 4 (5%) had given blood. See Baqi S., Nabi N., et al., HIV Antibody Seroprevalence and Associated Risk Factors in Sex Workers, Drug Users and Prisoners in Sindh, Pakistan, *Journal of AIDS and Human Retrovirology*, 1998, 18 (1), 73–79.

out of 272 tested, 18 were reactive to the VDRL and FTA-ABS tests.[70] Out of 3,441 male prisoners, only one infected by HIV-1 was found (0.03%) after having had intercourse with prostitutes; the same person donated blood twice in Karachi and often received therapeutic injections with previously used syringes. Out of 84 women prisoners, only one was found to be HIV-positive; of the sample, 12 had received blood transfusions and 4 had given blood.

The last survey quoted highlighted the low presence of HIV in three categories "at risk". The significant percentage of convicts and drug addicts who donate blood is to be underlined whilst the majority of the Pakistani blood banks buy it and few of them carry out the appropriate anti-HIV, hepatitis C and B tests. Also of concern is the number of drug addicts and convicts who are given "therapeutic" injections with syringes that are not sterile whilst drug addicts often reuse syringes thrown away by hospitals.

Morocco

According to the estimates of the WHO-EMRO and UNAIDS in 2002–2003, there were 14,000 people with HIV/AIDS. An element that can confirm the low degree of reliability of the official data is the worrying increase of the other STDs (syphilis, gonorrhoea, venereal ulcer, herpes genitalis, etc.) and estimated at about 100,000 cases every year[71]: it is objectively improbable that such a figure does not also include AIDS.

In a survey of 162 young Moroccans (100 resident in the country, 50 abroad and 12 emigrants) on AIDS, more than 40% raised the problem of the relationship between Islam and premarital relations[72]; as many young people are unemployed, illiterate and too poor to maintain a family, many would like to ask the *ulama* for consent to use condoms for protection in premartial relations, considered almost inevitable until marriage which is increasingly at a later age, especially for immigrants to the West where the visible manifestations of Islam are rare whilst temptations are stronger. However, this request clashes with the Muslim solution consisting of premarital abstinence and early marriage, a solution which seems accepted more easily by Moroccan girls than males. These young people often assign to sexuality functions and values that go against the Muslim model: a more pronounced "sexual liberalism" (precocious premarital sexuality, plurality of partners, etc.) can also be conceived as a sort of compensation (one of the rare areas of personal initiative) against the impossibility of marriage[73];

[70] Baqi et al., Nabi, art. cit., 75.
[71] See in Manhart L.E., Dialmy A., et al., Sexually Transmitted Diseases in Morocco: Gender Influences on Prevention and Health Care Seeking Behavior, *Social Science & Medicine*, 2000, 50, 1369–1383.
[72] Dialmy, Moroccan Youth, Sex and Islam, art. cit., 16–17.
[73] Dialmy A., Sexualité, Emigration et Sida au Maroc, in Observatoire Marocain des Mouvements Sociaux, *Emigration et Identité*, Actes du Colloque International, Fez, 24–25 November 1995, 155–200.

in addition, the absence of sex is sometimes interpreted as a probable cause of psychological imbalance.

Morocco, in the rear on the road towards modernisation, follows a faster pace in the field of sexuality, which tends to become a value in itself, independent of marriage and the family; this is accompanied by rapid urbanisation,[74] which has produced the breakup of the patriarchal family and the regression of the importance of religion in society. Even the publicised but sporadic acts of sexual repression by governments meet the need to present a good image to a public opinion that is potentially sensitive to the warnings of the "fundamentalists".

Contraceptive pills are practically on free sale whilst many private clinics perform illegal abortions. All this takes place even though the Moroccan Penal Code prohibits extramarital sexual relations, equating them with the offence of prostitution. The official statistics on population control do not include the consumption of contraceptives by unmarried women, but only by married women, which masks the reality of sexual permissiveness.

Islam, nevertheless, still maintains a very important role in managing sexuality even amongst young people. Tazi Saoud, Rector of the University of Qarawiyyin, believes it is meaningless to discuss the way of protecting the sinner from the risk of AIDS, as Islam prohibits fornication: the use of the condom encourages – in any case – fornication which is prohibited with and without the prophylactic.[75] Mohammed Issef, a professor at the same university, recalls the alternative offered by the Prophet: marriage or abstinence; bearing in mind the current difficulties, Muslim society has to make an effort to make marriage practicable. In the same way, recourse to the "principle of necessity" is inapplicable – i.e. to the lesser evil – to try to legitimise the use of the condom in order to avoid a possibly fatal infection. Indeed, the need that makes the unlawful lawful exists only to save lives and has to be evaluated on a case-by-case basis.

The situation in which young Moroccans find themselves is emblematic: on the one hand, the Minister of Health invites them to use condoms; on the

---

[74] Dialmy's interpretation of the consequences of the development of the modern city on the level of sexuality is instructive. The modern Western city, unlike the traditional Arab-Muslim one, is alleged to be characterised by anonymity and regression of group control. A mass without morals or restraint dominates. The spatial frontiers between man and woman and the spaces reserved for a single sex are rare. The mixing of the sexes does not obey the rule of the veil that allowed women to cross the male space without being a source of seduction. In contemporary cities, the two sexes live side by side, but in the Muslim world people are not yet ready for this mixing without frontiers, from either the historical or psychological point of view. The result is the irresistible necessity to "chase" women. In these cities the woman is by definition an outsider, as relationship is no longer an organising and protective principle of this space. The woman dressed in Western clothes becomes exciting, the absence of a veil and bonds of relationships transform her automatically into an object of immediate sexual consumption. The various places (neighbourhoods, schools, universities, workplaces, streets, cafés, cinemas, public gardens, beaches, hotels, etc.) become the privileged places of sexual conquest. See Dialmy, Sexualité, Emigration et SIDA au Maroc, op. cit. 162–163.

[75] Dialmy, Moroccan Youth, art. cit., 17.

other hand, the Minister of Religious Affairs asks them to abstain or to get married. At the same time, both political and religious power appear fearful of publicly presenting any statistics on lawful and unlawful sexual customs of the citizens, because this would be the equivalent of acknowledging that they are widespread to a certain extent, which is a politically risky fact in a Muslim country.

The development of premarital and extramarital sexuality in Morocco may paradoxically be shown precisely by the epidemiological situation of AIDS as at 30 June 1993 when it was noted that "overall, there exist as many single men with the infection as married people" (INFOSIDA Maroc).[76] Amongst the factors that facilitate the spread of AIDS, we can find:

1. The average age at the first marriage is increasing, which contributes to the precocity of sexual relations (as in Egypt, the practice of suturing the hymen to remedy lost virginity, in particular after rape, is not infrequent in Morocco) and the autonomy of sex with respect to marriage.
2. The attitude towards prostitution appears bivalent. On the one hand, it is condemned as a scourge of the family order but, at the same time, it is tolerated as it allows the young man to initiate his sexual life and also allows husbands to satisfy their erotic fantasies. Today Morocco is a destination for sexual tourism. Brothels are illegal but two phases can be distinguished: from 1956 to the early 1980s, there were women who sold themselves only to Moroccans, from the 1980s to 1995 women who prostituted themselves with foreigners, especially Arabs and men who prostitute themselves with Western men.

The Ministry of Health has disclosed some global statistics on the cases of AIDS in the country[77]: 1986–1; 1987–10; 1988–24; 1989–44; 1990–70; 1991–98; 1992–128; 1993–145; 1994–234 and 14,000 HIV-positive cases; 1995–290 up to September. Limitedly to the 145 cases of infection recorded in 1993, heterosexual transmission is the main cause (47% of cases) but poor women are three–four times more exposed; 13% of infections can be attributed to drugs (men are five times as numerous as women). In the same year, almost 80% of the women were infected in a heterosexual relationship.

Some considerations on the situation of women are in order. Moroccan women do not in general appear to be able to decide how the sexual act is performed to protect themselves or their husbands, or clients in the case of prostitution. The general condition of these women shows a strong social inequality compared to men and this makes them particularly vulnerable as they cannot obtain information, demand that the condom be used and play an active role in prevention. It is clear that the struggle for personal rights is inseparable from that for the right to health and prevention from AIDS.

---

[76] See Dialmy, Sexualité, Emigration et SIDA au Maroc, op. cit., 168–169.

[77] The Casablanca newspaper L'Opinion of 3 December 1995 defined these official figures (coming exclusively from public doctors) as ridiculous, similarly with the figures on HIV-positive cases. See Dialmy, Sexualité, Emigration et SIDA au Maroc, op. cit., 173–174.

Regarding the evolution of the ways of transmission in the periods 1986–1989 and 1990–1993, heterosexuality increased from 20% to 61%, i.e. the absolute majority, whilst all the other ways appear to have decreased.

In the 290 cases of 1995, heterosexual contagion reached 52% of cases against 14% due to homosexuality, 11% to intravenous injections and 5% to transfusions. In the same year, 70% were resident Moroccans, 13% Moroccans abroad and 17% foreigners. The cities are the most affected (82%) and women remain particularly unprotected as 78% of them were infected by their stable partner.[78]

A further serious element of infection is represented by child prostitution (which also concerns other Muslim countries such as Bangladesh, Pakistan, Indonesia, etc.). The factors triggering this off can be identified in abject poverty, in the disintegration of family models, unemployment and drug addiction. Street children[79] may spontaneously offer sexual favours to potential clients, very often tourists (mainly Arabs from the Persian Gulf and Westerners because they pay more). Moroccan law punishes sexual abuse on children under 15 with five-years imprisonment and this can go up to 20 years if accompanied by violence. Similar punishments are foreseen for the rape of girls under 15 but imprisonment can be up to 30 years if loss of virginity can be proven. Homosexuality is illegal and is punishable by three- to six-months imprisonment and consequently the practice tends to be clandestine.

A paradoxical obstacle to the control of the spread of infection in mercenary sexual relations is the behaviour of the police, who consider the possession of condoms as evidence of an unlawful relationship: this contributes to the refusal of "protected" relations by prostitutes.

In perspective, the attempts by the Moroccan authorities to minimise the risks of AIDS inevitably produce ignorance and the absence of prevention. This attitude was in part encouraged by the observation that until the 1980s the majority of infected cases in the country had lived in Europe. The result was a double prejudice with regard to Moroccan immigrants in European countries and Morocco: Morocco considered the emigrants who returned home as potential importers of the infection, whilst the European countries where the Moroccans immigrated considered them as categories "at risk". Paradoxically, according to the results of a survey, the immigrant tends to show greater attention to AIDS in Europe and so he is more likely to use condoms (compared to his fellow countrymen at home).

According to more recent data of the Ministry of Health, in December 2002 there were 1,085 cases of AIDS and 15,000 cases of HIV-infected individuals. The majority (about 70%) were infected through heterosexual relations. The reason for the relatively low number of infected people may depend on various reasons, such as[80]:

---

[78] Dialmy, Sexualité, Emigration et SIDA au Maroc, op. cit., 177–178.

[79] Kandela P., Marrakesh: Child Prostitution and the Spread of AIDS, *The Lancet*, 9 December 2000, 356.

[80] Elharti E., Alami M., et al., Some Characteristics of the HIV Epidemic in Morocco, *Eastern Mediterranean Health Journal*, 2002, 8 (6).

1. The HIV-1 subtype B virus is predominant in Morocco; this subtype is less transmissible than the other subtypes. This predominance is more similar to the situation of Europe and USA (whilst in sub-Saharan countries HIV-1 subtypes C, A, D and E are predominant). In addition, the HIV-1 subtype B prevails in northern Africa and this is not surprising because the cultural and economic relations (in particular tourism and immigration) are closer between these countries and Europe compared to sub-Saharan countries.
2. The diffusion of male circumcision, a practice which is often associated with a reduced risk of HIV infection.
3. The rules of Muslim life which forbid anal intercourse and sexual relations during menstruation (for example Koran 2.222).

Saudi Arabia

Article 23 of Ministerial Resolution no. 288/17/L of 23 January 1990 states that doctors have to keep secret what they learn in their professional practice except in five cases; no. 23 (2) specifies that disclosure is lawful to report a contagious disease.[81]

Initially, the three main sources of infection from HIV in the Kingdom were: (1) blood transfusions (the first cases in the country)[82] and the use of haemoderivatives; (2) mother–foetus transmission (for prevention, mothers are invited to follow health education on the pathology and the use of the condom in their marital relations); and (3) the infections procured during "pleasure" trips to south-eastern Asia. Since 1986, blood samples and haemoderivatives have been subjected to screening. In later years, self-sufficiency appears to have been reached for the blood supply essential for hospitals, thanks to an intense advertising campaign aimed at increasing donations. Citizens who donate blood at least ten times are given public recognition.

In Saudi hospitals, the preoperation anti-HIV test is not routine, due to the limited spread of the infection and the high costs of tests; therefore, they are performed only on categories "at risk", the members of which are difficult to identify for cultural reasons as their personal and intimate behaviour is unlikely to be communicated, especially by women.[83] However, the categories concerned include convicts, patients under dialysis and patients with STDs. HIV cases are isolated in their own homes.[84]

---

[81] The revelation is also lawful: (a) if it reveals a death caused by a crime and if it is useful to prevent other crimes; (b) disclosure protects the doctor from accusations against his professional capacity; (c) if the patient agrees (in writing) with disclosure or if disclosure to the family is therapeutically beneficial; (d) if ordered by a court. See *International Digest of Health Legislation*, 1992, 43 (1), 28–29.

[82] The first two cases in the Kingdom date back to the early 1980s. They were a 42-year-old man and a child of 5 and a half who had received transfusions of infected blood in Riyadh in 1981 imported from the USA. Both died a few years later. See Harfi H.A. and Fakhry B.M., Acquired Immunodeficiency Syndrome in Saudi Arabia, *JAMA*, 1986, 255 (3), 383–384.

[83] Zawawi T.H., Abdelaal M.A., et al., Routine Preoperative Screening for Human Immunodeficiency Virus in a General Hospital, Saudi Arabia, *Infection Control and Hospital Epidemiology*, 1997, 18 (3), 158–159.

[84] Kandela P., Gulf States test Foreigners for AIDS, *British Medical Journal*, 5 March 1994, 308, 617.

Between 1988 and September 1990, 96 infected people were identified at the King Faisal Specialist Hospital and Research Centre (KFSH & RC) of Riyadh.[85] Of these, 6 had already been blood donors, the other 90 were in hospital for a variety of reasons. The vast majority were Saudis, 9 were foreigners and were deported. Medical records were available for 71 patients, 52 adults and 19 children. Transfusions and coagulation factors (haemophilia) represented the probable cause of the infection in 51 cases out of 71; heterosexual transmission concerned 7 people, another 7 associated drug addition as a probable cause of infection and there were 2 cases of homosexual/bisexual origin.[86] A kidney transplant represented the cause in 6 patients, 3 had bought the organs in Bombay from LNRDs and 3 had been transplanted in the USA with two organs from corpses and one from a LNRD. Out of 128 multi-transfusion haemophiliacs treated at KFSH, 20 (16%) were HIV-positive.

In other words, there may exist a significant sample of HIV-positive cases amongst the Saudis. Estimates for the region of Riyadh (2.4 million in 1990) were of 2 cases every 100,000. For the Kingdom as a whole (at the time 11.6 million), the estimates of 0.7/100.000 were probably underestimated (limitedly to hospital patients; other figures speak of a rate of 0.56/1,000).[87] The spectrum and frequency of opportunistic infections is very similar to the cases described in Europe and North America.[88]

The measures of the Saudi Ministry of Health to identify the presence of active and latent syphilis (with tests such as VDRL, rapid plasma reagin (RPR), FTA-ABS and *Treponema Pallidum* microhemagglutination assay (TP-MHA), HIV and hepatitis B surface antigen (HBsAg) on all foreigners who enter the country for work are particularly strict. In early 1994, the Ministry for Finance and Trade estimated that at least 27% of the population of the Kingdom were foreigners, representing over 50% of the entire labour force. This situation places

---

[85] Ellis M.E., Halim M.A., et al., HIV Infection in Saudi Arabia: Occurrence, Pattern of Disease and Future Implications, *Saudi Medical Journal*, 1993, 14 (4), 325–333. We will also refer to the clarifications of the authors in *Saudi Medical Journal*, 1994, 15 (6), 462–463.

[86] Ellis, Halim, et al., *Saudi Medical Journal*, 1994, 15 (6), 462–463.

[87] Ellis, Halim, et al., art. cit. 331.

[88] The commonest include pneumonia from *Pneumocystis Carinii* and infection from Citomegalovirus whilst the Kaposi's sarcoma appears rare; the characteristics of the pathologies connected with HIV with the Saudis are clearly different from those described in the bordering African geographical areas. See Ellis, Halim, et al., art. cit., 331–332. To a questionnaire on the causes of transmission of AIDS answered by 61 patients with different infective diseases and 22 HIV-positive patients, 5 (6%) had never heard of AIDS; the remaining 78 expressed considerable uncertainty on the causes of transmission of HIV. For the sake of accuracy, of the 78, 16 and 13 (29% and 59%) knew of the existence of AIDS amongst Saudis; 15 (27%) and 1 (5%) deemed it an infection exclusive to homosexuals; 48 (79%) and 21 (95%) thought it could be spread by sexual intercourse, transfusions, syringes, etc. but 14 (25%) and 1 (5%) thought it could be spread by handling objects, 26 (47%) and 1 (14%) by shaking hands, 32 (57%) and 1 (9%) through kissing. Only 16 (29%) and 13 (59%) thought that the condom offered protection in sexual intercourse. See Ellis, Halim, et al., art. cit., 330–331.

the Saudi authorities in an explicitly "defensive" position with regard to these continuous arrivals. A census in the mid-1990s showed the presence of 4,624,459 foreign workers, 70.4% male and 29.6% female. The desire to perform systematic and periodic screenings on all these workers in Saudi Arabia inevitably has extremely high costs to which must be added those of the tests performed on Saudis.

The directives for screening foreign workers are as follows[89]:

1. A complete medical examination in the home country is required, including tests for the most important sexually transmittable pathologies.
2. On arrival in Saudi Arabia, a general examination and haematological screening is repeated within the first four weeks.
3. The anti-HIV test is compulsory every two years for the renewal of the work permit.
4. Foreign workers resulting positive to any of the tests are deported.

Between April 1987 and November 1994, 1,648 foreign workers, of whom 585 males and 1,063 females, were examined at the King Abdulaziz University Hospital of Jeddah.[90] These workers came from 22 countries, the majority from Indonesia (683, 194 males and 489 females), the Philippines (457, 83 males and 374 females) and Egypt (164, 115 males and 49 females). The commonest occupations were chauffeurs and domestic workers. Out of 1,648 individuals, 19% were HIV-positive, 23.8% had syphilis and 57% hepatitis B. These are very high figures even though they refer exclusively to the data of the King Abdulaziz UH. The figures provided by KFSH & RC of Riyadh from a five-year study (1985–1990) on 64,294 units of blood from donors of various nationalities[91] present in Saudi Arabia are, on the other hand, less worrying.

To avoid mother–foetus HIV transmission, the doctors of the Armed Forces Hospital of Riyadh offer advice in the event that the mothers are "at risk" and without forgetting that transmission is not automatic and will affect only some of the children[92]:

1. Infected women should avoid pregnancy.
2. Perform appropriate tests on blood and haemoderivatives for transfusions.
3. Children with coagulation defects must be treated with "heat-treated factor products".

---

[89] Hamdi S.A. and Ibrahim M.A., Sexually Transmitted Diseases in Domestic Expatriate Workers in Jeddah, Saudi Arabia, *Annals of Saudi Medicine* 1997, 17 (1), 29–31.

[90] Ibid., 31.

[91] The study revealed the presence of six HIV-positive individuals (by Western Blot+), i.e. 0.009%, specifically 0.005% amongst the Saudis (the number of Saudi samples was 19,775 units) and 0.011% amongst immigrants (with 44,519 samples). Only one Saudi Arabian was HIV-positive, the other five were Europeans/North Americans. Bernvil S.S., Sheth K., et al., HIV Antibody Screening in a Saudi Arabian Blood Donor Population: 5-Year Experience, *Vox Sanguinis*, 1991, 61, 71–73.

[92] See Isobar A.O., Fairclough D., et al., Maternal Transmission of Human Immunodeficiency Virus (HIV), *Saudi Medical Journal*, 1990, 11 (2), 125–129.

4. As screening on the prenatal condition of women is not yet recommended, pregnant women in "at risk" categories or who have received blood that has not been tested should be screened.
5. Babies born to HIV-positive mothers should not be breastfed by their mothers on condition that alternatives are available. In Muslim law, breastfeeding by another woman creates a "milk relationship" between the child and the nursing mother's family; in this case, recourse to dried milk is a valid alternative.
6. Babies born to HIV-positive mothers must be kept under close observation for at least five years to check whether signs of the infection appear.

Initially the victims of HIV-1 in the Kingdom were haemophiliacs, thalassemics and recipients of blood transfusions. Subsequently, the main method of infection became sexual intercourse; the categories at risk, such as injection drug users and, above all, those whose sexual practices were high risk are difficult to verify due to the social taboos concerning them.[93] The last category is the one that could really increase the number of infected people in the Kingdom.

Another recent study reveals that from 1984 to 2001, 6,046 cases of HIV infection were diagnosed in the Kingdom; 1,285 (21.3%) were Saudi citizens whilst 4,761 (78.7%) were foreign workers[94]. Amongst the Saudis, 945 (73.5%) were male whilst 340 (26.5%) were female. The means of transmission are: heterosexual relations in 487 cases (37.9%); blood transfusion in 322 (25%) cases; perinatal transmission in 82 (6.5%) cases; homosexual relations in 32 (2.5%) cases; intravenous drug use in 17 (1.3%); bisexual relations in 10 (0.8%) cases; unclarified reasons in 334 (26%) cases, probably due to the desire to hide illegal sexual behaviour. Of the 340 Saudi women infected, the main cause was identified in blood transfusions (25.3%); sex with their husbands (21.8%); in 41% of the cases, the causes of infection were unknown. The study notes that the causes include the decadence of religious values and the increase in poverty and unemployment that increase relations outside marriage and prostitution. Western strategies favourable to "safe sex" (promoting the condom) and "needle exchange programmes" (for drug addicts) are considered anti-Muslim. On the contrary, Muslim customs that can be beneficial in limiting extramarital relations include in particular marriage with up to four wives; the absence of age limits for marriage; compulsory (*zakat*) and voluntary (*sadaqa*) alms-giving which contribute to limiting prostitution; Islamic punishments; the use of the *hijab* for women in public, etc.

Lebanon

According to Jacques Mokhbaat, a Lebanese specialist on AIDS, Lebanese society expresses an ignorant hostility towards infected people, independently of

---

[93] Alrajhi A.A., Human Immunodeficiency Virus in Saudi Arabia, *Saudi Medical Journal*, 2004, 25 (11), 1559–1563.
[94] Madani T.A., Mazrou (Al) Y.Y., et al., Epidemiology of the Human Immunodeficiency Virus in Saudi Arabia; 18-Year Surveillance Results and Prevention from an Islamic Perspective, *BMC Infectious Diseases*, 2004, 4, 25, in www.biomedcentral.com/1471–2334/4/25

the origin of the disease; even if caused by contaminated blood imported by the government, citizens tend to consider all infected people as "deviants".[95]

Nevertheless, Lebanon is perhaps the Arab country where the discussion on AIDS is the most "open",[96] together with Egypt.

Circular no. 91 of 29 November 1991 of the Ministry of Health prohibits any discrimination in health care for patients with HIV/AIDS and recommends confidentiality for laboratory data. Subsequently, Law no. 334 of 18 May 1994 established the compulsory presentation of a premarital medical certificate with a period of validity of not more than three months from its issue. Sanctions are laid down for doctors and civil and religious authorities that fail to respect these measures in marrying the partners. Details can be found in Decision no. 857/1 of the Directorate-General of Health of 29 August 1994: the premarital medical certificate has been compulsory since 15 November 1994 and the essential tests include that for AIDS but tests for syphilis, toxoplasmosis, hepatitis B, etc. may be added. Circular no. 73 of 27 October 1994 of the Directorate-General of Health indicates how the anti-HIV test is to be performed: in the event of a positive result, a test of confirmation must be performed and the doctor informed of the result. The doctor has the duty to verbally inform the person concerned and to ascertain that the other future spouse has been informed; lastly, he must instruct the couples on methods of prevention.[97]

In November 2003, there were officially almost 800 patients (WHO-UNAIDS sources speak of 1951).[98] The main problem concerns the protection of the patient's rights. Many HIV-positive people are dismissed from their employment even when their activity does not expose other people to risks of infection. Infected workers cannot hide their illness because they are compelled to contact the administration of the company they work for, so that the Caisse Nationale de la Sécurité Sociale can reimburse the cost of medical care. In addition, there is the problem of admittance to the hospital of the HIV-positive person as many hospitals, including State-run ones, refuse to accept them, maintaining that they do not have the means to treat them. Antiviral treatment must never be interrupted; the doses and times of the drugs must be respected, as well as dietary rules. However, in Lebanon the chronic nature of the treatment is not respected due to the high cost of the drugs, about $1,200 a month. Many patients cannot obtain them on a regular basis; furthermore, the Ministry of Health provides them free of charge but there are continuous shortages of drugs (causing the interruption of the therapy) due to poor management of the stocks and purchases. A further problem is represented by the Ministry's inability to register new drugs, even if very useful for the treatment of the patient.[99]

---

[95] Kandela P., Arab Nations: Attitudes to AIDS, art. cit., 884.

[96] Kandela, art. cit., 884.

[97] *International Digest of Health Legislation*, 1995, 46 (2), 175.

[98] Merhi N., La grande misère des malades du Sida au Liban, *L'Orient Le Jour (Beirut)*, 4 December 2003. In survivreausida.net

[99] Merhi, art. cit.

Despite the great cultural and religious prejudice, there are six HIV/AIDS hotlines in the country, the first having been set up in 1994. There are no anti-HIV tests for tourists whilst it is required for foreigners who wish to work in the country.[100]

Libya

Article 1 of Decision no. 92 of the General Secretary for Health of 1987 lays down the deportation of any foreign individual who has been ascertained as infected. Article 2 specifies that an employer cannot take on an HIV-positive individual.

Oman

Article 4 of the Ministerial Resolution no. 1 of 19 January 1990 on the National Programme for the Prevention of AIDS says: "Foreigners residing in the country with proven AIDS infection shall be deported to their countries for completion and follow-up of medical treatment with the concurrence of the specialised authorities."[101]

Syria

Ordinance no. 17/T of 30 March 1991 abrogates the previous ordinance 36/T of 1987. The main measures contained in its 10 paragraphs are as follows[102]:

Paragraph 1. The following categories must have the anti-HIV tests: blood donors before each donation; foreigners between the ages of 12 and 70 who intend to take up residence in the Republic[103]; haemophiliacs; Syrian citizens who have been abroad for more than three months; any individual suspected of being infected; individuals who enter a country that requires a certificate of HIV negativity.

Foreigners residing in Syria for at least five years and who wish to marry a Syrian citizen are exonerated from the test on condition that they have not left Syria for a continuous period of more than 3 months in the past five years.

Paragraph 2. The tests are performed in authorised laboratories. The validity of the results is 3 months from the test. Experts from the UN and individuals with diplomatic immunity are exonerated from the test, independently of the duration of their stay in Syria.

Paragraph 3. With the exception of certain categories of residents, the tests mentioned in Paragraph 1 are free of charge if performed to protect public health.

Paragraph 4. The person in charge of the authorised laboratory must: make a confidential notification of the positive results of the first tests to the AIDS Reference Laboratory of the Ministry; notify the confirmed cases of HIV-positivity requesting further tests to be performed after a specified period; send a confidential monthly report on the activity of the laboratory to the Chief of the Reference Laboratory; guarantee the confidentiality of the data and of the test results.

---

[100] Wiessner and Lemmen, op. cit., 26.

[101] In *International Digest of Health Legislation*, 1992, 43 (1), 38.

[102] Syrian Arab Republic, Order no. 17/T of 30 March 1991, *International Digest of Health Legislation*, 1994, 45 (2), 174–175.

[103] On the current provisions, the US Department of State, Consular Information Sheet, Syria (8 May 2004) specifies: "AIDS tests are mandatory for foreigners aged 15 to 60 who wish to reside in Syria." No tests on entry are required; however, the person whose infection is certain will be deported from the country. See Wiessner and Lemmen, op. cit., 38.

Paragraph 7. In the case of confirmed infection, a group of epidemiologists will question the person concerned to establish the origin of the infection; a group of consultant psychologists will help the infected person to deal with the social problems connected with his or her condition, informing them how not to infect other people.

Paragraph 10. All the staff involved in the AIDS control programme are obliged to guarantee the confidentiality of the personal information and the information on the results; they must also give all the information exclusively to the people directly involved [without specifying who these are]. Any failure to do so is prosecutable as a criminal offence.

These rules are specified by Regulation no. 106/T of 3 July 1993 and no. 112/T of 13 July 1993.

Iran

The first meeting of Iranian AIDS patients took place on 21 November 2002 at the Research Centre for Gastroenterology and Liver Disease of Teheran. The aim was to examine their problems, taking as their starting point the great stigma they undergo. Only 17 of the 45 people invited were brave enough to attend.[104] Their analysis reports that only a few centres are willing to take care of them; society's approach to the AIDS patient is ignorant, discriminatory and based on false beliefs; even some doctors share this attitude. It is recalled that the best system to reduce the social stigma is interpersonal contact between infected people and healthy people; however, this strategy seems difficult to apply in Iran.[105]

Figures on the diffusion of the virus are uncertain: until 2002–2003 the figures oscillate between 4,000 and 40,000 infected people depending on the sources. The official data underline that 95% are men and 65% are drug addicts. The rate of infection increases especially amongst intravenous drug addicts infected in prison and in rehabilitation centres.[106]

Regarding health restrictions to enter the country, there are none for tourists and businessmen resident for less than three months. "Foreign nationals applying for a work or residence permit must present a negative HIV test result"[107]; the exceptions concern holders of diplomatic, service or special passports.

Malaysia

At the end of 2003, according to the UNAIDS-WHO[108] estimates, 52,000 people were HIV-infected or had symptoms of AIDS. The majority were infected

---

[104] Sherafat-Kazemzadeh R., Shahraz S., et al., Iranian Persons living with HIV/ AIDS unveil the Epidemic of Stigma, *Archives of Iranian Medicine*, 2003, 6 (2), 77–80.

[105] Ibid., 80.

[106] Mansoori S.-D., Zadsar M., et al., Immunological and Clinical Features of HIV in a Group of Hospitalized Iranian Patients, *Archives of Iranian Medicine*, 2003, 6 (1), 5–8. See also: Iran's HIV Cases Exceed 12,000, *IPPF News*, 6 January 2006 in www.ippf.org/

[107] Wiessner and Lemmen, op. cit., 23.

[108] UNAIDS-UNICEF-WHO, *Epidemiological Fact Sheets on HIV/AIDS and Sexually Transmitted Infections, 2004 Update*.

through intravenous drug use, followed by heterosexual transmission. At the end
of 2004, the Government announced, for 2005, that the highly active antiretroviral therapy (HAART) would be free for people infected with HIV. These drugs
were already free for mothers and children.[109]

On 13 November 2001, the State of Johor (one of the 13 states of Malaysia),
through the Johor Health Department, applied a *fatwa* of the local religious
authorities who considered the premarital HIV test on Muslim couples compulsory,[110] especially in order to avoid any infection of children. According to the
fatwa, the future spouses must do the test in State hospitals three months before
the wedding[111] and the result will be passed to the religious authorities. If they
are healthy, they will obtain a certificate that allows the marriage to take place.
If either is infected, it is up to the partners to decide whether to marry or not, as
it is impossible to prohibit the marriage against their wishes.

According to the Johor Chief Minister, A.G. Othman, the State was obliged
to take this measure due to the great increase in HIV infections in 2000.

The Malaysian AIDS Council opposed the measure because it is difficult to
apply and ineffective in preventing the spread of the infection, especially because
it overlooks the "window of time", i.e. the period of several months that passes
between infection and testing positive, invalidating the credibility of the test.

For tourists arriving in the country there are no controls. HIV tests are required,
on the other hand, for foreigners seeking permission to work as unskilled labour-
ers or domestic staff or construction workers. If the HIV test is positive, they are
prohibited from entering the country or they are deported.[112]

Indonesia

Law no. 9 of 6/6/1994 revealed that "the problem of HIV/AIDS infection in
Indonesia is still... insignificant".[113] Point III-A-2 recalls that "Any combating
effort shall reflect the religious and cultural values existing in Indonesia". In
general, the 1994 decree is strongly oriented towards defending the dignity of
the HIV-infected individual or AIDS patient and in protecting their families. In
principle, the test for HIV diagnosis must always be voluntary and consensual by
the citizen or infected person and the result is strictly confidential.

The cultural diversities present in the State (which has over 200 million inhabitants, the vast majority of whom are Muslims) are underlined where it says that

---

[109] Free AIDS Drug Therapy Next Year, *New Straits Times*, 30 November 2004, in *AIDS/HIV
News: Malaysia* in www.utopia-asia.com/unews/archive.htm

[110] *Malaysian AIDS Body against Plan to force Pre-marriage HIV Tests*, Agence France Press, 1
January 2001, in *AIDS/HIV News: Malaysia* in www.utopia-asia.com/unews/archive.htm

[111] Ng E., *Malaysia-Sex-AIDS: Forced HIV Tests spark Calls for Sex Education in Malaysian
Schools*, Agence France-Press, 28 November 2001.

[112] Wiessner and Lemmen, op. cit., 28.

[113] Decree No. 9/KEP/MENKO/KESRA/VI/1994 of 6 June 1994 Concerning the National Strategy of Combating AIDS in Indonesia, in *International Digest of Health Legislation*.

"the basic information [about HIV/AIDS, methods of prevention, etc.] should be varied in regard to the method and specific emphasis so as to be suitable for Indonesian people, who are varied in their social and cultural conditions" (III-B). Recourse to condoms is included as one of the possible means to be publicised to control the infection, despite the opposition of several Islamic groups.

Since 1994, Indonesia has seen a fast and unstoppable spread of the infection. The main causes include the very low use of the condom, even in commercial sex. It is estimated that less than 10% of the 7–10 million Indonesians who frequent sex workers use condoms.[114] However, the greatest cause of the infection is through needle exchange by injecting drug users, etc.

The 2002 government data estimates an average of 110,800 people infected with HIV. Of these, 42,749 are injecting drug users, 32,922 are clients of prostitutes, 10,021 are homosexuals, 8,851 are prisoners and 6,085 are clients of transvestites.[115]

According to the press, in 2004 the number of HIV-infected people is alleged to have reached 200,000 and this would explain why the WHO classifies Indonesia amongst the most serious cases of south-eastern Asia along with Thailand and China.[116]

Other African Countries

It is not infrequent that the positions on contraception of African imams, especially those operating in rural areas, are very rigid. They may maintain that Islam always prohibits contraception and population control, ignoring the traditional tolerant positions of Muslim law on the matter. The religious leaders of Mali, Niger and Burkina Faso seem to belong to this category.[117] For all, the remedy for the HIV infection is abstinence; fasting also decreases the libido and allows sexuality to be controlled.

At the same time, many other African imams operating in areas where AIDS has spread at alarming rates are very tolerant and realistic in the application of Muslim principles. They prefer to encourage the spread of contraceptives (without too many limits) in order to avoid a social catastrophe (the principle of the "lesser evil").

There are some characteristic social customs of several Muslim areas in the world that may increase the spread of AIDS[118]: polygamous marriages; non-sterile male circumcision; ablution of the dead and practices of female genital

---

[114] UNAIDS-WHO, *AIDS Epidemic Update*, December 2003, 20–21, in www.unaids.org

[115] Ministry of Health of the Republic of Indonesia, *National Estimates of Adult HIV Infection, Indonesia 2002*, Ministry of Health, Jakarta, 2003, 46.

[116] Prystay C. and Mapes T., In *Indonesia, AIDS Education Clashes with Islam*, 29 March 2004, in www.aidschannel.org/; see also Indonesia on Verge of AIDS Epidemic, *IPPF News*, 30 November 2005 in www.ippf.org/

[117] Touré B., La Croisade anti-Sida des Imams maliens, *TchadForum*, 15 April 2004 in www.syfia.info

[118] Kagimu M., Marum E., et al., Planning and Evaluating Strategies for AIDS Health Education Interventions in the Muslim Community in Uganda, *AIDS Education and Prevention*, 1995, 7 (1), 10–21.

mutilation (FGM) (see Chapter 11). In the polygamous Muslim marriage (up to four wives), it is probable that an infected partner can infect the other members. Male circumcision is traditionally carried out in the first year of life by a local circumciser, in the mosque and on groups of several children at the same time. Generally, the circumciser uses the same razor blade on different children until the blade becomes blunt. If one child is infected, the others are also likely to become infected. As a consequence, it is important to oblige the circumcisers to use a new blade for each child as well as gloves. Regarding the ritual ablution of a deceased Muslim, all the orifices of the body have to be cleaned, including clots of blood, by relatives of the same sex as the deceased, guided by an imam. In order to avoid possible infection, the operators should be provided with gloves.[119]

Senegal

The estimates of 1999 spoke of about 79,000 people infected (out of a population of about 9 million). The first cases of AIDS cases in the country date back to 1986. The State reacted by declaring the fight against AIDS a health priority.

The factors that encourage infection are common to many other African countries[120]: poverty, inefficient consulting services and prostitution. In addition, the condition of women is very weak due to sociocultural traditions: constraint to do household work; difficulty of access to specialist services; economic dependence; risk of stigmatisation when going to STD assistance centres; submission (including sexual) to men.

The percentage of women who are virgins when they get married is 89% in rural areas and 68% in urban areas, especially Dakar. The decision to use the condom is always the husband's. Two more social customs may encourage the spread of AIDS: levirate and sororate.[121]

---

[119] The concerns linked to the practice are shown by the instructions of the *Jamiatul Ulama* (Council of Muslim Theologians of South Africa) in the newspaper *Al-Jamiat* under the headline "Ghusl of Aids Victims", which are reproduced here in full:
"Important guidelines
(1) Avoid direct contact with fluid and mucus membranes by using gloves, goggles, and masks. It is important to use rubber (latex) gloves. People who have abrasions and cuts on their hands must use double rubber gloves.
(2) Always use double gloves when cleaning the oral cavity.
(3) Aprons and boots must be used if spillage of fluids is anticipated.
(4) Wash hands thoroughly after ghusl.
(5) Contaminated surface, linen, and clothing must be disinfected in a sodium hypochloride solution for one or two hours. (One in ten dilution). The HIV virus is stable at room temperature in both wet and dry conditions. Therefore, thorough disinfection and sterilisation are necessary.
It may be preferable to adopt these precautions generally, because in most instances the HIV status of the deceased may not be known". See www. jamiat.org.za/al-jamiat/
[120] Seck A., Le Sida au Sénégal, un défi mondial de santé publique, *Soins*, 2001, 657, 46–47.
[121] Levirate is an institution according to which a man is obliged, or is entitled, to marry his brother's widow. Sororate indicates the institution whereby the widower is entitled, or has the duty, to marry his late wife's sister.

Of the HIV-positive cases, 61% travelled abroad in recent years; 65% of the wives of emigrants are HIV-positive.[122] A study by the "Fonds des Nations Unies pour la femme" (UNIFEM) in 2003 showed that the number of infected women in Senegal, in the period between 1988 and 2002, quadrupled (35,947 cases in 2002); amongst men it doubled (41,326 cases in 2002).[123] Infected women are often victims of great discrimination and stigmatisation.

In 1986, the national programme against AIDS was launched, involving the religious authorities, even though the results are at times controversial.[124] The religious leaders, very strict in principle, appeared to be very moderate and realistic in practice. The most important actions taken by the government include sexual education in schools, extensive promotion and distribution of condoms and the creation of consulting centres for adolescents.

Uganda

In Uganda 1 adult out of 10 is HIV-positive. In the country, Muslims are a minority; however, the work of the Islamic Medical Association of Uganda (IMAU) for the prevention of the spread of AIDS has aroused great interest as it was chosen by ONUSIDA as a brilliant model to be imitated in other countries.[125] In 1989, the Grand *mufti* himself declared war on AIDS. From 1992, the IMAU began to train more than 8,000 religious leaders and volunteers on the disease; they regularly went to more than 100,000 families, in 11 districts of the country, to inform them of the nature of the infection and to invite the faithful to change their sexual behaviour. Only two years later, the citizens who had benefited from this service had a good knowledge of the risks and means of transmission of the virus. In addition, they showed a significant reduction in the number of sexual partners and an increase in the use of condoms.[126] The activity of the 300 doctors of the IMAU is mainly dedicated to Muslims but is also open to the members of other religions. In 1995, an educational programme was started for young Ugandans called "Madarasa Aids Education and Prevention", which operated from schools attached to the mosques.

The emancipation of women is also decisive in Uganda. When totally economically dependent on their husbands, neither do they dare oppose his wishes, even when they suspect he is unfaithful to them, nor do they dare insist on the man using a condom. The IMAU operates so that women can gain financial autonomy (including by working outside the home).[127]

---

[122] Seck, art. cit., 47.
[123] Faye A., Santé-Sénégal: Les femmes porteuses du VIH/SIDA réclament un meilleur accueil dans les hôpitaux, *IPS Inter Press Service*, 16 March 2004.
[124] Lagarde E, Enel C., et al., Religion and Protective Behaviours Towards AIDS in Rural Senegal, *AIDS*, 2000, 14 (13), 2027–2033.
[125] Islamic Medical Association of Uganda (IMAU), *Education Sida grâce aux Imams*, Geneva, ONUSIDA, 1999, 5.
[126] Islamic Medical Association of Uganda (IMAU), op. cit., 6.
[127] Ibid., 27.

Education on the use of condoms in the IMAU's anti-AIDS programme has raised considerable opposition amongst the religious leaders. Doctors, however, have succeeded in convincing them that condom can be used against AIDS when calls for abstinence and relations exclusively with the spouse are useless.[128] The doctors also recalled that the high number of pregnancies before marriage and the many cases of STDs require a certain flexibility.

Mali

In many mosques in Mali, the imams are directly involved in the fight against AIDS and frequently devote the Friday sermon to this issue. A sermon by the imam of the mosque of Djikoroni (a popular neighbourhood in the capital, Bamako) was very clear: "You must all know that AIDS is in our country, in our cities and in our families. It threatens the whole of society."[129] In the capital, 140 places of worship and hundreds of state mosques act in the same way.

The initiative came into being in April 2003 with the "Réseau National de Lutte contre le Sida" and the NGO Population Service International. The involvement of the religious leaders is useful in particular to raise awareness of the not so young, as the prevention campaigns address young people.

The sermons preach abstinence for bachelors and fidelity for the married. The use of condom is recommended when one spouse is infected. Amongst the traditional customs that may contribute to spreading AIDS are levirate and sororate; in this case preventive tests are required.

In Mauritania,[130] the government has asked the religious authorities to use the Friday sermons to warn the faithful about the danger of AIDS. Until recently, it was a taboo subject. However, the number of infected people is very low.

---

[128] Ibid., 30.
[129] Touré, art. cit.
[130] Mauritania seeks Imam's Help in Fighting AIDS, *Reuters NewMedia*, 9 March 2001.

THE OPINIONS ON GENETICS

PRINCIPLES AND VALUES   235
THE DEBATE ON GENETICS   237
HUMAN CLONING   241
POSITIONS TOLERATING HUMAN CLONING   245
RESEARCH ON STEM CELLS   248
THE ABORTION OF HANDICAPPED FOETUSES   250
CONSANGUINEOUS MARRIAGE   254
PRE-NATAL DIAGNOSIS   259
CONCLUSION   265

PRINCIPLES AND VALUES

Genetic engineering refers to all those techniques aimed at inserting new genetic information into the structure of the cell of a living individual. Some preliminary remarks must be made, with it being opportune to distinguish between at least four different reasons for intervention[1]:

(a) Diagnostic reasons, by means of pre-natal or post-natal tests to ascertain the presence of pathologies of genetic origin. In the post natal phase, for example, the adult can be diagnosed before marriage or a relationship aimed at conception (a subject at risk like a carrier of the gene of a recessive genetic disease, e.g. thalassaemia, could marry a partner who is also a carrier and run the risk of generating offspring affected by the disease in 25% of cases); in cases of civil law for paternity tests; in the context of criminal law to identify guilty parties in various criminal offences.

(b) Manufacturing reasons in the pharmacological field with the production of hormones such as human insulin, interferon, bacterial, viral, parasitic vaccines, etc., all with the techniques of recombinant DNA; biotechnologies in the agricultural and food contexts are also included.

(c) Manipulating reasons, i.e. aimed at producing modifications that are not therapeutic but elective and selective and already used in animal and plant contexts (with the problem of the impact on the ecosystem) but also hypothesised for man with "ameliorative" genetic engineering (e.g. taller or longer living individuals).

(d) Regarding therapeutic reasons, when the ultimate objective is recovery from a disease, gene therapy is talked of, which can be defined as the replacement of a gene (a fragment of DNA) that is missing or malfunctioning by a gene

---

[1] See Sgreccia E., *Manuale di Bioetica*, Milano, Vita e Pensiero, 1994, 243ff.

Dariusch Atighetchi, Islamic Bioethics: Problems and Perspectives.
© Springer Science+Business Media B.V. 2009

that can fully carry out its function, i.e. that of preventing and/or curing a pathological condition.

There are two fundamental levels of gene therapy with differing ethical consequences.

1. The therapy performed on germinal cells (e.g. gametes or precocious embryos) consists of the introduction of a gene into the fertilised egg (implying risks of damage and morphological anomalies caused by the intervention of a mechanical type – microinjection – and physical damage on the cell) or precocious embryos so that the modification affects the genome of the subject that will be conceived or is already conceived but also its descendants. At present, this is a technique the effects of which on the descendants are not well-known.

2. Gene therapy on somatic cells (i.e. of the lymphatic system, of the blood and of the bone marrow) can return normality to the defective cells of the patient, curing him, but without any effect on descendants. The pathologies that can be cured by this system include those caused by the defect of a single recessive structural gene and therefore susceptible to complete cure thanks to the introduction of a single copy of the healthy gene such as the deficiency of adenosine-deaminase (ADA), cystic fibrosis, Duchenne type muscular dystrophy, haemophilia, etc.

Developing countries as yet have little familiarity with technological innovations and the ethical problems connected with these. In Muslim countries, there have not been many specific debates to date on the threats deriving from the applications of genetic engineering. However, faith in the direct Word of God and in His Law has to provide the points of reference to deal with any new problem. Here are some principles and guiding values drawn from medical ethics and Muslim law that have been formulated for guidance in the application of genetics:

1. The principle of justice (*istihsan*).
2. Defending the collective interest (*maslaha*).
3. Respect for the physical integrity of man. Koran 95.4: "We have indeed created man in the best of moulds."
4. Respect for the psychic integrity of man. Koran 30.30: "No change (let there be) in the work (wrought) by God"; Koran 32.9: "But He fashioned him in due proportion, and breathed into him something of His spirit."
5. Protection of the genetic inheritance of families.

The *Shari'a* abhors any confusion or uncertainty in identifying natural and legitimate parents. The Prophet said: "Select your spouse carefully in the interest of your offspring because lineage is a crucial issue."[2] This also implies the refusal of adoption[3] as expressed by the Koran 33.4: "nor has He made your adopted sons your sons" and 33.5 "Call them by (the names of) their fathers."

6. Prohibition on manipulating creation.

---

[2] In Serour G.A., *Ethical Implications of Human Embryo Research*, Rabat, ISESCO, 2000, 15.

[3] See discussion, concerning adoption in Chapter 6 "Assisted Procreation".

Koran 4.119–120 is very frequently quoted, in which the demon says: " I will mislead them, and I will create in them false desires; I will order them to slit the ears of cattle, and to deface the (fair) nature created by God'. Whoever, forsaking God, takes Satan for a friend, hath of a surety suffered a loss that is manifest. (120) Satan makes them promises, and creates in them false desires; but Satan's promises are nothing but deception" and Koran 30.30 (quoted above). However, Muslim scholars warn that these passages do not prohibit every intervention on creation as this would mean prohibiting all surgical and therapeutic practices (e.g. appendectomy, tonsillectomy, etc.) which nevertheless represent a "positive" modification of creation.[4]

Amongst the applications of the sixth principle, we find respect for Nature and the environment.

On this subject, literature often refers to Koran 45.13: "and He has subjected to you, as from Him, all that is in the heavens and on earth". This passage has been interpreted as an invitation to respect the surrounding environment and human nature itself.

As neither the Koran nor the *Sunna* talk of genetic engineering, it is necessary to "balance" this "manipulative" concern for creation with other principles (7 e 8) guiding human action on it.

7. The therapeutic principles for the care and protection of human life. These are the passages that represent the fundamentals of Muslim medical ethics:
   (a) Koran 5.32 "if any one saved a life, it would be as if he saved the life of the whole people";
   (b) *Hadith* (Bukhari): "There is no disease that Allah has created, except that He also has created its treatment"[5];
   (c) Koran 2.185 says: "God intends every facility for you; He does not want to put you to difficulties."

8. There are further juridical rules: preventing evil has precedence over doing good (*dar'u al mafasid muqaddam ala jalb al-masalih*); protection against pain; abstaining from causing damage to oneself and to others (*la darar wa la dirar*).

In the light of these principles and criteria of action, all the applications that promote life, health and well-being are praiseworthy and rewarded by God but require strict control and precautions against manipulative intentions and contingencies. The principle remains vague and requires continuous concrete redefinitions.

THE DEBATE ON GENETICS

Whilst genetics is flourishing in the rich countries, in developing countries the situation appears difficult. Frequently considered the "luxury" branch of medicine, it is confused with research, therefore rarely seen as a social service and

---

[4] Hathout H. and Lustig B.A., *Bioethical Developments in Islam*, in Lustig B.A., Brody B.A., et al. (eds.), *Bioethics Yearbook – Theological Developments in Bioethics: 1990–1992*, Dordrecht, Kluwer Academic Publishers, 1993, 133–147.

[5] Bukhari, *Sahih, Medicine*, Vol. 7, Book 71, Number 582, in www.usc.edu/dept/MSA/index.html.

scientists deal with genetics more than doctors. According to geneticist Habiba Chaabouni, genetics in Tunisia in 1981 was above all a discipline taught by "foreign missionaries".[6]

In the highly differentiated Muslim world, alongside situations of clear difficulty, there are initiatives and successes in specific fields in several countries.[7]

Regarding the general principles to be followed in genetic research, a very clear opinion was that expressed by the Council of the Academy of Muslim Law (Fiqh), Jeddah (10th session, 28 June to 3 July 1997) in the Introduction to the Resolution on Human Cloning[8]: Islam does not place any limits on the freedom of research when it is implemented to study in further depth the laws of nature; however the religion demands that the doors not be opened without rules regarding the applications of the research. They must first be evaluated by Muslim Law, which can decree them licit, and prohibit those proceedings which appear illicit, as not everything that can be implemented must be done; it is necessary to approve the practices that are of use to the collective interest and reject those that are immoral. The document continues: science must preserve human dignity, it must not attack the personality of an individual or destabilise social order; it must not destroy the bonds of kinship and blood or the family structure.

The constant imitation of modern Western biological sciences is a reason for concern,[9] although reference to the precepts of the *Shari'a* represents an instrument to stem this trend. The cautious attitude reserved for these new technological developments was confirmed in Jeddah by the Council of the Academy of Muslim Law (10th session, 28 June to 3 July 1997), in a resolution on human cloning.[10] Point 9 quotes Koran 4.83: "When there comes to them some matter touching (public) safety or fear, they divulge it. If they had only referred it to the Apostle, or to those charged with authority among them, the proper investigators would have tested it from them (direct). Were it not for the Grace and Mercy of God unto you, all but a few of you would have fallen into the clutches of Satan."

---

[6] Chaabouni H., La Génétique dans les pays émergents: l'urgence de créer des réseaux, *L'Observatoire de la Génétique*, March–April 2003, no. 10, www.ircm.qc.ca/bioetique/obsgenetique.

[7] See amongst many others: Acharya T., Rab M.A., et al., Harnessing Genomics to Improve Health in the Eastern Mediterranean Region-An Executive Course in Genomics Policy, *Health Research Policy and Systems*, 2005, 3 (1) in Biomedcentral, www.health-policy-systems.com/content/3/1/1; Dasilva E.J., Biotechnology in the Islamic World, *Nature Biotechnology*, 1997, 15, 733–735.

[8] See the complete Arabic and Italian texts in De Caro A., *La clonazione secondo il diritto islamico*, Academic year 1999/2000, Roma, Università La Sapienza, 123–130, 125–126. (Degree thesis).

[9] *Recommendations of the 9th Fiqh-Medical Seminar*, Casablanca, 14–17 June 1997, point 7, at www.islamset.com

[10] In Belkhodja M.H., *L'Islam et la Biologie*, Allocution à la Rencontre Internationale sur la Bioéthique, Tunis, 23–25 October 1997.

Sheikh Jad al-Haq appeared somewhat sceptical of recourse to the new genetic techniques at the First International Conference on "Bioethics in Human Reproduction" in the Muslim world (Cairo, 1991). In his opinion, scientists forget the words of God "of knowledge it is only little that is communicated to you"; indeed, they try to correct hereditary psycho-physical malformations in man or even improve human genes by recourse to preventive and therapeutic measures. This represents a serious attack: "Hereditary traits such as intelligence, stupidity, stature, beauty, ugliness, sterility or fertility have persisted over generations and cannot be terminated in seconds by a scalpel or an injection."[11]

An articulated stance on these issues is to be found in the Final Recommendations of the 11th Seminar on "Genetics, Genetic Engineering, the Human Genes and Genetic Treatment – An Islamic Perspective"[12] (Kuwait, 13–15 October 1998) organised by the Academy of Muslim Law (Fiqh) of Jeddah (Saudi Arabia), by the WHO Regional Office, Alexandria (Egypt) and the Islamic Education, Science and Culture Organisation (ISESCO). A list follows of some of the more significant points included in the six paragraphs of the document. With reference to the guiding principles in recourse to genetics (paragraph I) the document states that Islam promotes knowledge (Koran 39.9) and does not obstacle any "constructive" scientific research; for these reasons, Islam must move to the fore in genetic research. Recourse to genetics to cure hereditary or acquired pathologies does not contradict the acceptance of divine will. Research, treatment and diagnoses on the genetic condition of a person must be performed with the previous and free consent of the individual concerned or, if incapable, of the guardian. If consent cannot be obtained, the research may be carried out if it brings about a clear benefit for the health of the person (I,5). Every genetic diagnosis must be treated with confidentiality. Genetic research must never have priority over the rules of the *Shari'a* and respect for human rights (I,9). In the second Paragraph of the document concerning the human gene, a favourable opinion is expressed on the mapping of genes as an effort aiming at knowledge of the human being and comprehension of the mechanisms of some hereditary pathologies. The third paragraph confirms that genetic engineering on germinal cells is prohibited by the *Shari'a*. Genetic engineering should not be used with offensive purposes or crossing genes of different species to improve the human race or to tamper with the personality of an individual (eugenic purposes). Similarly to other Muslim documents, the Recommendation cautions against scientific monopolies that deprive the poor of the possibility to have recourse to such technologies. In the opinion of the Seminar, Muslim law has no objections to the use of genetic engineering in agriculture and for animals, but calls for an evaluation of the long-term risks on humans, animals and the environment. The fifth paragraph on genetic consulting recommends it be available to citizens but without any obligation; the results of

---

[11] Haq (Al) J., Islam, a Religion of Ethics, in Serour G.I. (ed.), *Ethical Guidelines for Human Reproduction Research in the Muslim World*, Cairo, IICPSR, 1992, 11.

[12] www.islamset.com

consulting must remain confidential; recourse to consulting must be encouraged by the health institutions, the mass media and in the mosques. These are urgent actions as marriage between close relatives, widespread in Islam, is associated with a greater risk of physical defects, especially in families with a history of genetic pathologies.

In this document of 1998, the Academy of Muslim Law of Jeddah and the ISESCO invited Muslim countries to work on the mapping of genes. The scarce interest shown by these countries in the Human Genome Project (HGP) was shown by the presence of only 14 experts from only 4 Muslim countries in the Human Genome Organisation (HUGO).[13] Equally as serious is the low presence of Muslim experts in international scientific and technological literature: the Muslim world contributed 1.033% of international literature on the topics mentioned against 1.959% by Belgium and 1.64% by Switzerland alone.[14]

Regarding diagnostic reasons, recourse to elements of identification provided by "genetic fingerprints" or "genetic identity cards" (to be found in the blood, hair, nails, tissues, etc.) for the confirmation of paternity seems to be positively evaluated in the Muslim world. The same attitude prevails with regard to recourse to these tests for judicial-penal purposes, i.e. for the identification of the culprits of serious criminal offences.

In the light of the debates to date, it can be said that the introduction of a healthy gene into somatic cells, if performed for therapeutic reasons, is approved by Islam; some experts deem this technique equivalent to a transplant at molecular level and is accepted by those in favour of organ transplants (the majority of jurists).

According to other authors, the modification of an organ through a gene for therapeutic purposes is acceptable; vice versa the modification of an entire organism is formally prohibited as a divine creation is altered.[15] Respect for what has been created, for physical human integrity and the therapeutic principle impose that each modification that can be transmitted to descendants (i.e. of the genome) is prohibited even if the purpose is therapeutic, as an organism other than that created by God would be created.[16] The first attempts at genetic manipulation were made on "transgenic" animals which, if used as experimental models for the cure of human disease, would be tolerated by Islam.

There is a clear-cut prohibition of genetic engineering for eugenic reasons.

---

[13] Nasim A., *Ethical Issues of the Human Genome Project: An Islamic Perspective*, in Fujiki N. and Macer D.R.J. (eds.), *Bioethics in Asia*, Tsukuba, Eubios Ethics Institute, 2000, 209–214. One of the first approaches on the subject in the region is the article by Hazmi (El) M.A.F., Human Genome Organisation – Is there a Role for Middle Eastern Scientists?, *Saudi Medical Journal*, 1995, 16 (5), 371–377.

[14] Nasim, art. cit.

[15] Zribi A., Etica medica e Islam, *Dolentium Hominum*, 1996, XI (31), 82–85.

[16] Ben Hamida F., in AA.VV., *The Human Rights, Ethical and Moral Dimensions of Human Health Care*, Council of Europe Publishing, 1998, no. 32, Genetic Manipulation.

HUMAN CLONING

Cloning, i.e. the reproduction of genetically identical individuals can be obtained in two ways: with the transfer of the nucleus[17] or by embryo splitting.[18] A third type of cloning by human nuclear transfer must be added, "non-reproductive cloning", which is not aimed at producing persons or animals. Therapeutic cloning comes under this category: the procedure is carried out only on the in-vitro pre-embryo (which is not transferred again into a uterus) for the purpose of producing certain cell lines (e.g. stem cells) or tissues or organs.

The subject of human cloning seems to be taboo for the monotheistic religions. In Islam, the debate presents diversified positions that are not all against it; there are different religious standpoints that can give different answers. Each of the positions groups together different principles that may also be present in other positions. The first two positions are the commonest. The first position is the most rigid and is contrary to human cloning. The second is contrary but more moderate in its approach. The third and fourth position are, on the other hand, upheld by a minority who tolerate cloning and will be discussed in the next section. Here are the four positions:

1. God is the creator of all things. Cloning aims to imitate divine creation; at the same time it wants to modify it. The perfection and variety of creation is damaged by human cloning. Divine will is ignored.
2. The previous principles are accepted. However, cloning is successful only if God so wishes (Principle of predestination and divine omnipotence), similarly to contraception and IVF. Therefore, it is not an act of creation but only a "reshuffle" of what has already been created by God.
3. Human cloning is only a new tool of assisted medical procreation. Indeed, when cloning by transfer of the nucleus takes place between married partners, the practice is similar to a homologous reproductive practice, e.g. IVFET.
4. Cloning helps understand the miracle of the Resurrection at the end of time (and the miracle of the birth of Christ without a father).

It must also be noted that the first position tends to be self-sufficient and corresponds to an "a priori" refusal that does not require justification or explanation. In fact, maintaining that cloning aims to imitate divine creation and omnipotence clashes with the pillars of Muslim theology and cloning automatically becomes unsustainable. The second position adds a further step of reflection: cloning is successful only if God so wishes and does not correspond to a creative

---

[17] Asexual fertilisation in which the haploid nucleus (typical of the germinal cells where there has not yet been the fusion of the paternal and maternal chromosomes) is removed from a fertilised oocyte, replacing it with the diploid nucleus from the somatic cell of an adult of the same species. This nucleus, inserted into the cytoplasm of the fertilised oocyte, takes on the characteristic of totipotency, forming an individual that is identical to the one from which the somatic cell comes.

[18] This is natural cloning which takes place in the first phases of embryonic development when the embryo at the stage of a cell divides, generating two embryos from which two identical individuals are originated.

act as it only mixes together what has already been created. In this case, the pillar on which the first position was based is weakened: the refusal of human cloning must be justified and cannot be "a priori".

There now follows a list of some accounts and documents which come under the first two positions.

In a document, the third last Grand *mufti* of Egypt, N. Farid Wassel, had defined human cloning an immoral and Satanic practice. "There exists no error or sin greater than wanting to change the creation of God."[19] With cloning, man is an exact copy[20] obtained with a procedure other than that desired by God. The male–female union is the only one that ensures well-being and harmony for the nature of man as well as establishing rights and duties between individuals and communities. Human cloning will completely change what God has created, will bring ruin to the earth as well as changes in human reproduction and in social relations. This will affect the bonds of the individual with the family, of families amongst themselves and their bonds with the State and society. "Therefore Islam does not admit this practice . . . and closes the door to scientific experiments on it . . . it is in favour of laws aimed at considering such a practice as a criminal offence, and punishments against researchers and collaborators working on this . . . directly or indirectly."[21] Wassel is, however, in favour of partial cloning for treatments and therapies that are for the benefit of man.

The final decision of the Council of the Academy of Muslim Law of Jeddah (10th session, 1997)[22] prohibits every type of duplication of man (I); all cloning is prohibited especially when a third person is involved with respect to the married couple (III); it is licit to have recourse to cloning and genetic manipulation in the fields of microbiology, bacteriology, botany and zoology, but within the limits authorised by the *Shari'a*, namely for the collective interest (IV).

According to the same Academy of Muslim Law (Fiqh) of Jeddah (10th session, 1997), but above all for the Academy of Research of Al-Azhar,[23] the techniques of cloning must not be confused with an act of creation; indeed, only God is capable of creating from nothing, without previous examples (Koran 52.35–36; 13.16; 19.19; 22.73–74), which is not the case with cloning practised by men who combine elements that have already been created by God in different ways.

Before expressing its own negative opinion, the Al-Azhar document formulates the theses for and against cloning. There are at least two motives in favour of it: (1) The greater genetic knowledge obtained will allow greatly modifying the

---

[19] Wasil N.F., *Compendio sulla ricerca della clonazione umana e giudizio della medicina e della scienza nella Shari'a*, in De Caro, op. cit., 159–167.

[20] The Arabic term used both by journalists and jurists to indicate cloning is *istinsakh*, meaning "copy".

[21] Ibid., 165.

[22] De Caro, op. cit., 131–134; Belkhodja, art. cit.

[23] Academy of Research of the University of Al-Azhar, *Progetto di Dichiarazione sulla clonazione*, 1998, in De Caro, op. cit., 141–142.

applications of medicine and pharmacology; (2) It would be possible to clone a genius or a leader with a noble soul.

The motives against cloning are more elaborate[24]:

1. The technique (of the "Dolly" type) is far from being effective. As only one case out of 300 has been successful, it is impossible to apply a technique with so few possibilities of success to man.

2. The natural union between male and female by spermatozoon–ovule fertilisation promotes variety and the diffusion of stronger and more adaptable creatures; in addition, it limits diseases compared to other reproductive methods. Cloning only produces repeated forms without creativity and variations.

3. Cloning is too complicated a procedure compared to the natural union as desired by God (Koran 30.21).

4. Procreation by means of cloning deprives men of the quality of good feelings and noble emotions that characterise him with regard to a new life. With cloning, man degenerates "to a rather abject bestiality".

5. If to clone we remove the cell from one female, the ovule from a second female and if we implant all this into the uterus of a third female, who is the mother? The owner of the cell or of the ovule or of the uterus from which the clone is born?

6. The cloned copy has to face moral, social and health difficulties as it remains outside the community, leading therefore to psychological illness, etc.

7. If the aim is to clone a genius of science, the arts, etc., the copy only concerns the physical characteristics. Intelligence and feelings cannot be cloned. In fact, man remains the result of the combination of genetics with the environment and the agents surrounding him.

8. If many attempts at human cloning fail, what will happen to the deformed and mutilated beings that may be produced? Are they to be killed? Are their organs to be sold? Are they to be shut up in a zoo?

In conclusion, after having evaluated the positive and negative aspects of human cloning, Al-Azhar deems that the resultant harm is greater than the common good, therefore the practice is prohibited.

In fact, the aspects that have aroused greatest criticism on human cloning concern asexual reproduction and the social and family consequences of the practice. The rules on marriage and inheritance of the *Shari'a*, even when partially introduced into the positive laws currently in force, become inapplicable if there is no clear bond between relatives.

Sheikh Yusuf al-Qaradawi refuses the idea that recourse to cloning by man is the equivalent of infringing divine creation, or defying His will because, if the technique is successful, it is only thanks to the will of the Creator. Qaradawi is interested above all in the social and family consequences of cloning and he expresses his negative opinion on the basis of these considerations. The *sheikh*

---

[24] De Caro, op. cit., 146–149.

wonders what the consequences would be of having several identical copies of the same individual.[25] The main consequence concerns the family based on the paternal and maternal roles, as marriage becomes useless in view of procreation by cloning; the egg and the uterus belong to the woman and are sufficient for procreation making the presence of man superfluous. A similar imbalance, maintains Qaradawi, would lead to illicit relations between men and women as is the case in some Western countries. In addition, the "photocopy" effect would lead to serious consequences in marriage as the partner might no longer recognise the "original" spouse, giving rise to adulterous relationships. Lastly the *sheikh* identifies in the Koran a further reason for rejecting human cloning and that is, the variety of human beings and creatures as the distinctive sign of divine creation. Despite these criticisms, recourse to the techniques of cloning remains tolerable, according to Qaradawi, for the purpose of protecting the health of the foetus; eliminating some hereditary pathologies; producing specific parts of the human body such as hearts and kidneys for therapeutic purposes.

According to Muslim family law, "the genetic material of the offspring had to be inherited from both parents".[26] This leads to a further reason to refuse a cloned baby: it does not have the maternal and paternal DNA because it is the copy of one of them. It becomes impossible to determine the baby's precise relationship with its parents. Furthermore, it would be possible to transfer the nucleus of a somatic cell of a woman to the enunucleated ovum of the same woman; procreation takes place with a single sex and all this is unacceptable for Islam.

There exists considerable concern by some Muslim countries and by religious authorities that they may become ground for experimentation and dissemination of cloning techniques practised by foreign organisations. In this regard, point 5 of the Resolution on Human Cloning by the Academy of Muslim Law (Jeddah, 1997) invited Muslim countries to pass laws and rules to close their doors to such foreign organisations and researchers. The Tunisian National Committee of Medical Ethics called for legislation aimed at prohibiting every sort of "human cloning, of any modification of the genetic code or programme".[27]

One of the most authoritative Muslim bio-ethicists, Gamal Serour, totally agrees with the prohibition of producing genetically identical human individuals by the transfer of a nucleus from an adult or a child. However, he deems that human cloning may at times be performed by a totally sterile couple, during the validity of the marriage contract, with the nuclear transfer from the husband's somatic cell to the enunucleated ovum of his wife.[28]

---

[25] Sachedina A., *Cloning in the Quran and Tradition*, www.people.virginia.edu/~aas/article/article4.htm; Qaradawi (Al) Y., *Cloning and Its Dangerous Impacts*, www.Islamonline.net.

[26] Serour, *Ethical Implications of Human Embryo Research*, op. cit., 22–23.

[27] Comité National d'Éthique Médicale, Avis no 3, *Le Clonage*, CNEM, Tunis, 1997.

[28] Serour, op. cit., 23.

The Recommendations of the Workshop on "Ethical Implications of ART for the Treatment of Infertility" (Al-Azhar, Cairo, 2000) also condemned reproductive cloning to produce a new person (point 8). However, non-reproductive cloning was accepted to produce stem cells. The possibility of reproductive cloning between married partners when the husband is totally sterile was also discussed, but no common ethical positions were found.[29]

Does a cloned man have a soul? Amongst the opponents of reproductive cloning, some maintain that a cloned person does not have a soul (*ruh*) "and will therefore not be a human being as we know. The product of cloning will have all the biological properties of the ordinary human being but will not have the spiritual qualities. Thus the life of the cloned product will be of little or no quality".[30] We will find the opposite position (I.B. Syed) amongst those who tolerate reproductive cloning in the next section.

Alongside balanced analyses, there are also strongly controversial political and religious approaches towards cloning and the West. In these cases, human cloning may represent the tip of the iceberg of the threat of Western scientific "ideology" aimed at infringing the honour and dignity of man and of religious cultures of all peoples, in particular of Islam.[31] In the case of the *mufti* M. Saheb, the refusal of cloning is accompanied by a refusal of genetic engineering. Interviewed by the France Presse news agency (AFP), the Saudi Arabian *sheikh* Mohammed Ibn Saleh al-Othimin approved of the death penalty for whoever did research on cloning, more specifically: "I think that the lowest penalty imposed on those who invented cloning should be amputation of their hands and feet. Otherwise, they should be executed."[32]

POSITIONS TOLERATING HUMAN CLONING

As shown, the opponents of cloning refer to multiple motivations, with different priorities. The minority in favour of cloning also base their arguments on a variety of reasons. In general, they deny that the procedure is against nature and divine will precisely because the omnipotence of the Creator makes the act possible only if God so wants (as is the case with IVF and contraception). In addition, those in favour of cloning accept it only within married couples, for example, with the DNA taken from the husband and implanted into the wife's

---

[29] Serour G.I. (ed.), *Proceeding of the Workshop on Ethical Implications of Assisted Reproductive Technology for the Treatment of Infertility*, November 22–25, 2000, Cairo, Al-Ahram Press, 2002, 107 and 102–103.

[30] Kasule O., *Preservation of Progeny: A Medical Perspective* in www.e-imj.com/Vol3-No1/Vol3-No1-H3.htm#_ednref1.

[31] Saheb M.S.M., Human Cloning, *Al-Jamiat*, May 1998, 3 (7), in www.al-jamiat/_private/jam_navbar.html.

[32] See Youssef R., Mary had a Little Cloned Lamb, *Middle East Times*, 21 March 1997, in www.metimes.com/.

egg. According to another approach, human cloning is proof of human resurrection by God on the Day of Judgement.

In the opinion of the Nobel Prize-winner for Literature (1988), the Egyptian Naguib Mahfouz,[33] God remains the only Creator whilst cloning is a medical procedure, no different from "in vitro fertilisation"; both these procreative acts remain miraculous and divine. In his opinion, cloning will produce unimaginable beneficial effects.

Shaker Helmi, Professor of Genetics at the University of Alexandria, refuses the idea that human cloning can imitate divine creation because the human being is not God. He also adds: "I think there should be no limit to freedom for scientists to find out about life, nature, etc. If we find a way through science to help mankind, cure diseases and find solutions, then why not go ahead with research?"[34]

The authoritative Lebanese *Shi'ite ayatollah* S.M.H. Fadlallah deemed human cloning as the equivalent of a new procreative technique.[35] It is the equivalent of the discovery of IVF, currently a practice accepted (if homologous) by jurists. This discovery has been successful only thanks to divine will. Creation must be understood as "the invention of a new law", namely a law that has never existed before. This has never happened before in the history of human science as man is only now discovering the secrets of what has been created and the laws that govern existence. According to Fadlallah, the real problem lies in the possible negative use of this information. If the effects of human cloning are more negative than positive, the practice should be prohibited. Therefore, the opinion of Fadlallah is only moderately positive. According to the *ayatollah*, human cloning is not destined to be a great success because it is too expensive as well as requiring enormous scientific efforts. Sex is the natural reproductive instrument; vice versa, cloning only interests some scientists and a small number of people. To those who object that the legitimisation of cloning by the *ayatollah* is the equivalent of submission to the words of Satan (Koran 4.119) "I will order them . . . to deface the (fair) nature created by God," the *Shi'ite* religious leader answers that the passage, taken literally, would oblige us not to touch anything, not even mountains or the earth. Cloning does not modify creation; it is only one of the ways of human reproduction desired by God.[36]

Another *Shi'ite*, the Iraqi *ayatollah* Sa'id at-Tabataba'i al-Hakim, refers to Jesus who, in the Koran, does not have a biological father. By analogy, this proves

---

[33] Mahfouz N., Cloning! *Al-Ahram Weekly*, Online, 12–18 December 2002, no. 616.

[34] Youssef, *Mary had a Little Cloned Lamb*, art. cit.

[35] *Cloning, Interview with ayatollah Fadlallah*, www.bayynat.org.lb/www/english/newsletters . . . .

[36] To the question on what he thought of the hypothesis that Jews who want to rule the world are behind the invention of cloning, the *ayatollah* replied: We know that from the book *The Protocols of the Learned Elders of Zion* that in their history the Jews have always tried to perturb the values of humanity in the interest of International Judaism or Zionism. However, we cannot base our opinion on this assumption, even if the hypothesis is not to be excluded.

that other methods of reproduction exist. Therefore cloning does not represent a deviation from reproductive methods.[37]

For some experts, human cloning is not explicitly prohibited by the Koran, the *Sunna* or the *Shari'a*. According to Ibrahim B. Syed, however, a clone is comparable to a photocopy of the original or to an identical twin but younger. If this twin has a soul, the human clone has one too.[38] Often, continues the author, the Koran, in referring to the truth of the Resurrection, highlights the example of the creation 22.5 "O mankind! If ye have a doubt about the Resurrection, (consider) that We created you out of dust, then out of sperm, then out of a leech-like clot, then out of a morsel of flesh, partly formed and partly unformed . . . ." To those who do not believe in the Resurrection of corpses with the recomposition of the dispersed chemical and physical compounds, Koran 36.78–81 replies: " 'Who can give life to (dry) bones and decomposed ones (at that)?' – Say, 'He will give them life Who created them for the first time! For He is well-versed in every kind of creation!' 'Is not He Who created the heavens and the earth able to create the like thereof?'." The Koran underlines the infinite capacity of God to restore every detail of the human body as stated in 75.3–4: "Does man think that We cannot assemble his bones? Nay, We are able to put together in perfect order the very tips of his fingers."[39] As a consequence, cloning allows us to understand that the restoration of human life with the Resurrection is no more difficult than the original creation. In short, cloning becomes proof of the Resurrection and allows us to understand it, even proving it scientifically.[40]

Yusuf al-Qaradawi, quoted above amongst the opponents of cloning, also agrees with the idea that the phenomenon of human cloning helps to have a better understanding of the re-creation of man on the day of Resurrection.

Similarly, *sheikh* M. Shihabuddin Nadvi (1931–2002) appears enthusiastic about the consequences brought by cloning on the theological level: "Cloning has successfully produced a replica of an animal through a process which does not involve any sexual intercourse. This development of a cell into a fully-fledged living entity does provide irrefutable proof of the claim of resurrection by religion. Whatever might be the purpose and aim of such an experiment, none can dare dismiss the happening of the Resurrection Day now!"[41] Man does not become God (Koran 22.73), but only limits himself to imitating Him, replicating the natural processes established by the Creator.

---

[37] Eich T., Muslim Voices on Cloning, *ISIM Newsletter*, 12 June 2003, 38–39.

[38] Syed I.B., The Human Cloning: An Islamic Perspective, *The Muslim World League Journal*, 1997, 25, 2, 43–48; O. Kasule refutes the presence of a soul in a cloned man, see previous section.

[39] According to many commentators, the peculiarity of the fingerprints of the individual was known to the Koran whilst the English did not discover fingerprints until after 1880.

[40] Syed, art. cit., 48.

[41] Nadvi M.S., *Cloning Testifies Resurrection*, in www.2muslims.com/directory.

RESEARCH ON STEM CELLS

In general, many juridical and religious bodies and authorities approve search on stem cells, especially in the first few days of life of the embryo, to protect and increase the public good.

The position in favour of research on embryonic stem cells is influenced by the very widespread perception of the benefits that will be obtained thanks to their use. This research will satisfy the principle of the public good (*maslaha*), with the treatment of a great number of pathologies. The reason for the attention raised by this research is fostered by the fact that it is practised on creatures without a soul and outside the mother's uterus. Amongst the multiple distinctions that reinforce this willingness for research, some distinguish between actual life and potential life (the pre-embryo is believed to be only a potential life); or the ovule fertilised in a test-tube greatly differs from the one in the uterus; others maintain that the destruction of supernumerary embryos is not an abortion as they are not yet real human beings. Indeed, the use of these embryos to save mankind is an obligation (*fard kifayah*).[42] Obviously dangerous excesses have to be avoided. For many experts it is unlawful to produce embryos for the exclusive purpose of research whilst it is lawful to use only embryos already produced in excess for IVFET (as an alternative to their destruction), but there are discordant opinions on this. Lastly, there must always be the consent of the partners from whom the embryo comes for it to be used.

The Recommendations of the Workshop on "Ethical Implications of ART for the Treatment of Infertility" (Al-Azhar, Cairo, 2000) accepted non-reproductive cloning to produce stem cells that would be of benefit to others (point 8).

In 2001, the Muslim Law (Fiqh) Council of North America stated it was in favour of the use of unused supernumerary embryos "provided that this is done in the first few days after fertilisation" instead of destroying them.[43] The Ethics Committee of the Islamic Medical Association of North America (IMANA) maintains that "genetic research using stem cells from products of miscarriages or surplus ova after IVF procedures is permissible"; it is prohibited to conceive in order to abort and use the foetus and its stem cells.[44]

Gamal Serour, Professor of Obstetrics and Gynaecology at the University of Al-Azhar, tolerates non-reproductive cloning and believes it right to use precocious supernumerary embryos – but only within the first 14 days – for research for the benefit of others instead of leaving them to die.[45] On the contrary, Hamdy

---

[42] Siddiqi M., *An Islamic Perspective on Stem Cells Research*, 27 February 2002, in www.IslamiCity.com.

[43] Fiqh Council of North America, *Embryonic Stem-Cell Research*, 2001, on the Council's web site. The opinion of Sachedina A. is even more favourable to the use of stem cells in *Islamic Perspectives on Research with Human Embryonic Stem Cells*, in National Bioethics Advisory Commission (USA), *Ethical Issues in Human Stem Cell Research*, Vol. III, Maryland, Rockville, 2000, 29–33 in www.bioethics.gov.

[44] Imana Ethics Committee, *Islamic Medical Ethics: The IMANA Perspective* (PDF format), in www.imana.org, 1–12.

[45] Serour, *Ethical Implications of Human Embryo Research*, op. cit., 25.

al-Sayed, President of the Egyptian Medical Syndicate, appears to be against this position as the conceived being is already a human being; for this reason, only the use of stem cells from umbilical cord blood (i.e. non-embryonic stem cells) is legal in Egypt.[46]

A *fatwa* of January 2003 by the Egyptian *mufti* and professor at the University of Al-Azhar, Ahmad al-Tayyeb, maintained[47] that it is lawful to clone parts of the human body to replace diseased parts or as a therapeutic instrument to cure some pathologies; early embryos and five-day-old blastocysts created by means of transfer of the nucleus (nucleus transfer) are not equivalent to human beings to be protected, but are a useful source of cells to cure others.

On 9 January 2003, the Malaysian National Fatwa Council issued a *fatwa* allowing the therapeutic cloning of human embryos[48] for research purposes on condition that the foetus is destroyed within 120 days. On the other hand, the reproductive cloning of humans was prohibited because, according to the Koran, a child must be the product of the legal union between a man and a woman. The Chairman of the Council, Ismail Ibrahim, maintained that scientific therapeutic research on embryos within 120 days is in agreement with the opinion of Islamic sectors for whom the foetus does not possess a soul in the first 4 months and can be aborted. The Malaysian government is moving in the same direction, i.e. prohibiting reproductive cloning but accepting therapeutic cloning to produce stem cells.

In March 2003, the official Iranian press agency (IRNA) announced that the country was one of the first 10 in the world capable of producing, cultivating and freezing embryonic human stem cells. An Iranian survey on the opinion of 26 *Shi'ite* scientists, scholars and lawyers, connected with the activity of an assisted reproductive technology (ART) laboratory, expressed great willingness to use embryos before implantation as these are deemed to be without rights similar to those possessed by a foetus or child.[49]

These juridical-religious legitimisations push many Muslim countries to devote considerable resources to developing this research. For example, Saudi Arabia is one of the countries in the world with the highest rate of IVF births (4% of births against 1% in the USA). Thanks to these practical experiments, the Kingdom immediately started a stem cell research programme under the Jeddah BioCity private venture. This plan would allow the Saudi kingdom to become the biotechnological capital of the Middle East.[50] However, Saudi researchers

---

[46] Dabu C., Stem-Cell Science stirs Debate in the Muslim World, *The Christian Science Monitor*, 22 June 2005, in www.csmonitor.com/world/globalIssues.html.

[47] Walters L., Human Embryonic Stem Cell Research: An Intercultural Perspective, *Kennedy Institute of Ethics Journal*, 2004, 14 (1), 3–38.

[48] Mahmood K., *Malaysia Taking Steps to ban Reproductive Cloning: Report*, 11 January 2003, in www.IslamOnline.net; *Malaysia Muslims approve Cloning, Abortion up to 120 Days*, in www.lifesite.net.

[49] Ziaei S. and Farokhi M., The Ethical Challenge of Stem Cell Research and Tissue Transplantation, *Eubios Journal of Asian and International Bioethics*, 2004, 14, 97–99.

[50] McKay Duncan M., An International View of the Regulation and Funding of Embryonic Stem Cells in Therapeutic Research, *Texas Transnational Law Quarterly*, 2003, 17 (1), 19–26.

have to use stem cells from legally aborted foetuses and with the consent of the parents. Creating embryos for the purpose of research is illegal. The possibility of using supernumerary IVF embryos for research and therapeutic cloning is the subject of debate.[51]

The uncertainty of the religious positions is reflected in the policies of Muslim countries. On 18th February 2005, the United Nations' Legal Committee adopted the text of a declaration of intent on "human cloning", which is not legally binding and also concerns research on embryonic stem cells. The final text, which comes at the end of four years of discussion, was approved by the Legal Committee of the United Nations with 71 votes in favour (e.g. Morocco, Saudi Arabia, Sudan); 35 against (no Muslim States) and 43 countries abstaining, including the majority of Muslim countries (e.g. Algeria, Egypt, Indonesia, Iran, Iraq, Jordan, Kuwait, Lebanon, Malaysia, Senegal, Oman, Tunisia, Turkey). On 8th March 2005 the General Assembly approved the same document (84–34–37), with the value of a Recommendation. However, the final text asks countries to prohibit all forms of human cloning. This is a declaration of intent that various countries have declared not wishing to respect, especially for the limits on research on embryonic stem cells.

THE ABORTION OF HANDICAPPED FOETUSES

The diagnostic use of genetic tests in the pre-natal phase (which started in 1967) allows the early identification of the state of the foetus or the chromosome complement, in other words it allows ascertaining whether the foetus is affected by malformations or defects such as trisomy 21, which is responsible for the Down syndrome. This use arouses strong ethical problems deriving, more often than not, from the non-existence of treatment for the foetus after the identification of the defect, therefore the mother's only choice is to abort or accept a handicapped child. In Saudi Arabia, out of a sample of 30 families expecting a child who would probably be affected by a pathology of genetic origin,[52] 12 took abortion into consideration if the doctor so consented; 3 fathers were willing to have an abortion performed abroad whilst 16 parents refused on religious grounds and as it was not certain that the child would be born handicapped.

At present, thousands of illnesses caused by congenital malformations or incurable pathologies have been identified. These illnesses are caused by the alteration, amplification or absence of one or more genes (e.g. haemophilia, mucoviscidosis, sickle cell anaemia, beta-thalassaemia). For Islam, abortion for eugenic purposes is prohibited whilst if performed to avoid the birth of a gravely handicapped and incurable child the positions of the jurists appear highly varied.[53]

---

[51] Ibid.

[52] Panter-Brick C., Parental Responses to Consanguinity and Genetic Disease in Saudi Arabia, *Social Science and Medicine*, 1991, 33 (11), 1295–1302.

[53] See Chapter 5 "Abortion".

Some Tolerant Positions

After having ascertained the presence of serious handicaps in the foetus, a considerable number of jurists (both *Sunni* and *Shi'ite*) tolerate a voluntary interruption of pregnancy before the infusion of the soul into the foetus. The consent of the parents is required. Kuwait law (1981 and 1984) and the local Committee of Fatwa (1984) lay down the possibility of aborting with the consent of the parents if the foetus shows serious physical or cerebral handicaps before infusion of the soul.

Sheikh Qaradawi (February 1996) was in favour of the abortion of a deformed foetus exclusively within the first 40 days of pregnancy[54] (this means that abortions cannot be performed after pre-natal tests done after this limit). The position of the authoritative *Hanafite* Syrian M. al-Buti appears very similar.[55]

Jad al-Haq (former *sheikh* of Al-Azhar and *mufti* of Egypt) allowed the abortion of such a foetus within 120 days but not for all types of handicaps because, for example, a normal life can be lived by the blind or people born without a hand. After 120 days, however, it is no longer licit to abort a handicapped foetus (*fatwa* December 1980).[56] The next *mufti* of Egypt, Nasser Farid Wassel, also agreed on this limit.

The Muslim Law Academy of the Muslim World League (Mecca), 12th session, 10–17 February 1990, approved, by a majority, a document that sanctioned abortion within 120 days from conception if it is proven that the foetus is affected by incurable handicaps and that life, after birth, would become miserable for him or her and the family. To abort, the consent of both parents is required.[57]

In 1994, before opening the first department of pre-natal diagnosis in Lahore (Pakistan), two authoritative national religious leaders (M.T. Othmani and M.G. Murtaza) were asked for their opinion on the lawfulness of abortions performed due to serious genetic pathologies.[58] Both declared abortion before 120 days lawful if the seriousness of the pathology is confirmed; after 120 days it is illicit.

The Islamic Medical Association of South Africa, in a letter of comment[59] on the new South African law of 27 September 1996 with which abortion within the first 12 weeks was liberalised, limited to three situations the possibility of abortion in Islam: (1) when the mother's health would be seriously damaged by the

---

[54] Rispler-Chaim V., The Right not to be Born: Abortion of the Disadvantaged Fetus in Contemporary Fatwas, *The Muslim World*, 1999, 89 (2), 130–143.

[55] Omran A.R., *Family Planning in the Legacy of Islam*, London and New York, Routledge, 1994, 192–193.

[56] Rispler-Chaim V., *Islamic Medical Ethics in the Twentieth Century*, Leiden, Brill, 1993, 14–15.

[57] Albar M.A., Counselling about Genetic Disease: An Islamic Perspective, *Eastern Mediterranean Health Journal*, 1999, 5 (6), 1129–1133.

[58] Ahmed S., Saleem M., et al., Prenatal Diagnosis of Beta-Thalassaemia in Pakistan: Experience in a Muslim Country, *Prenatal Diagnosis*, 2000, 20, 378–383.

[59] Islamic Medical Association of South Africa, *Objection to the Enactment of Termination of Pregnancy Legislation*, 15 October 1997, in www.ima.org.za/abortion.html.

continuation of the pregnancy; (2) in the presence of serious foetal anomalies which would make life impossible; (3) when the pregnancy is the result of incest or rape.

A *fatwa* (September 2000) by the spiritual leader of the Islamic Republic of Iran, Ali Khamenei, judges lawful the abortion of thalassaemic foetuses within 10 weeks of pregnancy; in any case the abortion of similar foetuses is recommended before the infusion of the soul. This *fatwa*, however, would take on full legal value only if approved by the Iranian Parliament. In April 2005 the Iranian Parliament approved a new law that was more favourable to the abortion of handicapped foetuses.[60]

The Saudi expert M. Albar tolerates the voluntary interruption of pregnancy within 120 days if the foetus has a serious malformation whilst it is possible afterwards only to protect the mother's health[61]; similarly, the Algerian doctor Aroua is in favour of abortion when there is the certainty that the child will be born in conditions incompatible with a normal life.[62] The Tunisian Ben Hamida deems abortion possible for Huntington's disease, a dominant genetically transmitted chromosome anomaly, i.e. each individual affected may transmit it to half of their children who will transmit it to half of their children, etc.; the individual that is not suppressed at the foetal stage generally lives until the age of 40. If only the husband desires the abortion of his child and the wife is against it, the abortion is illicit.[63]

In a *fatwa* in October 2002 the Malaysian National *Fatwa* Council specified that abortion after 120 days is murder except to save the life of the mother. Before 120 days it is also licit when the foetus is handicapped.

The Ethics Committee of the IMANA has recently stated[64] that the foetal congenital malformations in which abortion is permitted before the 120th day include lethal malformations such as bilateral renal aplasia, trisomy 13 and 18. If medical experts and Islamic scholars give their consent, abortion is allowed in the case of non-lethal malformations such as hydrocephaly, cervical meningomyelocele, chromosomal aneuploidies, etc.

Some Rigid Positions

The Islamic Medical Association of South Africa asked the *Dar al-Fatwa* in Riyadh (Saudi Arabia) for an answer on the lawfulness of the abortion of seriously handicapped foetuses. *Fatwa* no. 2484 of 1982 denied the lawfulness of the

---

[60] See Chapter 5 "Abortion".

[61] Albar M.A., Ethical Considerations in the Prevention and Management of Genetic Disorders with Special Emphasis on Religious Considerations, *Saudi Medical Journal*, 2002, 23 (6), 627–632.

[62] Aroua A., *Islam et Contraception*, Algiers, OPU, 1993, 66–68.

[63] Ben Hamida, in AA.VV., *The Human Rights, Ethical and Moral Dimensions of Human Health Care*, op. cit., Genetic Manipulation, case 33.

[64] Imana Ethics Committee, *Islamic Medical Ethics: The IMANA Perspective*, op. cit., 9.

abortion of a foetus with a transmittable genetic pathology or with a suspected congenital defect or handicap.[65]

Sheikh M. Shaltut advised parents who are carriers of hereditary pathologies to limit pregnancies but without going as far as abortion. Shaltut also advanced doubts on the possibility of understanding whether a retarded child is unhappy: "Who knows whether the retarded child or Down syndrome child is unhappy as he is? It is often the projection of the healthy on the life of the handicapped, but not necessarily the truth."[66]

In 1994, the Islamic Research Academy of Al-Azhar, prohiited abortion for any reason except when the mother's life was in danger.

Ebrahim indicates as preferable for Islam every preventive measure aimed at avoiding the birth of handicapped children, rather than recourse to abortion.[67]

The majority of jurists appear to be against the abortion of deformed foetuses and foetuses which are the result of rape: they tend to underestimate the psycho-physical damage on the woman except when her life is at risk. In the West, the debate on the damage caused by the handicapped foetus also strongly involves the financial burden on the parents, their disappointment, the emotional stress and the violation of the right to privacy and that of choice in the case that the mother has not been informed of a deformed foetus, preventing her from a possible abortion.

The real problem for Muslim jurists remains, however, abortion, i.e. the induced and violent death of an innocent, whilst all the other problems remain contingent.

Many doctors of the Law fear that approving abortion for deformed foetuses or from rape would lead to approving it in commoner cases.

Positions of Different Countries

Only a small number of legislations in Muslim countries and countries with a Muslim majority include the abortion of handicapped foetuses. The main reason for this silence is of a socio-economic nature. In developing countries, the mortality of handicapped foetuses is far higher and is proportional to the limited possibilities of screening the condition of the foetus and of the pregnant woman. The more widespread the possibility of screenings, the more deformations are identified. This opens up the question of the lawfulness of abortion, a question associated with a growing sensitivity to the problem. The result is the need for a law that regulates the practice.

However, there are legislations which deal with the issue tolerating, to various degrees, voluntary interruptions of pregnancies of seriously damaged foetuses: Iran, Kuwait, Bahrain, Burkina Faso, Saudi Arabia, Tunisia, Qatar, Turkey, Iraq (art. 63 of the Penal Code under Saddam Hussain) and all the former Soviet

[65] Ebrahim A.F.M., *Biomedical Issues-Islamic Perspective*, Mobeni, Islamic Medical Association of South Africa (IMASA), 1988, 126.

[66] Rispler-Chaim, art. cit., 136.

[67] Ebrahim, op. cit., 129.

Muslim states (Azerbaijan, Kazakhstan, Kyrgyzstan, Tajikistan, Turkmenistan, Uzbekistan). It is unnecessary to recall that they mainly enjoy the support of the local official religious authorities, as in countries with opposing positions.

The legislations that to date do not appear to have settled the issue include Egypt, Libya, Syria, Nigeria, Mauritania, Senegal, Sudan, Pakistan, Mali, Malaysia, Lebanon, Jordan, United Arab Emirates, Niger, Oman, etc.

However, it should be noted that in many of these (e.g. in Egypt), in the event that a serious condition is detected, it is possible to have recourse to abortion on the basis of an extensive application of the principle of the necessity to save the mother's life, a classic principle in Muslim law and which is always introduced into the positive laws in force. This extensive application is all the easier when the law also allows abortion to protect the psycho-physical health of the pregnant woman (e.g. Morocco, Algeria, Indonesia).

CONSANGUINEOUS MARRIAGE

Pre-natal tests and genetic consulting take on particular importance for Muslim families (in particular in the Arab world) due to the high percentage of marriages between first and second cousins or between consanguineous relations, giving rise to a higher rate of neuro-metabolic disorders, genetic pathologies, congenital malformations, precocious mortality and neural tube defects.[68] Such unions represent an important genetic phenomenon producing an increase in the homozygous genotypes to the detriment of the corresponding heterozygotes and a higher percentage of disorders caused by recessive autosomal genes (e.g. phenylketonuria, cystic fibrosis, Cooley's disease, Tay-Sachs disease, etc.)[69] as well as a higher rate of morbidity and mortality.

The recourse to consanguineous marriage is not only the result of local habits but also of beliefs, cultural values and the desire to strengthen family bonds, to facilitate agreement between the couple, to keep the property in the family, etc. The possible knowledge of risks for the offspring is not sufficient to dissuade people from following the tradition[70] which has pre-Islamic origins as related by Omar Ibn Khattab, a companion of Prophet Muhammad, who

---

[68] Mazrou (Al) Y.Y., Farid S.M., et al., Changing Marriage Age and Consanguineous Marriage in Saudi Females, *Annals of Saudi Medicine*, 1995, 15 (5), 481–485.

[69] Teebi A.S., Autosomal Recessive Disorders among Arabs: An Overview from Kuwait, *Journal of Medical Genetics*, 1994, 31, 224–233.

[70] Panter-Brick, art. cit., 1298–1299. Out of a sample of 36 families (1988) with children suffering from metabolic and neurological pathologies, the attitudes of parents Towards marriage between close relatives was divided almost equally between those in favour and those against. However, this is a very deep-rooted tradition in society as shown by a husband, who after genetic consulting at the King Faisal Specialist Hospital, divorced his first wife (his first cousin on his father's side) only to marry a cousin on his mother's side, or the case of three daughters betrothed to the three sons of a maternal uncle; one of the sons had a child with phenylketonuria but this did not make the parents of the girls change the marriage arrangements.

invited the members of the clan of the Sa'ib to marry out of the clan to avoid the risk of a weakened offspring. In a *hadith* in Ibn Maja, Prophet Muhammad declared: choose where you deposit your sperm for the line of descent is conducive. For exactly the same reason, Ghazali (d. 1111) dissuaded first cousins from marrying. In fact, Koran 4.23 prohibits marriage between certain degrees of kinship[71] but not between cousins; Prophet Muhammad's own wives included a first cousin whilst the daughter of the Prophet, Fatima, married Ali, the Prophet's cousin.

The Holy Text also prohibits marriage of a person with his milk sister or mother. The milk relationship is determined when a wet nurse feeds the child of another woman; the milk mother and father (i.e. the husband of the wet nurse besides the "owner" of the milk) become the equivalent of the natural parents whilst the breastfed child becomes the equivalent of a brother or sister of the other natural children of the "helping" couple, with the same rules and prohibitions on marriage for the brothers and sisters applying for the milk child. At the same time, however, the milk relatives do not inherit from one another whilst the milk parents do not have obligations of protection or maintenance over the breastfed child.

The institution of the milk relationship has pre-Islamic origins. Prophet Muhammad himself was breastfed by his own mother Amina for three days, by a servant for 40 or 120 days and, lastly, by Halima, a wet nurse belonging to the Banu Sa'ad tribe, until he was five-years-old.[72] This is, however, an institution that today has almost disappeared due to the possibility of using dried milk. The use of milk banks, thanks to the donations of different women, is not lawful as the babies end up being related to unknown "mothers".

However, the conclusive Recommendations of the Seminar on "Human Reproduction in Islam" (IOMS, Kuwait, 1983) specified the conditions for the use of the milk banks for the purpose of helping premature babies. Point 1 states[73]: "The setting up of banks of mixed human milk is to be discouraged. If medical need calls for them, banks of human milk may be set up for premature babies. A group of participants believes, on the basis of the opinion of the majority of Fiqh scholars, that the collection of milk should be done in a way that guarantees the identification of each donor and each baby receiver. The nursing should

---

[71] Koran 4.23–24: "23. Prohibited to you (for marriage) are: Your mothers, daughters, sisters; father's sisters, mother's sisters; brother's daughters, sister's daughters; foster-mothers (who gave you suck), foster-sisters; your wives' mothers; your step-daughters under your guardianship, born of your wives to whom ye have gone in, no prohibition if ye have not gone in; (those who have been) wives of your sons proceeding from your loins; and two sisters in wedlock at one and the same time, except for what is past; for God is Oft-forgiving, Most Merciful; 24. Also (prohibited are) women already married, except those whom your right hands possess...".

[72] Koçtürk T., Foetal Development and Breastfeeding in Early Texts of the Islamic Tradition, *Acta Paediatrica*, 2003, 92, 617–620.

[73] *The Seminar on Reproduction in Islam. Minutes of the Recommendation Committee*, in Gindi (Al) A.R. (ed.), *Human Reproduction in Islam (1983)*, Kuwait, IOMS, 1989, 274–277.

be written down in records that are kept, and everyone involved should be notified to avoid the marriage of persons who have a milk relationship entailing the prohibition of their marriage. Others, however, believe there is no need to identify the donors and receivers, on the basis of the opinion of Al-Laith Ibn Sa'd and the scholars of Al-Zhahiriyah School . . . who believe that milk relationships result only when a baby sucks the breast of a milk mother."

Similar indications have been applied by the Neonatal Intensive Care Unit of the Adan Hospital (Kuwait) to feed "very low birth weight" (VLBW) babies, whose natural mothers did not have enough milk, with that of other nursing mothers who had surplus milk. Maternal milk is particularly beneficial for preterm children. The donating and receiving mothers (often friends or relatives) met beforehand (with their respective families) in hospital after having understood all the legal consequences linked to the existence of milk children. Both the donors and the mothers of the babies receiving the milk signed a written consent form in which they accept the legal consequences. Before collecting the milk to be stored in refrigerators, the donors must undergo clinical tests (including HIV I and II, Hepatitis B and C, Cytomegalovirus and Syphilis).[74]

The prohibitions resulting from acquired kinship through breastfeeding may be recognised by legislation in countries such as, inter alia, Tunisia[75] and Morocco.[76]

In Egypt consanguineous marriages[77] represent 28.96% of the total (of which 14.10% of the total of marriages unite relatives of the first degree, 5.40% of the second degree and 9.46% of marriages are between couples of other degrees of kinship); in Kuwait 54.3% of marriages (of which 32.2% of the total are consanguineous to the first degree); in Iraq 57.9% (30% between first degree relatives); in Bahrain 39.4%; in Lebanon 29.6% of marriages between Muslims and 16.5% of those between Christians are believed to be consanguineous; in the United Arab Emirates, 61.6% of legal unions are consanguineous; in Jordan 50.23% of

---

[74] For the positive aspects and the problems related to this type of donation, see Naqueeb (Al) N.A., Azab A., et al., The Introduction of Breast Milk Donation in a Muslim Country, *Journal of Human Lactation*, 2000, 16 (4), 346–350.

[75] République Tunisienne, *Code du Statut Personnel*, Tunis, 1997, art. 17, page 9: "Breast feeding produces the same impediments as kinship and alliance. The child who has been breastfed, with the exclusion of his brothers and sisters, is alone considered the child of the wet nurse and her husband. Breastfeeding only prohibits marriage when it takes place during the first 2 years of the infant's life."

[76] *Moudawana* or Code of Personal Status, art. 38: "The prohibitions resulting from kinship by breastfeeding are the same as those of kinship or alliance. Only the child who has been breastfed, with the exclusion of his brothers and sisters, is considered as the child of the mother by breastfeeding and her husband. Breastfeeding forms an impediment to marriage only if it has taken place in the first 2 years before weaning", Law no. 70.03, Family Code (3 February 2004), in Mounir O., *La Moudawana. Le nouveau Droit de la Famille au Maroc*, Editions Marsam, Rabat, 2005, 137–179.

[77] The data on the various countries is mainly from Hazmi (El) M.A.F., Swailem (Al) A.R., et al., Consanguinity among the Saudi Arabian Population, *Journal of Medical Genetics*, 1995, 32, 623–626.

the total are consanguineous marriages (more than 35% between consanguineous partners of the first degree).[78] In Algeria these marriages concern 22–25% of unions. In Oman, marriage between first cousins accounts for 33.6% of unions.[79] In Turkey, consanguineous marriages applied to 20.9% of couples.[80]

There is a great deal of information from Saudi Arabia where 58.5% of marriages are consanguineous. Pathologies of genetic origin are the result of two factors in particular: the high number of consanguineous marriages (in particular amongst the Bedouin where the percentage touches on 90% of unions) and the large number of children in each family. The most characteristic pathologies in Saudi society include the "hereditary hearing impairment", a disease caused by recessive autosomal genes very frequent in consanguineous unions in the Kingdom, as well as the "maple syrup urine disease" (MSUD), a disorder resulting from a defective metabolism of the amino acids and far more widespread in Saudi Arabia than in Europe or North America. A study was carried out by the Saudi National Child Health Survey (NCHS) in 1987–1988 on 8,482 married Saudi women[81] of whom 5,974 in cities and 2,508 in rural areas. Regarding the first 5,974, 2,031 (34%) married a first cousin, 1,434 (24%) married another relative and 2,509 (42%) married a non-relative. Regarding the 2,508 in a rural context, 1,028 (41%) married a first cousin, 477 (19%) another relative and 1,003 (40%) an unrelated man; it is to be noted that amongst the women with a certain degree of education (at least able to read and write) namely 2,078, the number of them married to unrelated men was 51% compared to the 37% of illiterate women. The low female schooling rate increases the difficulty in understanding the complicated hereditary mechanisms in the case of genetic consulting.

In a study of the early 1990s, out of 3,212 Saudi families in the 5 provinces of the Kingdom, 57.7% of the unions (i.e. 1852) were based on consanguineous relationships; the most common relationship was between first cousins (28.4%), followed by distant relations (15.2%) and by marriage between second cousins (i.e. children of cousins, 14.6%).[82] In the five provinces, the rate of consanguinity varied from 52% to 67% whilst the union between first cousins was almost always prevalent, compared to the unions between second cousins.

A total of 1,307 married Saudi Arabians, including 994 women, took part in a more recent survey in Dammam[83] and province; 52.3% of the marriages (679)

---

[78] Shanti (El) H., The Impact of Genetic Diseases on Jordanians: Strategies Towards Prevention, *Journal of Biomedicine and Biotechnology*, 2001, 1 (1), 45–47.

[79] Riyami (Al) A.A., Suleiman A.J., et al., A Community-Based Study of Common Hereditary Blood Disorders in Oman, *Eastern Mediterranean Health Journal*, 2001, 7 (6), 1004–1011.

[80] Bökesoy I. and Göksel F., *Ethics and Medical Genetics in Turkey*, in Wertz D.C. and Fletcher J.C. (eds.), *Ethics and Human Genetics*, Berlin, Springer-Verlag, 1989, 380–387.

[81] Mazrou and Farid, art. cit., 484.

[82] Hazmi and Swailem, art. cit., 623–625.

[83] Abdulkareem (Al) A.A. and Ballal S.G., Consanguineous Marriage in an Urban Area of Saudi Arabia: Rates and Adverse Health Effects on the Offspring, *Journal of Community Health*, 1998, 23 (1), 75–83.

were consanguineous; the most frequent union was with a first cousin (20.4%) whilst non-consanguineous unions represented 48% of the couples. The rate of consanguineous marriages in the sample of Dammam increases up to 1979 and then decreases: in 1950, 51.2% of marriages; from 1950 to 1959, 54.3%; from 1960 to 1969, 52.4%; from 1970 to 1979 56.3%; after 1980 the average becomes stabilised at around 48.9%. In fact, the frequency of the commonest patholo-gies found amongst the offspring of consanguineous marriages, namely sickle cell anaemia, bronchial asthma, diabetes mellitus, mental retardation, cardiac anomalies, etc., was very similar to that of the offspring of non-consanguineous parents. This study has not shown significant relationships between consanguin-eous marriages and congenital or hereditary pathologies[84] of the offspring.

Amongst Saudi parents, reference to divine will as responsible for their chil-dren's illnesses of genetic origin appears deep rooted and this conception helps tolerate the pathology in the respect of Koran 57.22: "Do misfortune can hap-pen on earth or in your souls but is recorded in a decree before we bring it into existence: that is truly easy for God" independently of the knowledge or not of hereditary mechanisms. Reliance on divine will offers a more acceptable and less uncertain explanation as it avoids attributing the illness of a child and the suffering of the parents to simple chance or "blind probability". However, this acceptance does not mean that cures are not sought.

Recourse to readers of the Koran and local healers, whose actions are often associated by the parents with treatment by official medicine, still seems to enjoy a certain degree of popularity in the Middle East and the Arabian Peninsula. To explain sudden illnesses that affect babies, there is also recourse to the "evil eye" cast by the look of an envious or jealous woman.[85] The possibility for a husband to marry up to four wives or divorce and remarry, in the event that the mar-riage in course has produced or risks producing a handicapped baby, is judged ambivalently by Saudi wives: they can only live with this choice of their hus-band in suffering or resignation as it is considered the husband's right to marry other women to generate healthy children.[86] By way of prevention, from the end of the 1980s, the King Faisal Specialist Hospital made screening for all babies compulsory to identify phenylketonuria whilst preventive diagnosis by means of amniocentesis was recommended for mothers deemed "at risk".

However, the rate of consanguineous marriages in Saudi Arabia is tending to decrease in the educated classes; amongst the inherited pathologies there is con-genital syphilis (*bijel*) in some areas of the Kingdom. The syndromes associated with thalassaemia seem fairly widespread, the region of Qatif has a high diffu-sion of haemoglobinopathies in particular, in addition to thalassaemia (espe-cially beta-thalassaemia) and sickle cell anaemia.[87]

---

[84] Abdulkareem and Ballal, 79–80 and 82.
[85] Panter-Brick, art. cit., 1297–1298.
[86] Ibid., 1298.
[87] Nasserallah Z., Patients with Beta Thalassaemia Major, *Saudi Medical Journal*, 1995, 16 (1), 13–16.

Consanguineous marriages inevitably also concern immigrant couples. Great Britain is the country where the phenomenon has been analysed best, in Pakistani,[88] Indian and Arab immigrants. Amongst the Pakistani immigrants,[89] current estimates show that 75% of marriages are consanguineous and 50% are marriages between first cousins. Curiously enough, these high percentages represent an increase compared to the percentage of the parents, amongst whom only 30% married first cousins. Marriage in this context is to be understood as a union between two families rather than between two individuals. Amongst the advantages of consanguineous marriages there is the direct knowledge of the partner for one's son or daughter by the parents, a fact which is important in certain Muslim minority groups when the traditional social network which facilitates the search for a suitable partner is missing. In addition, the parents consider a marriage that reinforces the family bonds as a guarantee for their old age.

Many malformed and handicapped babies do not survive in under-developed countries; as a consequence, professional care for these children is relatively new amongst immigrant Muslims in Great Britain. Psychologically, culturally and religiously, these parents tend to accept the deformity as an act of God, rationalising it as an "ordeal from God" or a punishment for previous sins. The latter can take on the characteristics of a blessing because it is better to be punished in this life rather than for eternity after death.

In short, the state of privation, consanguinity and general reluctance by Muslims to abort foetuses with congenital anomalies are some of the reasons for the high rate of handicaps in the British Muslim community.[90]

PRE-NATAL DIAGNOSIS

As far as pre-implantation diagnostic techniques are concerned (tests on "in-vitro" embryos within three days of insemination, i.e. when the embryo is at the stage of eight cells and without a soul therefore only healthy embryos are transferred into the mother) in order to identify genetic problems, these represent an alternative to pre-natal diagnosis which allows avoiding selective abortion. Muslim jurists would appear to approve pre-implantation diagnosis as IVF is not against the will of God nor does it modify creation.[91] In fact, post-implantation

---

[88] See amongst others: Honeyman M.M., Bahl L., et al., Consanguinity and Fetal Growth in Pakistani Moslems, *Archives of Disease in Childhood*, 1987, 62, 231–235; Darr A. and Modell B., The Frequency of Consanguineous Marriage among British Pakistanis, *Journal of Medical Genetics*, 1988, 25, 186–190; Roberts A., Cullen R., et al., The Representation of Ethnic Minorities at Genetic Clinics in Birmingham, *Journal of Medical Genetics*, 1996, 35, 56–58.

[89] Dhami S. and sheikh A., The Family: Predicament and Promise, in sheikh A. and Gatrad A.R. (eds.), *Caring for Muslim Patients*, Oxon, Radcliffe Medical Press, 2000, 43–55.

[90] Gatrad A.R. and sheikh A., Birth Customs: Meaning and Significance, in sheikh A. and Gatrad A.R. (eds.), *Caring for Muslim Patients*, op. cit., 57–71.

[91] Hashemite (El) N., The Islamic View in Genetic Preventive Procedures, *The Lancet*, 19 July 1997, 350, 223 (Letter).

practices (e.g. amniocentesis, taking a sample of the chorionic villus) should be accepted within the first 120 days (in any case, before the infusion of the soul) also because the absence of disorders of genetic origin in the pre-implantation phase does not guarantee a healthy foetus; in addition, other serious malformations of extra-genetic origin in the foetus may be revealed later through, for example, ultrasounds.

A Conference at the University of Al-Azhar in November 2000 dealt with preimplantation genetic diagnosis (PGD). The prevailing thesis upheld the lawfulness of not implanting embryos produced *in vitro* when they show serious chromosomic or genetic anomalies identified through PGD. More controversial was the debate on the non-medical use of PGD. Participants condemned recourse to PGD for discriminatory purposes against females, but specified that total refusal of the practice implies serious risks for mothers in Middle Eastern societies, especially when the birth of a male remains fundamental for the psycho-physical and social well-being of the mother.[92]

Regarding recourse to pre-natal diagnosis (little used in the majority of Muslim countries, especially for economic and organisational reasons) the problems shown in Turkish society appear representative. Pre-natal diagnosis is carried out in Turkey but the indications given by similar tests should be evaluated in the light of their technological level and the qualifications of medical staff.[93] The main ethical problem in many emerging countries concerns the impossibility of providing certain health services to many of those who need them; at the same time, more problematic is the fact that in some areas there is a tendency to have recourse to pre-natal diagnosis for anti-ethical purposes. The most common technique in Turkey would seem to be ultrasonography, used above all to identify the sex of the foetus in particular in the eastern parts of the country where males are preferred. This encourages the attitude of aborting a healthy but female foetus when discovered.[94] A study on 524 doctors in the Ankara area in 1995–1996 showed that 64% of them were against establishing the sex of a foetus without medical reasons.[95] Of these, 190 were against it for social reasons; 135 feared that the practice would have aggravated sexual discrimination; 113 feared that the practice would have anticipated the selection of other characteristics of the infant.

The revelation of an "unfavourable" pre-natal diagnosis requires particular caution in cultural areas where marriage and procreation take on a

---

[92] Serour, *Proceedings of the Workshop on Ethical Implications of Assisted Reproductive Technology for the Treatment of Infertility*, op. cit., 96–97; Serour G.I. and Dickens B.M., Assisted Reproduction Developments in the Islamic World, *International Journal of Gynaecology and Obstetrics*, 2001, 74, 187–193.

[93] Arda B. and Cangirkayi A., Les problèmes éthiques relatifs au diagnostic prénatal et à la consultation génétique, *International Journal of Bioethics*, 1995, 6 (2), 160–162.

[94] Arda and Cangirkayi, art. cit., 161.

[95] Pelin S.S. and Arda B., Physicians' Attitudes Towards Medical Ethics Issues in Turkey, *International Journal of Bioethics*, 2000, 11 (2), 57–67.

fundamental value for social life. In addition, in contexts where the female role is inferior or less protected, the woman with genetic problems is socially more vulnerable than a man. In Turkey, a large number of consanguineous marriages take place despite the limitations placed on such unions first by the Islamic legislation concerning marriage (in force until 1926) and subsequently by the Turkish Civil Code which takes up the previous rules of the *Shari'a*. In both systems, however, marriage between children of siblings was and continues to be lawful.[96]

In Muslim countries there appears to exist a close relationship between religion, the perception of pre-natal diagnosis and abortion. However, the high rate of illiteracy often makes it difficult to distinguish between religious, traditional and personal beliefs; this leads to ambiguity in determining the licit or the illicit in religion.

In a study from Lebanon published in 1999, out of a sample of 90 couples (80% Muslim and 20% Christian) at risk of genetic pathologies,[97] after an explanation about genetic risks and pre-natal procedures, 56% of the couples asked for pre-natal diagnosis in their next pregnancy, 19% were uncertain whilst 26% were absolutely against pre-natal diagnosis. In 87.5% of the cases, this last position was motivated by religious opposition to abortion which was linked to the refusal of the pre-natal diagnosis. Although this is a very limited sample, 61% of Christian couples were in favour of pre-natal diagnosis and 28% were against it. Of the Muslim couples (80% of the total) 53% were in favour of pre-natal diagnosis, 28% were against it and 19% uncertain. The *Shi'ite* component formed the majority of the Muslims and of these, most of them (42%) were against pre-natal diagnosis compared to 40% of *Shi'ites* in favour. On the contrary, 71% of the *Sunnis* were in favour and 6.5% against pre-natal diagnosis.[98]

In a Saudi survey, the inclination of accepting pre-natal diagnosis and/or abortion was studied, after having learnt of the *fatwa* of the Islamic Jurisprudence Council (linked with the Muslim World League), Mecca, 10–17 February 1990, which approved abortion in the first 120 days of conception if the foetus is affected by incurable handicaps.[99] This was a limited sample of 32 families with children suffering from sickle cell anaemia, thalassaemia or both; consanguinity was present in 23 families (71.9% against the 57.7% of the Saudi population

---

[96] Bökesoy–Göksel, art. cit., 385.

[97] Zahed L., Nabulsi M., et al., Acceptance of Prenatal Diagnosis for Genetic Disorders in Lebanon, *Prenatal Diagnosis*, 1999, 19, 1109–1112; See also Zahed L. and Bou-Dames J., Acceptance of First-Trimester Prenatal Diagnosis for the Haemoglobinopathies in Lebanon, *Prenatal Diagnosis*, 1997, 17 (5), 423–428.

[98] Zahed and Nabulsi, art. cit., 1110–1112.

[99] Alkuraya F.S. and Kilani R.A., Attitude of Saudi Families Affected with Hemoglobinopathies Towards Prenatal Screening and Abortion and the Influence of Religious Ruling (Fatwa), *Prenatal Diagnosis*, 2001, 21, 448–451; See also Hazmi (El) MAF, The Natural History and the National Pre-Marital Screening Program in Saudi Arabia, *Saudi Medical Journal*, 2004, 25 (11), 1549–1554.

as a whole). In this sample, knowledge of the religious document increased the number of those in favour of abortion but not of those in favour of pre-natal diagnosis.[100]

In May 1994 in Pakistan the first department of pre-natal diagnosis for thalassaemia (it is the commonest single gene disorder) was opened in the country. Before it opened, two well-known religious authorities, M.T. Othmani and M.G. Murtaza, were asked about the position of Islam on the abortion of foetuses with serious genetic pathologies.[101] Both declared abortion before 120 days lawful. In the sample of 141 couples studied at the Fatimid Thalassaemia Centre of Lahore that already had children who were carriers of genetic pathologies, 124 (87.4%) were in favour of the abortion of a thalassaemic foetus, 2% against and the remainder uncertain. In the experience of the Centre, the couples that are uncertain about pre-natal diagnosis were comforted by knowing that Islam allows abortion in certain situations. Other couples, after learning the limit of 120 days, although disappointed, refused pre-natal diagnosis after that period of time.[102] The majority of the couples with a genetically diseased foetus would abort without great hesitation.

A comparative study investigated the attitudes of two culturally very different samples of families with haemophilia in Iran and Italy towards information on PND (pre-natal diagnosis) and abortion.[103] The samples included 59 Iranian (38 haemophilia A patients of more than 16 and 21 mothers) and 50 Italians (27 haemophilia A patients of more than 16, 16 mothers and 7 fathers). The quality of life of 84.7% of the Iranians was severely influenced by haemophilia A, against 50% of the Italians. The main reason for malaise for both samples was the psychological distress which affected 69% of the Iranians against 28% of the Italians. Amongst the Iranians, only 15.3% knew about PND of haemophilia A compared to 64.6% of the Italians. Regarding the willingness to abort a future male child affected by haemophilia,[104] 58.2% of Iranians were in favour against 16.7% of the Italians. The reasons why there was such a strong inclination for abortion by the Iranians comes from a lower level of care with respect to Italian standards and very little confidence in the appearance of effective care with respect to the Italian sample.[105]

Beta-thalassaemia represents an important and very costly problem for the Iranian health service. To cope with this and similar pathologies, Iran, like many other Muslim countries (see below), imposes mandatory premarital blood tests. In an extensive study on Thalassaemia – which lasted from 1997 to 2001 – more

---

[100] Alkuraya and Kilani, art. cit., 449–450.
[101] Ahmed S., Saleem M., et al., Prenatal Diagnosis of Beta-Thalassaemia in Pakistan: Experience in a Muslim Country, *Prenatal Diagnosis*, 2000, 20, 378–383.
[102] Ahmed and Saleem, art. cit., 382.
[103] Karimi M., Peyvandi F., et al., Comparison of Attitudes Towards Prenatal Diagnosis and Termination of Pregnancy for Haemophilia in Iran and Italy, *Haemophilia*, 2004, 10, 367–369.
[104] On Iranian legislation see Chapter 5 "Abortion".
[105] Karimi and Peyvandi, art. cit., 368.

than 2.7 million prospective couples were screened and 10,298 couples at risk were identified.[106] The future couples at risk attended genetic counselling with relatives. Subsequently, about half of the couples married in any case; many of the others split up and many others were uncertain. Despite the social and marriage obstacles placed by the pathology, widespread willingness to have pre-natal diagnosis was evident, thanks to a greater possibility of legal abortion which has emerged in recent years.[107]

Where genetic tests are easily accessible, as in Western countries, immigrant parents may appear to be more willing to interrupt pregnancies of seriously handicapped and incurable foetuses[108] although Muslim parents, e.g. in Great Britain, are reluctant to do such tests.

Changing continents, a different situation can be found in the Muslim community of Cameroon (about 20% of the population) as – according to the Minister for Health, the Muslim H.M. Salihu – it is difficult to provide help for couples with genetic problems expecting children.[109] In any case, the local Imams tend to refuse abortion before implantation (some also seem opposed to coitus interruptus) and before the infusion of the soul.

The problem of transmittable pathologies is increasingly important for the health authorities of Muslim countries and this is a reason for the legislator to intervene. In the Seminar mentioned above[110] on "Genetics – An Islamic Perspective" organised in Kuwait (13–15 October 1998) by the Academy of Muslim Law of Jeddah, paragraph VI of the conclusive document concerned the reasons why genetic diagnosis should be compulsory or voluntary: (1) To promote public awareness on the risks of genetic pathologies. (2) To encourage couples to have pre-marital genetic tests and to educate the public through the information of the mass media, public meetings and sermons in mosques.

Public and specialised meetings are increasingly frequent in these countries to examine the possibility of making specific tests compulsory, for example, to identify the gene of thalassaemia as a prerequisite to obtain authorisation for marriage.[111] Many ethical problems emerge. The majority of Muslim governments have other priorities due to the poor health resources available. Poverty makes the test prohibitive for individuals without state aid. Furthermore, making such tests compulsory would hurt the autonomy of the individual. If imposed by law, a medical certificate can always be procured illegally. In addition, if the test is

---

[106] Samavat A. and Modell B., Iranian National Thalassaemia Screening Programme, *British Medical Journal*, 13 November 2004, Vol. 329, 1134–1137.

[107] Samavat and Modell, art. cit., 1136. See also Christianson A., Streetly A., et al., Lessons from Thalassaemia Screening in Iran, *British Medical Journal*, 13 November 2004, Vol. 329, 1115–1117.

[108] Black J., Broaden Your Mind About Death and Bereavement in Certain Ethnic Groups in Britain, *British Medical Journal*, 29 August 1987, Vol. 295, 536–539.

[109] Salihu H.M., Genetic Counselling among Muslims: Questions Remain Unanswered, *The Lancet*, 4 October 1997, 350, 1035–1036.

[110] See the section on "The Debate on Genetics" above.

[111] Albar, Counselling about Genetic Disease: An Islamic Perspective, art. cit., 1129–1133.

positive, would it be licit to legally prevent the marriage? In this case who would be appointed to have this power?

Apart from the different interpretations of Muslim law on all these problems, even after having found a common interpretation, this would be translated in different ways into the positive law of the individual countries.

Here are some examples.

In Tunisia, genetic consulting is increasingly requested, as is the observation of the evolution of the foetus, by couples of high socio-economic level. Although increasing, the request for genetic consulting remains weak compared to the frequency of hereditary diseases and the strong degree of consanguinity in Tunisia. In any case, according to the law, a pre-nuptial medical certificate is required from the couple before marriage[112]; the doctor is required to give genetic advice (as well as on birth control).

A study presented to the Council of Ministers of Saudi Arabia in 2002 reported that at least 1.5 million Saudis are carriers of congenital diseases of the blood, in particular thalassaemia and anaemia.[113] These illnesses kill many Saudi Arabians every day. Despite the serious risks for their children, parents do not stop procreating. Prince Talal Ibn Abdul Aziz has officially requested the introduction of the obligation of similar tests before marriage and this obligation would not contradict the principles of Islam. Lastly, since February 2004, pre-marital medical check-ups have been compulsory in the Kingdom, after having been approved by Royal Decree passed on 30 December 2003.[114] The person performing a marriage must ask the bride and groom to show the results of the test for hereditary and contagious diseases (AIDS, hepatitis, etc.), before concluding the marriage contract. Nobody has the power to prevent the marriage if either or both the partners have or are carriers of any hereditary or contagious disease, against the wishes of the partners. The procedure followed in the analysis has one particularity: after genetic counselling the result will be issued to the person (carrier or diseased) concerned (man or woman) but to the woman in the presence of her guardian.[115]

Identical concerns are experienced by the Jordanian authorities. In July 2002, the Government approved a draft law making a premarital blood test compulsory, in order to reduce the incidence of thalassaemia (4% of Jordanians are carriers).[116] Until final approval by King Abdullah and the subsequent publication in the Official Journal, the measure in force lays down that couples resulting positive to the thalassaemia test are advised on what to do, in order to understand the risks

---

[112] République Tunisienne, *Code du Statut Personnel*, op. cit., 68.

[113] Arabnews, *Talal wants Compulsory Premarital Tests*, 11 September 2003.

[114] Hazmi, *The Natural History and the National Pre-marital Screening Program in Saudi Arabia*, art. cit., 1553. Ghafour P.K.A., Premarital Checkups Mandatory Now, *ArabNews*, 24 Feburary 2004

[115] Hazmi, *The Natural History and the National Pre-marital Screening Program in Saudi Arabia*, art. cit., 1551.

[116] Wakeel (Al) D., Thalassemia Screening required before Marriage – Draft Law, *Jordan Times*, 14 July 2002.

of procreation, but without impediments to the marriage. On 9 March 2004 the Council of Ministers approved a new Pre-Marital Blood Test Law which makes the test compulsory for all couples before their marriage.[117]

In the United Arab Emirates, neonatal screening of hypothyroidism and phenylketonuria has been present in the national programme for a few years. In July 2002 the President, *sheikh* Sultan al-Nahyan, established a Marriage Fund to help the young people of the Emirates get married without having to cope with the enormous costs involved. The Fund is for unmarried men and women, widows and widowers and divorcees.[118] Projects supporting the Fund include the *tawafuq* (matchmaking) project to help young people find a suitable partner. The project, supported by the local Muslim authorities, encourages social stability, solves the problem of marriage between partners of an advanced age and encourages marriage for divorcees and widows/widowers. All this would apply the teaching of Muhammad as it encourages cooperation between Muslims for marriage. To obtain the funding, there is a compulsory medical test, the main aim of which is to identify thalassaemia, anaemia and sexual transmitted diseases. The motivation is due to the awareness that 10–15% of the handicapped children in the UAE are due to hereditary disease. The project guarantees absolute confidentiality of the data collected and the partners are free to marry after having learnt of the risks.

CONCLUSION

The absence of the soul before infusion (120th or 40th day or any other date) can help or legitimise the use of the embryo and foetus for the benefit of others. This absence can foster therapeutic research (but also for non-therapeutic reasons) in the phases preceding infusion. Similarly, it can encourage the abortion of foetuses showing lesions or handicaps before infusion.

Pre-natal pre-implantation and/or post-implantation diagnosis can also be easily accepted in the period preceding infusion; as a consequence, the possibility of aborting a handicapped foetus becomes easier.

Answers on all these problems are diversified amongst Muslim jurists, patients and national laws. As on many of the issues of Muslim bioethics, it is a question of identifying the commonest position with respect to minority tendencies.

---

[117] Dalal K., Cabinet Approves Draft Pre-Marital Blood Test Law, *Jordan Times*, 10 March 2004.

[118] Mussallam N.S., Marriage Fund to take up Matchmaking, *Khaleej Times – Online*, 26 June 2002.

THE END OF LIFE

SUFFERING AND ILLNESS   267
THE INCURABLY AND TERMINALLY ILL PATIENT   271
PALLIATIVE CARE   272
INFORMATION AND CONSENT OF THE
SERIOUSLY AND TERMINALLY ILL PATIENT   274
THE LIVING WILL   283
EUTHANASIA   285
SUICIDE AND MARTYRDOM   290
DEATH   296
POST-MORTEMS   297
CONCLUSION   303

SUFFERING AND ILLNESS

Amongst the consequences of the belief in absolute divine omnipotence, the concept of "causae secundae" is generally refused by the *Ash'arite* theological school of thought (still the majority in *Sunnite* Islam), according to which every relationship of cause and effect is produced by the sole divine motor. What we define "causality" is, in fact, the observation of the habit with which God makes the things of the world exist.[1] Ghazali specifies in his "The Incoherence of Philosophers" (*Tahafut al-Falasifa*): "It is the habit of God to make certain events succeed certain other events, a habit that is unchangeable"; Ghazali explains natural causality as a "divine habit" whilst the Scottish philosopher David Hume explained it by the habit of the human mind to see a constant succession between certain events.[2] According to the *Ash'arites* neither nature nor human will has the power to produce real effects; the will remains void without God therefore causality is a word without meaning.[3] In the theological context, not even immutable natural laws such as to influence or limit divine freedom would exist.

All suffering and death therefore exist only by the will of the Creator according to the explanations given by Koran 57.22: "Do misfortune can happen on

---

[1] Baffioni C., *Storia della filosofia islamica*, Milano, Mondadori, 1992, 90–92.

[2] Rahman F., Islam and Health/ Medicine: A Historical Perspective, in Sullivan L.E. (ed.), *Healing and Restoring*, New York, Macmillan Publishing Company, 1989, 156 and note 17.

[3] The constant reference to God is curiously attested by the attitude of the Prophet to the use of amulets to obtain recovery. Initially, the Prophet forbade their use but when his Companions maintained that they had used them with a benefit, the Prophet accepted their use on condition that the writing on the amulets was only Koranic verses to avoid anyone invoking powers other than those of the only God. See Rahman F., *Islam and Health/Medicine: A Historical Perspective*, art. cit., 150.

Dariusch Atighetchi, Islamic Bioethics: Problems and Perspectives.
© Springer Science+Business Media B.V. 2009

earth or in your souls but is recorded in a decree before We bring it into existence: that is truly easy for God" and 2.155–156: "155. Be sure we shall test you with something of fear and hunger, some loss in goods or lives or the fruits (of your toil), but give glad tidings to those who patiently persevere; 156. Who say, when afflicted with calamity: 'To God we belong, and to Him is our return' ". Too literal an application of this concept risks leading to a passive attitude by man in the fight against disease and illness.[4]

In fact, Islam does not exalt suffering; several times God stresses this concept as in Koran 20.2 "We have not sent down the Quran to thee to be (an occasion) for thy distress" or it states that God never places unbearable burdens on human shoulders. In addition, both faith and the application of the *Shari'a* are traditionally considered a vehicle of beneficial effects at all levels of individual and social life, save the divine freedom of deciding otherwise. Islam has mystical but not ascetic tendencies and formally disapproves of excess austerity which debilitates the body and suppresses natural instincts. The purpose of all this is to observe measure and obey the precepts of the *Shari'a* which are not very burdensome.

The conduct of the Prophet and of his Companions in the face of illness, pain and suffering represents the models of behaviour to be imitated. Life is a test in which men are summoned to face adversity (Koran 2.153–157) with self-control (*sabr*), courage, strength of mind and resignation to divine will. However, Islam does not preach impassibility nor does it refuse the expression of pain, although prohibiting abandon to excessive expression as death is predestined. For example, Muhammad himself wept on the death of his son Ibrahim pronouncing these words: "Ibrahim, our eyes shed tears and our hearts are filled with grief, but we do not say anything except that by which Allah is pleased."[5] Thus pious Muslims will at times rebuke people who complain due to excessively long suffering, either physical or spiritual, or that mourn a deceased person for too long; nevertheless, traditional behaviour in the various communities have often betrayed these rules in reactions to grief and death. Men suppress their emotions more whilst women are more explicit.

Illness and suffering contribute to the expiation of sins, giving rise to a more positive and serene attitude towards suffering no longer understood as a curse or experience without meaning, before which there exists only despair. A delicate *hadith* recites: "I visited the Prophet during his ailments and he was suffering from a high fever. I said, 'You have a high fever. Is it because you will have a double reward for it?' He said, 'Yes, for no Muslim is afflicted with any

---

[4] This is the position initially held by the Egyptian sheikh Sha'rawi when, considering that procrastinating the death of a person already condemned by a pathology would be contrary to the will of God, ended up by deeming unlawful (*haram*) recourse to transplants, renal dialysis and resuscitation. *Controverse religieuse en Egypte à propos des transplantations d'organes*, AFP, 8 February 1989.

[5] Muslim, *Sahih, The Book Pertaining to the Excellent Qualities of the Holy Prophet*, Book 30, Number 5733, in www.usc.edu/dept/MSA/index.html.

harm but that Allah will remove his sins as the leaves of a tree fall down' "
(Bukhari).[6]

Islam exhorts the faithful to be sensitive to the ill and to take care of them in
order to avoid the possible desire for death. Society has the moral and religious
duty to contribute to the costs of caring for a patient, stepping in for relatives
who are unable to bear them; this action represents a post-mortem investment
and virtue has no price. The obligation of help by society is highlighted in the
following *hadith* "The faithful in their mutual love and compassion are like the
body; if one member complains of an ailment all other members will rally in
response."[7] Apologetically it is recalled that in Muslim societies, the seriously ill
or terminal patient can still enjoy, socially and psychologically, a network that
does not leave him unprotected and having to fend for himself as is the case in
the West. However, if the Koran recommends perseverance in the face of pain,
the absence of perseverance is not punished as Koran 2.45 recognises that each
individual possesses different capacities of suffering and therefore also a different
resistance to pain.[8]

A "saying" in which Prophet Muhammad said "If one organ complains, all
others share its complaint, suffering sleeplessness and fever"[9] takes on considerable importance, especially for the consequences. In line with the Old Testament, the Koran does not accept a radical dualism between soul and body in
the human being although distinguishing between the physical and the spiritual
aspects. Man is unity of body and soul, spirit and matter, reflecting in his image
the unity of divine creation and divine unity. This unitary vision of man is confirmed by the approach of Muslim "scholars" to the treatment of pain and illness, aiming to protect both spheres of the person, the spiritual and the material
one, giving precedence to the former. The material sphere has recourse to drugs,
surgery, etc.; in the spiritual one there is an attempt to recover the person's bond
with the Creator through prayer and reading the Koran.[10] This is believed to
reinforce the capacity of resistance, whilst the patient's attention is distracted
from his illness.

---

[6] Bukhari, *Sahih*, *Patients*, Vol. 7, Book 70, Number 550, in www.usc.edu/ dept/MSA/fundamen-
tals/hadithsunnah/bukhari7sbtintro.html.

[7] First International Conference on Islamic Medicine, *Islamic Code of Medical Ethics*, Kuwait,
1981, Chapter 10.

[8] Masoodi G.S. and Dhar L., Euthanasia at Western and Islamic Legal Systems: Trends and
Developments, *Islamic and Comparative Law Review*, 1995–1996, XV–XVI (1), 1–36.

[9] Sherbini (Al) I., Life and Death between Physicians and Fiqh Scholars, in Mazkur (Al) K.,
Saif (Al) A., et al. (eds.), *Human Life. Its Inception and End as Viewed by Islam*, Kuwait, Islamic
Organisation for Medical Sciences (IOMS), 1989, 321–333.

[10] Spiritual healing still appears to be fairly widespread in Saudi Arabia, especially in the cases of
incurable illness, and it can be found alongside modern medical practice. The main practices
consist of reciting passages from the Koran and the *ahadith*; other "curative" agents can be the
water from Zamzam, honey and black cumin. The practices aim in particular to free the patient
from possession by the *jinn* and the effects of the evil eye. See Shahri (Al) M.Z., Culturally Sensitive Caring for Saudi Patients, *Journal of Transcultural Nursing*, 2002, 13 (2), 133–138.

It is essential to make one clarification. The unitary conception of the person is present in the medical field and in the more properly religious context whilst a clear dualism characterises the Muslim philosophic tradition and mystical speculation (Sufism) due to the influence of Western philosophy, in particular Platonic. For the philosophers, the body and soul were irreducible from one another; the soul is the fire of the body from which it remains separate and distinct.

Regarding treatment, it is useful to clarify that the Sacred Text does not make explicit mention of medical care. Passages 10.57, 17.82 and 41.44 make exclusive reference to the spiritual sphere therefore it seems correct to believe that the Koran was intended to represent "medicine for the spirit" whilst it was only at a later date that the Islamic tradition formulated a literature that presented both physical and spiritual curative qualities in the Koran.[11]

However, the unitary conception of the human being has become the central idea of Muslim medical anthropology. This is an inevitable consequence: both illness and health are not exclusive to the physical–biological sphere but also include the psychological–spiritual sphere according to the *hadith* quoted above. Both well-being and physical illness are eventually, in some way, well-being and illness of the whole organism and the whole person. Indeed, the comparison between the Muslim community and the human body is not infrequent, in line with what is expressed by the aforementioned *hadith* according to which "The faithful in their mutual love and compassion are like the body; if one member complains of an ailment all other members will rally in response." It is not by chance that this is one of the most quoted "sayings" of the Prophet to legitimise organ transplants in Islam.

As healing is, to some extent, simultaneously physical and spiritual and as the mind and body are complementary to one another, it is no surprise to find psychosomatic medicine theorised and exercised by the greatest names of Islamic medicine, such as al-Razi (d. 925) and Avicenna (d. 1037).

The primacy in the medical field of the Muslim world from the 9th to the 14th centuries is confirmed in very modern practices and conceptions formulated in those centuries concerning total and partial anaesthesia, the treatment of pain, the first resuscitation teams and the management of upper airway obstruction[12] in a primacy far from present-day levels as shown by the many problems that will emerge in the course of this chapter.

---

[11] Rahman, *Islam and Health/ Medicine: A Historical Perspective*, op. cit., 154.

[12] See Mazrooa (Al) A.A. and Abdel-Halim R.E.S., Anaesthesia 1000 Years Ago, *Saudi Medical Journal*, 1991, 12 (5), 351–353. For example, total anaesthesia for operations was also obtained through the use of cannabis, opium and mandrake and the anaesthetic was administered by ingestion, inhalation and rectally. The exact dose of the drug to be administered to obtain general anaesthesia for a precise number of hours was determined (e.g. Avicenna). At the beginning of the 14th century the need was described (e.g. Ibn al-Quff) for the presence of an anaesthetist alongside the surgeon; this laid the bases for a resuscitation team. For resuscitation, the patient was fed externally by tubes in the throat. In those centuries, there are many in-depth descriptions of the use of endotracheal tubes for the management of upper airway obstructions, etc.

THE INCURABLY AND TERMINALLY ILL PATIENT

The Muslim doctor has the task of making every effort to give the incurable patient moral support and relieve physical and psychological pain, i.e. acting on the whole sphere of the person (reference to the classic *hakim*[13]). Every attempt to save a dying person, if the probabilities of success appear significant, should be made. On the contrary, artificially prolonging the vegetative state of a patient who is definitively unconscious is unlawful as it is consciousness that makes an individual responsible; otherwise dignity is violated beyond the divine will.[14] If it is certain that a person will not live, it is useless keeping them in a vegetative state with heroic attempts at resuscitation or other artificial equipment. However, the patient's life must not be deliberately terminated.[15]

In all developing countries, the majority of patients with serious pathologies (e.g. tumours) seek treatment when the pathology is too advanced and incurable. In Muslim countries it is still likely that the incurable patient will die in his own home, where several generations may live together and where the impression is that the terminal patient can tolerate his suffering better. This inclination is not without problematic consequences, when hospitals have modern clinical equipment. In fact, "letting the patients go home" often means forgoing direct care by Doctors and medical staff and hospital equipment and, at the same time placing a considerable "burden" on families that are not always in a position to take on a similar commitment, either on the material or psychological level. It has been asked whether behind this trend to "discard" the patient an insidious form of passive euthanasia can be identified[16] which can be associated with the refusal of the death of the patient. Some witnesses of the situation in the hospitals of Tunis confirm the widespread fear amongst Doctors of death and the reactions of the relatives (especially women) of the deceased: screams, fainting fits, hysteria, etc., with the result that in order to avoid similar embarrassment, the Doctors try to convince the relatives to take the dying patient home. In this, they seem comforted by Decree 1634 of November 1981; its art. 24 allows transferring a patient whose life is in danger to his home if he himself or his family so request. Similarly to what happens in the West, people prefer to conceal dying and the death of a relative from children.

In the case of the use of analgesics which reduce the psychological and physical suffering of the patient but which can, at the same time, accelerate death, the decisive element in Muslim reflection is the doctor's intention.[17] If his intention

---

[13] See section "Principles and Characteristics" in Chapter 3.

[14] Abdul-Rauf M., Contemporary Muslim Perspective, in Reich W. (ed.), *Encyclopedia of Bioethics*, New York, Free Press, 1978, 894.

[15] *Islamic Code of Medical Ethics*, Chapter 8.

[16] Labidi L. and Nacef T., *Deuil Impossible*, Tunis, Edition Sahar, 1993, 12.

[17] Hathout H. and Lustig B.A., *Bioethical Developments in Islam*, in Lustig B.A., Brody B.A., et al. (eds.), *Bioethics Yearbook. Theological Developments in Bioethics: 1990–1992*, Dordrecht, Kluwer Academic Publishers, 1993, 133–147.

is not to kill, but only to help the patient, then the doctor should not be subject to civil or penal prosecution.

Relatives frequently request extreme treatment in the hope of the recovery of a dying patient. Some Muslim "scholars" believe that every possible effort should be made for the survival of the patient whilst others introduce the categories of unnecessary, ineffective or inappropriate measures when the patient is in the final throes in order to avoid invasive, painful or, in any case useless interventions. At the King Faisal Specialist Hospital (KFSH) in Riyadh, recourse is made to a protocol similar to the Western "Do Not Resuscitate" Orders[18] which allows avoiding resuscitation practices in dying patients so that their death is less painful. Many Saudi hospitals do not have similar protocols to manage terminal patients whilst the Doctors themselves seem reluctant to use the concept of "appropriate cures" deeming this a limit to their possibilities of action.

PALLIATIVE CARE

Speaking of care for the terminally ill patient means speaking of palliative care, i.e. what to do when no more can be done for the patient.[19] Recourse to hospices (or similar structures) in the Middle East and North Africa is hindered by multiple reasons: in the first place those of an economic and social order as they require a complex and expensive structure; the inefficiency of nurses and the poor qualification of health personnel; the low salaries of the latter and, lastly, the general preference for home care.

Since the early 1990s, a palliative care department has been active in Saudi Arabia at the KFSH in Riyadh for the purpose of providing physical, psychological and social care to the terminally ill patient and his or her relatives either at home or in hospital (Inpatient Unit) by an inter-disciplinary team of professionals and expert nurses.[20] This is believed to be the most complete Palliative Care Service in the whole of the Arab world.[21]

In Europe and North America, care for the patient for whom therapy is useless is increasingly evaluated as a specialisation exercised by health personnel competent in clinical pharmacology and the specific pathologies. In the dominant social environment in Saudi Arabia, the strong family unity and the strong religious bond are some of the factors that help the patient to die at home, accompanied

---

[18] Younge D., Moreau P., et al., Communicating with Cancer Patients in Saudi Arabia, *Annals of the New York Academy of Sciences*, 1997, 809, 309–316.

[19] The definition of the WHO, in *Cancer Pain Relief and Palliative Care*, Geneva Technical Report Series, 1990 says: "The active total care of patients whose disease is not responsive to curative treatment. Control of pain, of other symptoms, and of psychological, social and spiritual problems is paramount. The goal of palliative care is the achievement of the best possible quality of life for patients and their families."

[20] Gray A.J., Ezzat A., et al., Developing Palliative Care Services for Terminally Ill Patients in Saudi Arabia, *Annals of Saudi Medicine*, 1995, 15 (4), 370–377.

[21] Isbister W.H. and Bonifant J., Implementation of the World Health Organization "Analgesic Ladder" in Saudi Arabia, *Palliative Medicine*, 2001, 15, 135–140.

by relatives. Numerous socio-cultural obstacles oppose the programmes of palliative care begun to date, which are also found in other Middle Eastern and North African countries. In the first place, there is a widespread lack of knowledge on modern pharmacological instruments to control and reduce pain. In particular, there is an underestimation of when and how to efficiently use morphine, an attitude associated with a general reluctance to use it independently of the methods of administration (orally, rectally, etc.) especially due to an unreasonable fear of creating an inevitable dependence. Both palliative care and the use of morphine are at times seen as equivalent to a form of euthanasia as they can shorten lives (on the contrary, it is intolerable pain that unquestionably contributes to shortening life). Other difficulties connected with the introduction of palliative care in Saudi Arabia are similar to those in many developed countries and can be summarised as follows: emphasis on the "cure" when recovery is no longer possible, scarcity of Doctors interested in this type of approach, general refusal to inform the patient of the diagnosis of cancer and the fatal prognosis,[22] absence of adequate methods of pain control, reluctance to discuss the problems of dying and death.

In 1989, when the administration of the KFSH began to study the possibility of launching a programme of home care for very seriously ill patients, two problems in particular emerged in connection with the social context (in addition to those already mentioned): the language barrier and the possible negative reaction from Saudi families to the possibility that non-Saudi (and non-Muslim) experts (male and female) could enter their homes to care for kin. The Kingdom had 14,500 Doctors, of whom only 12% were Saudi nationals, whilst virtually all nurses were foreigners. In this regard, the Home Care Programme approved in May 1991 included the use of chauffeurs and Saudi interpreters to allow the foreign specialists (Doctors and nurses) to provide adequate palliative care.[23]

In 1999 the European School of Oncology sponsored a symposium at the KFSH on "The Modern Management of Advanced Cancer: How to help your patients"; there were about 150 attendees, the majority health professionals in the Riyadh area, who analysed the availability and distribution of opioids in the Kingdom.[24] Eight main problems were identified: (1) Too many administrative problems hindered the availability of the drugs. (2) General lack of awareness of the importance of palliative care and scarce support from the community. (3) Lack of preparation on the issue between Doctors, nurses, patients and relatives, with consequent mistaken ideas and fears such as that of interfering with the patient's religious duties; a strong lack of experience on the "analgesic ladder"[25] is underlined. (4)–(5) All fear that the use of opioid analgesics goes against

---

[22] In Turkey and many Middle Eastern countries, the population tends to identify a cancer diagnosis with a death sentence therefore doctors are reluctant to reveal it. See Sen M., Communication with Cancer Patients, *Annals of the New York Academy of Sciences*, 1997, 809, 514–524.

[23] Gray and Ezzat, art. cit., 373–375; Younge and Moreau, art. cit., 315.

[24] Isbister and Bonifant, art. cit., 135–140.

[25] In 1986 the WHO introduced the concept of "analgesic ladder" for pain control in advanced cancer patients, passing from non-opioid analgesics to weak opioids and finally to the strongest ones.

religious principles; the recommendation is to instruct the religious authorities on the necessities and methods of care for the terminally ill in order to include these very authorities in the palliative care staff. (6) Many people fear side effects (dependence) caused by the use of strong opioids. The suggestion was to have recourse to the mass media to promote a national policy of education on coping with the terminally ill patient. (7) There exists a very restrictive government policy on the use of opioid analgesics. (8) There is disorienting bureaucracy that hinders the use of the drugs and the conditions of prescription.

In Egypt, the Egyptian Society for Pain Management was created in 1980 and three years later the Pain Clinic at the National Cancer Institute (NCI) of Cairo was opened. Until the early 1990s, there had been no pre-established protocols for the management of pain with the exception of an experimental programme implemented at the NCI: the home use of morphine tablets with a lasting effect requires the authorisation of the Ministry for Health and an attentive weekly medical examination. In fact, the pharmaceutical companies and chemists prefer not to deal with morphine out of fear of the responsibility involved: the only form available for taking it is in tablets.[26] Home management requires the consent of a relative with responsibility for the use of the drug and its return at the end of the treatment (i.e. on the death of the patient). In this context the NCI started up a pilot project which, using close family bonds, aims to specifically train a selected relative of the terminally ill patient who is instructed and informed, as far as possible, of the patient's needs (treatment of pain, side effects, treatment of symptoms, mobility, nutrition, sleep, etc.). In fact, these volunteers proved to be up to the task taken on.

In 1995–1996 in Indonesia, out of a population of 203 million inhabitants, 203–365 thousand new patients with tumours were reported each year.[27] The first pilot centre for palliative care and the therapy of pain was set up in 1990 at the Dr. Soetomo Hospital (DSH), in eastern Indonesia. To care for pain, in the majority of cases recourse was made in Indonesia to non-opioids and weak opioids due to the limited availability of oral morphine. Codeine, non-opioid analgesics and other drugs on the other hand are easily available. Since January 1995 oral morphine has become available at the DSH.

INFORMATION AND CONSENT OF THE SERIOUSLY AND TERMINALLY
ILL PATIENT

All over the world Doctors frequently conceal the truth from patients to avoid potential negative consequences. This attitude was dominant until a few decades ago, even in the USA. Subsequently, in the USA and Northern

---

[26] Tawfik M.O., Egypt: Status of Cancer Pain and Palliative Care, *Journal of Pain and Symptom Management*, 1993, 8 (6), 409–411.

[27] Soebadi R.D. and Tejawinata S., Indonesia: Status of Cancer Pain and Palliative Care, *Journal of Pain and Symptom Management*, 1996, 12 (2), 112–115; Lickiss J.N., Indonesia: Status of Cancer Pain and Palliative Cancer, *Journal of Pain and Symptom Management*, 1993, 8 (6).

Europe this approach began to be considered paternalistic and disrespectful of the patient and this was accompanied by increasingly open disclosure of the truth.

There are very great differences in the world in the willingness to communicate the truth, and these differences are linked to the evaluation of the autonomy of the patient in different cultures and also in a same culture with the passing of time.[28] A study of 1984 compared countries in which the cultural tradition is paternalistic with others where the autonomy of the patient was increasing[29]: oncologists deemed that the percentage of those who told the truth in Africa, France, Hungary, Iran, Panama, Portugal, Spain was less than 40% whilst other estimates indicate a willingness to tell the truth by more than 80% amongst oncologists in Austria, Denmark, Finland, Netherlands, New Zealand, Norway, Switzerland and Sweden.

The *hadith* quoted ("If one organ complains, all others share its complaint, suffering sleeplessness and fever") underlines a vision that tends to be "unitary" of the person. At the same time however, it can give rise to contradictory consequences according to the context in which it is applied. In fact the *hadith* offers a possible instrument to diminish or annul the validity of the requests of a terminally ill patient in the case that this request is not "in the norm" with respect to dominant ethics and values.[30] For example, if the terminal patient expresses requests "within the norm" they are approved and respected, otherwise, as in the request for some form of euthanasia, it becomes easy to reply that his condition of psychophysical inferiority (as the serious illness of one part is always, to some extent, the illness of the whole individual) does not allow him to formulate rational requests.[31]

The origin of this conception has its roots in Muslim law where the subject *optimo iure* of rights and duties is the free, pubescent male who is healthy in mind and body and with an irreproachable life. Only the possession of these requisites gives full juridical capacity. The absence of one or more of them entails the diminution of the right of disposal (*tasarruf*). These limitations are called "impediments" or "restraints" (*hagr*).[32] *Malikite* and *Shafi'ite* doctrines organise these limitations of capacity to act into five categories: sex; age; illness and certain pathological, physical and mental conditions; special and transitory circumstances (insolvency, marriage for the woman, etc.); apostasy and other deprivations that suppress

---

[28] See, amongst others: Dalla-Vorgia P., Katsouyanni K., et al., Attitudes of a Mediterranean Population to the Truth-Telling Issue, *Journal of Medical Ethics*, 1992, 18, 67–74; Surbone A., Truth Telling to the Patient, *JAMA*, 7 October 1992, 268 (13), 1661–1662; Pellegrino E.D., Is Truth Telling to the Patient a Cultural Artefact?, *JAMA*, 7 October 1992, 268 (13), 1734–1735.

[29] Holland J.C., Geary N., et al., An International Survey of Physician Attitudes and Practice in Regard to Revealing the Diagnosis of Cancer, *Cancer Investigation*, 1987, 5 (2), 151–154.

[30] See section "Autonomy and Consent of the Patient" in Chapter 3.

[31] See under the section "Euthanasia".

[32] Santillana D., *Istituzioni di diritto musulmano malichita con riguardo anche al sistema sciafiita*, Roma, Istituto per l'Oriente, 1926–1938, Vol. I, 121–122

juridical capacity. Therefore, the psychophysical infirmities cause limitations of the capacity to act. Some infirmities are temporary, others are chronic. The individual affected by a fatal illness is "bound" or "impeded" (*mahgur*), i.e. he has a limited capacity because it is believed that his psychophysical state does not allow him to be fully aware in providing for his own interests.

In Muslim law, the individual affected by a fatal disease is seen as equivalent to:

(a) A woman close to childbirth, i.e. who has reached the sixth month of pregnancy or, according to more recent opinion, the woman who is suffering labour pains.

(b) The soldier in battle.

(c) The prisoner condemned to death or to mutilation implying a danger of death. This is one of the factors that helps to explain why, in the case of explantation of organs from a person condemned to death (obviously after death), the request of his prior consent is not foreseen (e.g. 1972 Syrian law on transplants).

The classic doctrine of the limitation of juridical capacity of a person affected by serious or fatal illness is confirmed by the widespread behaviour in Muslim societies towards the seriously ill patient as shown by several contemporary studies. Following the opposite trend, in the Islamic Law of Succession, disadvantageous obligations such as donations, sales of goods at prices lower than their value, etc., made during a fatal illness are considered as *wasiya* (testamentary dispositions) and are valid even if the patient does not die.[33]

A study on 369 members of the Turkish Society of Anaesthesiology and Reanimation[34] on the use of do-not-resuscitate (DNR) orders to limit (or interrupt) treatments in patients in intensive care units shows that 65.9% had made verbal use of them. However, before the decision, the Doctors had discussed the case with colleagues in 82.7% of cases, with relatives in 14.4% of cases, with the Ethical Committee in 1.7% of cases and with the patient in 1.2% of cases. The dominant role of Doctors is also evident here in relation to the role of the family. On the contrary, the majority (58%) of the Doctors – when asked who, in theory, should be involved in a DNR order – indicated that the decision should be taken collectively by the patient, relatives, Doctors, hospital administrators and Ethical Committee; 31.4% say that the patient, relatives and doctor must decide jointly; 10% stated that only the doctor is responsible.

A study was carried out in six Saudi hospitals in 1993 on a sample of 249 Doctors.[35] Seventy-five per cent of them preferred to speak to relatives about the

---

[33] Schacht J., *Introduction au droit musulman*, Paris, Maisonneuve et Larose, 1983, 146.

[34] Iyilikçi L., Erbayraktar S., et al., Practices of Anaesthesiologists with regard to Withholding and Withdrawal of Life Support from the Critically Ill in Turkey, *Acta Anaesthesiologica Scandinavica*, 2004, 48, 457–462.

[35] Mobeireek A.F., Kassimi (Al) F.A., et al., Communication with the Seriously Ill: Physicians' Attitudes in Saudi Arabia, *Journal of Medical Ethics*, 1996, 22 (5), 282–285.

very seriously ill patient's conditions, even when the latter was judged of being of sound mind. This is the most interesting element affected by the strong family bond, a characteristic of the Arab world, where the patient himself seems to spontaneously delegate to relatives the faculty of deciding on the best treatment or measures to be adopted for him. Furthermore, the tendency of the Arab-Muslim family to instinctively take on the wishes (even those that are not expressed) of the patient is complementary to the principle of Muslim medical ethics ("If one organ complains, all others share its complaint, suffering sleeplessness and fever") as well as the approach of the Muslim law mentioned above, both judging a serious illness as the cause of the state of psychophysical inferiority of the patient.[36]

The situation in Saudi Arabia highlights at least two contradictory but representative trends of the present situation in the variegated Arab-Muslim world. Imitating the laws of the more developed Western countries, oriented towards guaranteeing an increasingly greater balance in the relationship between patient and doctor and following the line of some principles of Muslim medical ethics, the first "theoretical" trend emphasises the autonomy of the patient and his right to know the truth. For example, the Code of Kuwait (1981) specifies that each patient is a particular case and the doctor has the task of respecting the patient's right to know his clinical situation, reconciling this duty with that of finding the most suitable words to keep the patient hopeful and calm. These "Westernised" models of rules appear somewhat far removed from concrete social behaviour and the collective mentality.

The second tendency is that shown by a daily problematic reality far removed from the "theoretical" criteria, a situation distinguished by the Doctors' fear of death, by the reduced direct communication with the terminally ill patient, by the reluctance to reveal the truth, by the communication of the diagnosis to relatives, etc. This last situation is not however automatically deemed as an infringement of the autonomy and respect of the patient, precisely due to the possibility – offered by tradition – to address relatives as the natural "long hand" or "extension" of an ill relative, presumably no longer capable of taking the best decisions for himself and accepting this "proxy". We could speak of a "broader" conception of the

---

[36] Another study on the possibility of avoiding recourse to cardio-pulmonary resuscitation was carried out in 1992 on 100 doctors (85% Muslims) in two large hospitals in Riyadh (the King Khalid University Hospital and the Riyadh Central Hospital) in Saudi Arabia. An overall analysis of the data showed the following positions:

1. 39% of the doctors recognise the autonomy of the patient in the decision but the last limit must be established by the doctor.
2. 32% are in favour of full medical "paternalism" (71% are in favour of full and moderate "paternalism").

For 19% of doctors, the patient should enjoy total autonomy. See Mobeireek A., The Do-Not-Resuscitate Order: Indications on the Current Practice in Riyadh, *Annals of Saudi Medicine*, 1995, 15 (1), 6–9.

individual (i.e. incorporating the close relatives) with respect to the current Western one, which is inevitably translated into personal autonomy the "borders" or "limits" of which are diversified according to cultural and local contexts.

Training and specialisation abroad (e.g. USA where communication is considered the rule) of Doctors from contexts where the psychological protection of the patient seems a priority (e.g. Lebanese Doctors) does not appear to imply that they are any more willing to communicate the truth after they return home.[37] In other words, independently of what they have learned abroad, many of these Doctors remain influenced by the dominant trends and values in the culture where they practise (e.g. first the USA then Lebanon) and they adapt consequently. The most positive aspect is that these Doctors express a pragmatic approach open to modifications to handle different contexts, far removed from rigid ideological positions. With regard to Lebanon, Hamadeh and Adib observe that Lebanese Doctors continue to avoid the direct and complete revelation of a cancer diagnosis; this may be avoided only if the patients themselves seek greater empowerment and a desire to know the truth.[38]

In Muslim countries the patient is first and foremost a member of the family which feels responsible for him; the consent of the seriously ill and/or incurable patient is easily taken over by that of the family in order to avoid emotional problems for him.[39] In other words, avoiding any negative consequence for the patient at the psychophysical level caused by learning the truth is preferred, even at the price of lies and concealing the truth.[40] For the latter approach, both medical paternalism and the protective role of the family are indispensable instruments. In emergencies, it is opportune to involve the male members of the family because taking a decision alone can place a woman in a situation of embarrassment.[41]

Iranian society is strongly protective towards a seriously ill or incurable patient as the serious illness of one member of the family tends to emotionally involve all the others.[42] For this reason the Doctors prefer to communicate with chosen members of the family (son-in-law, daughter-in-law, nephew of the patient, etc.).

[37] Hamadeh G.N. and Adib S.M., Cancer Truth Disclosure by Lebanese Doctors, *Social Science and Medicine*, 1998, 47 (9), 1289–1294.

[38] Ibid., art. cit., 1293.

[39] Shetty P., Rizk K., et al., Supportive Care of Cancer Patients in Dubai, *Support Care Cancer*, 1997, 5, 196–199.

[40] Families often ask doctors not to inform the patient of the need for a radical operation, preferring to tell the truth directly but more slowly to the patient. However, this risks being translated into a systematic concealment of the truth, which is why, at the King Faisal Specialist Hospital and Research Centre in Riyadh, there is a reluctance to accept patients except when they are aware of their situation and have given their consent to the operation. See Younge and Moreau, art. cit., 311–312.

[41] Lawrence P. and Rozmus C., Culturally Sensitive Care of the Muslim Patient, *Journal of Transcultural Nursing*, 2001, 12 (3), 228–233.

[42] Montazeri A., Vahdani M., et al., Cancer Patient Education in Iran: A Descriptive Study, *Support Care Cancer*, 2002, 10, 169–173.

For example, at the Department of Oncology and Haematology and in the Marrow Transplant Centre of the Shari'ati Hospital of Teheran[43] the diagnosis is preferably disclosed only to one member of the family to avoid confusion and different versions between various relatives and negative effects on the patient. Subsequently, if the family allows, it is given to the patient. On going into hospital, a general informed consent form must be signed by the patient or by his legal representative. In the case of chemotherapy, some oncologists request the signature of another consent form in which the side effects of the treatment are explained. This consent may also be signed by the legal representative in order to avoid the patient learning the diagnosis.

The Egyptian situation is very similar. One hundred male surgeons divided into two groups of 50 took part in a survey carried out at a University Hospital in Cairo; the first group consisted of surgeons with more than 20 years of experience in relations with patients with tumours, the second group had less than 10 years of experience.[44] Amongst the answers to the questions, several agreed with what has been shown so far. For example, to the question on which person they preferred to inform first about the diagnosis of a tumour choosing between the patient, the spouse or relatives, only 3 surgeons out of 100 chose to inform the person directly concerned (none of the group with less than 10 years experience). How much information do you wish to give the patient on his real clinical condition? Out of 100 surgeons, 17 were willing to tell him the whole truth (15 in the group of 50 Doctors with more than 20 years experience but only two in the second group); 25 Doctors were willing to reveal very little (5 in the first group of 50, 20 in the second group), 58 Doctors out of 100 said they were willing to tell the patient something. Do you tell the terminally ill patient that some hope may exist? The 100 surgeons answered unanimously in the affirmative that the patient must never lose all hope. Similarly all the Doctors avoid informing the terminally ill patient that there is no hope left.

In Pakistan too, both Doctors and family members opt to conceal the truth, if considered psychologically damaging, from the patient.[45] Doctors are not very willing to discuss the pathology, therapy and diagnosis with a patient, especially if seriously ill. Amongst the causes taken into consideration are the importance of numerous families with strong internal bonds, the fear of creating incomprehension, a general sense of impotency and despair in the case of a tumour and anti-tumour therapies and the cultural level of the patient. In the Pakistani

[43] Ghavamzadeh A. and Bahar B., Communication with the Cancer Patient in Iran, *Annals of the New York Academy of Sciences*, 1997, 809, 261–265.

[44] Ghazali (El) S., Is it Wise to tell the Truth, the Whole Truth and Nothing but the Truth to a Cancer Patient?, *Annals of the New York Academy of Sciences*, 1997, 809, 97–108.

[45] Moazam F., Families, Patients and Physicians in Medical Decisionmaking: A Pakistani Perspective, *Hasting Center Report*, 2000, 30 (6), 28–37; Malik I.A. and Qureshi A.F., Communication with Cancer Patients, *Annals of the New York Academy of Sciences*, 1997, 809, 300–308.

context, there is frequent recourse to unorthodox or alternative therapies[46] which may replace or accompany official treatment (but delaying recourse to the latter can cause irreparable damage). The request of alternative treatment appears linked to the size of the family, as it reflects the range of opinions present in a large group, with the consequent recourse to different therapies by the patient. The more serious the pathologies appear, the more the therapeutic decisions may be taken by the family. It is always the doctor who makes the final decision. As already mentioned, in the case of a fatal diagnosis reported to relatives, they prefer not to communicate the truth to the patient and want the Doctors to do the same. On the contrary, when the patient is the first to know the truth, he often avoids facing up to the problem with the family, concealing his illness until some effect becomes visible. The reasons for this behaviour can be identified in the lack of comprehension, in fear of surgery and chemotherapy, the lack of confidence in Doctors and in the strong desire not to be a burden on the family.

The authority and the unity of the family also exceed the importance of the autonomy of the single individual in Saudi culture. A patient's decisions are easily changed by the family's desire to protect him; similarly, the family expects to be the first to know the bad news concerning the patient. Subsequently, the family decides if and how to tell the patient.[47] This can obviously create difficulties for a doctor wishing to disclose the serious truth to the patient.

In Turkey, with some exceptions,[48] it appears widespread practice not to give the patient directly specific information on negative diagnoses and prognoses. Article 19 of the 1998 Regulations of the Ministry of Health of Patients' Rights states: "The diagnosis may be kept from the patient in order not to jeopardise the mental state if revealing the diagnosis is likely to aggravate his disease or if the development or effects of the disease are considered to be serious."[49] According to the Regulations, a fatal diagnosis may be revealed to the patient only with great caution; if the patient is not opposed, the family will be informed of the diagnosis.

In Turkey it is usually a relative who makes the decisions on the treatment of a patient, especially in the face of a diagnosis of cancer (an illness considered synonymous with death). The "do no harm" and "beneficence" principles appear to have priority in Turkey over the principles of autonomy and justice; this may represent the basis for hiding the truth from the patient.[50] In the country, medical paternalism is based on social structures and values deeply rooted in society. Requests for information by the patient and the desire to actively participate in

---

[46] Including homeopathy; therapies based on the humours of Hippocrates administered with herbs by *hakims* (sages); recourse to "holy men" for spiritual therapies by means of amulets or reading holy verses, etc.

[47] Shahri, art. cit., 137.

[48] See, for example, Sen M., Communication with Cancer Patients, *Annals of the New York Academy of Sciences*, 1997, 809, 514–524.

[49] In *International Digest of Health Legislation*, 1999, 50 (3), 364–369.

[50] Ornek Buken N., Truth-Telling Information and Communication with Cancer Patients in Turkey, *Journal of the International Society for the History of Islamic Medicine*, 2003, 2 (4), 31–36.

decisions are limited.[51] The patient wants the doctor to be an expert and competent authority; this attitude can be accompanied by a "fatalistic world view" by patients, an approach generally associated with Islam. Nevertheless, it would appear that the attitude of Turkish Doctors is increasingly open to direct disclosure to the patient.[52]

Generally speaking, the protective attitude of society and the family to the seriously ill or terminal patient is accompanied by the paternalistic role of the doctor. Nevertheless, the trend by Doctors to inform the patient of his state of health is gaining ground everywhere.

An interesting study was carried out in the Arab Emirates in 1995 with the purpose of examining the information to be given to a terminally ill patient on a sample of 100 citizens and 50 local Doctors.[53] Two cases were analysed:

1. The death of the patient within six months is almost certain: Should the doctor tell him?
2. A patient has a 50% chance of dying in the next six months: Should the doctor tell him? There were three possible answers: Yes, No, It depends.

Case 1. Forty-two per cent of citizens and 38% of Doctors answered NO; 27% and 8%, respectively YES; and 31% and 54%, IT DEPENDS. If the patient asks expressly for the truth then 38% of citizens and 36% of Doctors answer NO, and 59% and 42%, respectively, answer YES.

Case 2. Seventy per cent of Doctors answer NO, 0% YES and 30% IT DEPENDS. If the patient expressly asks for the truth only 28% of Doctors say YES and 54% continue to say NO. The majority of citizens are also contrary: NO 43%, YES 30%, IT DEPENDS 27%. In the case of explicit request, 39% of citizens answer YES and 36% NO.

Comment. When asked for the reasons for their choices, no citizen mentioned the patient's right to know the truth about his health or the duty of the doctor to tell him, but practical reasons were given, e.g. making a will or allowing the patient to seek treatment abroad. Religious reasons justified both silence and revealing the truth. In general, the Doctors seem much less in favour than the citizens of telling the truth. Amongst the consequences, 90% of citizens revealed that they did not completely trust what Doctors tell them about their state of health. Lastly, the majority of Doctors maintain that Western patients seem more inclined to want to know the truth about their condition.

In Algeria, the revelation of a diagnosis of cancer "is an ordeal that many Doctors try to avoid, leaving it to another colleague",[54] similarly to what hap-

---

[51] Ornek Buken, 34–35.

[52] Pelin S.S. and Arda B., Physicians' Attitudes Towards Medical Ethics Issues in Turkey, *International Journal of Medical Ethics*, 2000, 11 (2), 57–67.

[53] Harrison A., Saadi (Al) A.M.H., et al., Should Doctors Inform Terminally Ill Patients? The Opinions of Nationals and Doctors in the United Arab Emirates, *Journal of Medical Ethics*, 1997, 23, 101–107.

[54] Abid L., *Consensus et règles de bonnes pratiques médicales. La concertation pluridisciplinaire à propos de la cancérologie*, in www.santemaghreb.com, 1 March 2003.

pens for major surgery. Many healthy people prefer to know their real clinical situation, whilst many other healthy people prefer the doctor to say nothing. Patients are often less willing to know and it is no surprise that even oncologists with clear signs of neoplasia are seen who delay diagnostic tests not to discover anything.[55] Obviously the cultural context plays a decisive role in the approach to the problem. The right approach, maintains Abid, is to prolong the disclosure in a context of great caution as the doctor–patient relationship becomes consolidated. In Algeria there is a certain fatalism (*mektoub*, i.e. "what is written") therefore the population generally thinks that God has sent us an incurable disease[56] to punish us. The truth should not be told bluntly "as in the West"; it is opportune to go by degrees with the aim of involving the patient in the therapy. For example, the doctor can begin by announcing a lesion that has to be evaluated as quickly as possible, before any worse consequences, etc. Article 51 of the Code of Medical Ethics of Algeria (1992) states that a serious or fatal prognosis can be concealed from the patient for reasons that the doctor evaluates with full awareness ("en toute conscience"); in this case, the family must be informed unless forbidden by the patient or the same has named a third person to be told. In the case of communication, "the serious diagnosis or the fatal prognosis must be revealed only with extreme caution".[57]

Article 44 of the Lebanese Code of Medical Ethics (1994)[58] advises the doctor to dissimulate a very serious diagnosis from the patient, especially in cases where death is certain. At the same time, the Code recognises that the patient has the right to know. As a result, the preferred attitude by Doctors is to avoid telling the seriously ill patient the truth and opting for a close relative.[59]

In a study of 1995 on a sample of 212 Doctors (especially oncologists) working in Beirut and the surrounding areas, Doctors who habitually told the truth (tellers) were confronted with those who habitually concealed it (non-tellers). The result was that 47% of the 212 usually told the truth whilst 53% (i.e. 112) usually did not disclose it.[60] The younger Doctors were less inclined to tell compared to the older Doctors; the tellers seem to have greater clinical experience compared to the non-tellers. Fifty-nine per cent admitted occasionally making an exception to their rule. Whether Doctors tell the truth or not appears to be conditioned by some characteristics of the patient[61]; the most important in decreasing order are the emotional stability of the patient, his intelligence, if the patient is a doctor,

---

[55] Abid L., *Cancer: doit-on ou non dire la vérité?*, in www.santemaghreb.com, 3 February 2004.

[56] It is curious to note that in everyday language a malignant tumour is defined a "female tumour" (tumeur femelle) as it can reappear after treatment or produce "little ones", i.e. metastasis, whilst a benign tumour is called a male tumour.

[57] Code de Déontologie Médicale, *Décret Exécutif* no. 92–276, 6 July 1992.

[58] Code Libanais de Déontologie Médicale, *Journal Officiel du Liban*, no. 9, 3 March 1994.

[59] Adib S.M. and Hamadeh G.N., Attitudes of the Lebanese Public Regarding Disclosure of Serious Illness, *Journal of Medical Ethics*, 1999, 25, 399–403.

[60] Hamadeh and Adib, art. cit., 1290.

[61] Ibid., 1290–1291 and 1293.

the patient's wish to know the truth, etc. Regarding the professional characteristics of the doctor which influence their personal willingness to tell the truth, in
the sample, telling the truth is closely linked to the length of clinical experience
and to the type of medical specialisation: concealing the truth decreases with
longer experience. Concealing the truth was more frequent amongst primary
health care practitioners compared to specialists and surgeons, etc.

In conclusion, further elements of "medical paternalism" can be identified in
other laws and regulations currently in force. For example, art. 20 of Ministerial
Resolution 288/17/L of 23 January 1990 of Saudi Arabia states that in the case of
incurable pathologies, or ones that endanger the life of the patient, the doctor will
decide whether to tell the patient or his relatives, or to say nothing, according to
the indications of his conscience, except where the patient himself has forbidden
him beforehand to inform others or explicitly asks him to inform specific people.
Regarding the therapy, the following art. 21 requires the consent of the patient
or that of his legal representative in cases where the capacity of judgement of the
patient is diminished. In the event of accidents or emergencies, the doctor can provide the essential treatment to save a patient's life or an organ without waiting for
the patient's consent. The Committee of Senior Ulama of Saudi Arabia expressed
themselves similarly in Resolution 119 of 27 February 1984.

Article 36 of the Tunisian Code of Medical Ethics in 1993 maintains that a
serious or fatal prognosis may be concealed from the patient. He may be told
only with the greatest circumspection, but in general close family members may
be informed except when the patient has forbidden this beforehand or has desig-
nated third parties to be told.

In Law no. 17 of 3 November 1986 of the Libyan *Jamahiriyya* on the doctor's
responsibility, art. 5 (f) states that the doctor must inform the patient in the case
of an incurable disease "if it is in his best interests to do so, and if his psychological condition permits it" and inform the close relatives, unless the patient is
opposed to this or has indicated the persons he wishes to be told.

THE LIVING WILL

The expression *living will* indicates a declaration made by a person in possession
of his mental faculties, in which he specifies the limits within which he wants to
be treated in the event of being in an extremely critical condition, without the
possibility of recovery.

The term *al-wasiya* in Muslim law indicates the last wishes and, by extension,
also a will (even if the latter is not strictly correct). Their regulation is present
in Koran 5.106: "O ye who believe! When death approaches any of you, (take)
witnesses among yourselves when making bequests, two just men of your own
(brotherhood) or others from outside if ye are journeying through the earth, and
the chance of death befalls you (thus)."

The division of the inheritance is fixed by the complicated rules of succession in
Muslim law and only depends for one-third on the wishes (*wasiya*) of the deceased.

The *living will* cannot be included in the will (*wasiya*) as what is included in a will can be executed only after the death of the testate; vice versa, the *living will* refers what to do in the phases preceding death. Consequently, if a Muslim signs a *living will*, it is considered without legal value.[62] In addition, the *wasiya* has as its object only the property or enjoyment of assets. It should be noted that the laws of contemporary states generally regulate the last wishes (*wasiya*) according to criteria of the *Shari'a*. For example, art. 184 of the Algerian Law on the Family no. 84–11 (9 June 1984) states that the *wasiya* is a transfer of property free of charge that takes effect after death.[63]

However, according to Ebrahim, a Muslim can draw up an alternative *living will*, without a binding value and which can be defined a *wasiya mubahah* (admissible document) and includes the following elements[64]:

(a) Request to suspend treatment if this does not improve the quality of life according to the principle *"la darar wa la dirar"* (no harm and no harassment). In this case death is not accelerated as only excessive treatment is refused. However, hygiene and nutrition should be maintained.

(b) Instructions to switch off life-supporting equipment after brain stem death has been diagnosed. This was allowed by the Council of the Islamic Jurisprudence Academy of the OIC in Amman, 1996, 5th Resolution.

(c) Inclusion of the wish to donate organs according to the principle of public benefit (*maslaha*).

(d) It is worthwhile appointing an attorney, Arabic *wakalah* (*pl.*), to be expressly mentioned. When the patient's mental faculties are compromised, his legal representative (*wakil*) is morally obliged to communicate the dying patient's wishes to his relatives and Doctors.

Lastly, the *living will* must be signed by the author, by his legal representative (*wakil*) and by two witnesses, according to Koran 2.282: "The witnesses should not refuse when they are called on (for evidence)". If none of the requests contained in the *living will* contradicts the teachings of the *Shari'a*, there is no justification for ignoring the directives it contains.[65]

The Islamic Medical Association of North America (IMANA) recommends that all Muslims sign a kind of "Living Will" or "Advance Directive". On the point of interest here, the sample declaration presented by the IMANA is similar to the content shown by Ebrahim. In the event of fatal illness or injury where

---

[62] Ebrahim A.F.M., The Living Will (wasiyat al-hayy): A Study of its Legality in the Light of Islamic Jurisprudence, *Medicine and Law*, 2000, 19, 147–160.

[63] Aluffi Beck-Peccoz R. (ed.), *Le leggi del diritto di famiglia negli stati arabi del Nord-Africa*, Torino, Edizioni della Fondazione Giovanni Agnelli, 1997, 57. For Egypt, see art. 37 of law no. 71 of 1946 on last wishes (Aluffi, 93); for Libya the Provisions on the last wishes (Law no. 7 of 1993), Aluffi, op. cit., 125–134.

[64] Ebrahim, art. cit., 157–158.

[65] Ibid., 160.

the doctor certifies in writing that the use of life-prolonging procedures only artificially prolongs the agony, the patient asks that "such procedures be withheld or withdrawn, . . . with only the provision of appropriate nutrition and hydration and the administration of essential medications and the performances of any medical procedures . . . to alleviate pain".[66]

A model of "health care proxy and *living will*" was drawn up by the Connecticut (USA) Council of Masajid (Mosques) in November 2000 and differs from the previous two. In this,[67] the future dying patient explicitly authorises his agent "to direct that no treatment be conducted or withheld from me if to do so is against the teachings of Islam. . . . I direct that medication be judiciously administered to me to alleviate pain". The subject refuses the possibility of anyone voluntarily interrupting his life and requests that "the 'life support systems' [may] be used in a judicious manner and their use discontinued . . . if it becomes reasonably apparent that they have no curative value". Furthermore, the "life support systems" must "include but are not limited to artificial respiration, cardiopulmonary resuscitation, artificial means of providing nutrition and hydration, and any pharmaceutical drugs".

The two positions do not appear identical as the second seems more "open". IMANA and Ebrahim state that "hygiene and nutrition should be maintained", otherwise death is caused whilst only further excessive treatments may be suspended if they "do not improve the quality of life". In the second document (Council of Mosques) the suspension of "life support systems" (which include artificial nutrition and hydration, cardiopulmonary resuscitation, etc.) seems lawful "if it becomes reasonably apparent that they have no curative value".

EUTHANASIA

In this section, the positions tend to oscillate between two polarities. On the one hand theoretical statements which in general condemn euthanasia (not well defined); on the other several medical–clinical pronouncements that seem to leave room for "interventions" aimed at terminating or shortening the life of the patient who is in hopeless and very undignified conditions.

It is opportune to remember that the definitions and distinctions between suicide, murder and euthanasia are sometimes confused and/or different depending on authors and documents. To put it simply, euthanasia could be defined as an act which hastens the death of a human being or removes the

---

[66] Imana Ethics Committee, *Islamic Medical Ethics: The IMANA Perspective* (PDF format), in www.imana.org, 1–12.

[67] Connecticut Council of Masajid, *Last Will and Testament . . .* , Hamden, Mahtab 20 November 2000, 8.

obstacles that slow it down. On the evaluation of the act itself, other variables have a strong incidence, for example, if an act is performed actively or passively on an individual; if the act is performed with the consent of the dying person or without this consent; if the act is performed by the dying person himself or by someone else.

For Islam, man is not his own master, therefore the termination of the treatment of a living individual on his request or on the request of his relatives in the event that the patient is unable to distinguish between right and wrong is prohibited. From a religious perspective, the patient who shows strength of mind in facing up to, accepting and overcoming his suffering earns credit with God; euthanasia – amongst other things – would prevent him from obtaining this credit. Again from a religious standpoint, it must be remembered that where there is life there is hope and a miracle can always happen.

There were no specific distinctions between the different cases of euthanasia in classic Muslim law. In any case, the modern *fatwas* do not dwell on the different types of euthanasia and generally do not go beyond the distinction between active and passive euthanasia, condemning them both as murder. However, many differences can be perceived within this condemnation.

Following the traditional apologetic approach, contemporary juridical–religious documentation asserts that the spread of euthanasia in the West is the result of rife materialism,[68] the dissolution of the family accompanied by the absence of affection between relatives therefore the seriously and terminally ill patient risks being left to fend for himself. Chapter 8 of the Islamic Code of Medical Ethics (1981) states that suicide and mercy killing "find no support except in the atheistic way of thinking that believes that our life on this earth is followed by void"; on the other hand, there is no human pain that cannot be overcome by medical treatment or neurosurgery.

In the case of the suspension of treatment aimed exclusively at delaying inevitable death by means of a decision of the patient himself or of the Doctors or relatives, this attitude does not seem contrary to the principles of the *Shari'a*. Artificially delaying an inevitable death by means of support therapies is contrary to the interest of the patient, hindering the natural process of dying.

This principle is clear in the *fatwa* issued by the *Fatwa* Department of the South African *Jamiatul Ulama* (Council of Muslim Jurists), which states as prohibited active euthanasia where patients can end their lives, for example, by means of a lethal injection; vice versa, "passive euthanasia where patients may withhold treatment or artificial life-support is only permissible if a trustworthy, reliable opinion and specialist feels that there is no hope of survival".[69]

---

[68] Boubakeur H., Position de l'Islam sur l'euthanasie, *Ethique*, April 1992 to January 1993, no. 6–7, 97–101.

[69] See the website of the Fatwa Department – Jamiatul Ulama (KZN), *The Islamic Ruling on Euthanasia*, Bishopsgate, South Africa, last modified 12 February 1999, mailto:info@jamiat.org.za.

In the opinion of another expert, artificially prolonging the vegetative state of a patient who has definitively lost consciousness is unlawful as it is consciousness that makes an individual responsible, therefore human dignity would be violated beyond divine will.[70] If it is certain that a person cannot continue to live, it is useless to endeavour to keep them in a vegetative state with heroic attempts at reanimation or other artificial equipment. The doctor must lengthen life, not hasten death. According to the Ethical Committee of the IMANA, there nevertheless remains the duty to continue hydrating, nourishing and limiting pain (the so-called Ordinary Care).[71]

The Islamic Code for Medical and Health Ethics (IOMS, 2004) in art. 62 does not consider as euthanasia the termination of treatment "when its continuation is . . . useless, and this includes artificial respirators . . . "; similarly, giving drugs to reduce severe pain does not come within the scope of "mercy killing" even if this drug may accelerate the patient's death.[72]

Article 24.1 of the Pakistani Code of Ethics (2001–2002) appears prudent in its language: "Futile treatment need neither be offered to patients nor be provided if demanded"; the futile treatment is a treatment that "merely preserves permanent unconsciousness or fails to end total dependence on intensive medical care or when physicians conclude . . . that a medical treatment has been useless".

Muslim law does not prohibit the refusal of useless and disproportionate treatment following the consent of relatives. Death would not be attributed to the termination of the "therapies" but to the inexorable progress of the disease. Therefore, any comprehension seems reserved for the so-called passive euthanasia (not better defined) which may not be considered a crime if not considered equivalent to suicide; according to some authors passive euthanasia contributes to protecting the dignity of the life and death of the suffering patient.[73] Koran 2.45 recalls that men show different capacities of perseverance even in the face of suffering in life. It is therefore conceivable that in specific cases a person suffering beyond every limit may choose to die to defend his dignity as compelling him to live has no positive effect on society or on the individual.

An additional difficulty regarding euthanasia comes from what was mentioned earlier: the value to be given to refusal by a terminally ill patient of care or treatment (food, oxygen supply, pharmacology, etc.) may be challenged, in

---

[70] Abdul-Rauf art. cit., 894.

[71] Imana Ethics Committee, art. cit. The document states: "Imana does not believe in prolonging the misery of dying patients who are terminally ill or in a persistent vegetative state (PVS) . . . When death becomes inevitable, . . . the patient should be allowed to die without unnecessary procedures. While all ongoing medical treatments can be continued, no further or new attempts should be made to sustain artificial life support. If the patient is on mechanical support, this can be withdrawn. . . . No attempt should be made to withhold nutrition and hydration" (page 5).

[72] Islamic Organization for Medical Sciences (IOMS), 2004, in www.islamset.com/ioms/Code2004/index.html, 1–61.

[73] Masoodi and Dhar, art. cit., 25–26.

principle, objecting that his psychophysical equilibrium is altered by the pathology in course.[74]

However, according to a rigid approach, the Muslim doctor should not intervene directly to voluntarily take the life of the patient, not even out of pity (Islamic Code of Medical Ethics, Kuwait 1981); he must see whether the patient is curable or not, not whether he must continue to live. Similarly, he must not administer drugs that accelerate death, even after an explicit request by relatives; acceleration of this kind would correspond to murder. Koran 3.145 states: "Nor can a soul die except by God's leave, the term being fixed as by writing"; Koran 3.156 continues "It is God that gives Life and Death, and God sees well all that ye do", resulting that God has fixed the length of each life, but leaves room for human efforts to save it when some hope exists.

The patient's request for his life to be ended has in part been evaluated by juridical doctrine in some aspects. The four "canonical" *Sunnite* juridical schools (*Hanafite, Malikite, Shafi'ite* and *Hanbalite*) were not unanimous in their pronouncements. For all, the request or permission to be killed does not make the action, which remains a murder, lawful; however, the disagreement concerns the possibility of applying punishments to those that cause death: the *Hanafites* are in favour; *the Hanbalites*, the *Shafi''ites* and the *Malikites* are partly in favour and partly contrary to penal sanctions. In turn, those in favour do not agree on whether the punishment should consist of the "blood price" or of the "lex talionis".[75] In the specific case of the responsibility of relatives, the *Shari'a* imposes that each heir

---

[74] Some commented clinical cases illustrate this situation: see Ben Hamida F., *L'Islam e la bioetica*, in Massué J.-P. and Gerin G. (ed.), *Diritti umani e bioetica*, Roma, Sapere, 2000, 90 and cases 55, 27, 14 and 49:

1. Case 55. Patient aged 42 with two children, following a viral infection, he suffered cardio-respiratory failure in a terminal phase. With his consent, he was put on to a list of patients who must have, extremely urgently, a heart-lung transplant. On the day of the transplant, when he was still conscious, he refused the operation. His wife and children asked the doctors to operate in any case. In a Muslim context, in the face of similar cases, the seriously ill patient is said not to be capable of expressing himself in full consciousness, and so his refusal cannot be accepted. This is a consideration that can frequently be found: the more seriously ill a patient is, the more the value of his request is diminished if deemed incoherent with traditional principles and those of Muslim medical ethics (e.g. a request for euthanasia), even if expressed by a person formally of sound mind. The classic principle of the unity of the person can contribute to understanding how illness invades every aspect of the personality.

2. Case 27. Woman aged 43, cancer of the ovaries, metastasis, gives her consent for the injection of substances without any relation with the pathology in order to ascertain the post-mortem effects. According to Muslims, this experimentation is prohibited as the free and conscious consent of the patient is not valid due to her condition as a terminally ill patient.

3. Case 14. Man aged 45 with amyotrophic lateral sclerosis, disorders of phonation and deglutition; the psychological analysis is normal. He requests active euthanasia. For Islam, nobody can put an end to life; in addition experience shows that a prognosis is never certain; lastly, the patient's request cannot be taken into consideration due to his weakness.

[75] Keskin A., *La Morale musulmane*, in AA.VV., *Le Médecin face aux droits de l'homme*, Padova, CEDAM, 1990, 1251–1261.

who has played any role in the death of the person from whom he should have inherited, can no longer inherit anything.

In June 1995 the Muslim Medical Doctors Conference in Malaysia (Kuala Lumpur) reasserted that euthanasia (not better defined) goes against the principles of Islam; this is also valid in the military context, prohibiting a seriously wounded soldier from committing suicide or asking other soldiers to kill him out of pity or to avoid falling into enemy hands.[76]

The law in some Muslim countries explicitly states what has been said so far about euthanasia. On the other hand, the social debate is at times more articulated.

The Penal Code of Turkey describes euthanasia as a crime. Article 13 of the Regulations of 1998 of the Ministry of Health on Patients' Rights (or Patients' Bill of Rights) prohibits euthanasia and specifies: "It shall be prohibited to take life, by medical methods or in any other manner whatsoever"; this is prohibited both on the request of the patient and of others.[77] In Turkey, the debate on euthanasia started in the mid-1980s thanks to technological development in medicine and in care for the dying; in those years the vast majority of deaths took place at home, whilst in the mid-1990s, 40% of deaths were in hospital, in a technological context. Passive euthanasia seems to be a widespread practice. It is easy to discharge from hospital a terminally ill patient who is not informed of his condition and without his consent (that of his relatives is sufficient); the prescription of fatal doses of anaesthetics seems common.[78] A study of 1995–1996 on 524 Doctors operating in the Ankara area showed that more than half (61.5%, i.e. 320 Doctors) were in favour of some form of passive euthanasia.[79] The majority of those in favour limited this consent to those patients who consciously ask for it; moreover, precise rules and protocols on the subject were requested. 38.9% (204 out of 524 Doctors) were against passive euthanasia as the duty of a doctor is to save lives or because there can never be certainty as to whether the patient's request is really conscious or because life belongs only to God.

Article 552 of the Lebanese Penal Law of 1943 (still in force) can punish by up to 10 years imprisonment those who, out of compassion, put an end to a person's life on the latter's request. Article 27 (10) of the Code of Ethics (1994) specifies[80] that, if the patient suffers from an incurable disease, the doctor must limit himself to relieving psychophysical suffering with treatment compatible with keeping the patient alive (if possible). The doctor does not have the right to voluntarily cause death but, it is added, it is preferable not to use excessively technical methods which prolong the patient's agony. The doctor must treat the patient until his death but without violating his dignity. In fact, there are indications of Doctors who have

---

[76] Masoodi and Dhar, art. cit., 23–24.
[77] In *International Digest of Health Legislation*, 1999, 50 (3), 364–369.
[78] Yasmin Oguz N., Euthanasia in Turkey: Cultural and Religious Perspectives, *Eubios Journal of Asian and International Bioethics*, 1996, 6, 170–171.
[79] Pelin and Arda, art. cit., 61 and 66–67.
[80] *Code Libanais de Déontologie Médicale*, art. cit.

performed forms of active or passive assisted suicide with or without the consent
of relatives or the patient. This seems to respond to a less rigid attitude of society
to the problem of euthanasia, an attitude shown by the approach of 135 judges
in Beirut (31 were women, 103 were already judges or prosecutors, 32 were lawyers training for appointment as judges, 61% were Muslims, 39% non-Muslims) to
hypothetical cases concerning withdrawal or withholding of life-sustaining devices
(passive euthanasia) and physician-assisted suicide (voluntary euthanasia).[81] Fifty-
six per cent of the judges express a different opinion from art. 552 of the Penal law
and are in favour of the doctor assisting the terminal patient if the latter asks for
death. The younger judges and the lawyers training to become judges seem to be
the most in favour, together with the women.[82]

If in Lebanon a debate has started, it has yet to begin in Sudan. In the first
study carried out in the African country (in April 2000) on the attitude to
euthanasia and assisted suicide[83] on 248 Doctors in the two main hospitals of
Khartoum, all the Doctors who graduated in Sudan were strongly against both
solutions, especially for religious reasons (90%) or as actions contrary to their
duties as a doctor (85%) and similar considerations.

Ministerial Resolution no. 288/17/L of Saudi Arabia of 23 January 1990 in
art. 21, dealing with the patient's consent, states that under no circumstances
must a doctor put an end to a terminally ill patient's life even if requested by the
patient himself or by his relatives.

Law no. 17 of the Libyan *Jamahiriyya* of 3 November 1986 on the responsibility of the doctor maintains that no patient's life may be taken, not even on his
request, for the reason of deformity, incurable or terminal disease or acute pain,
even when the patient's life depends on artificial life-support equipment.

SUICIDE AND MARTYRDOM

The Koran does not contain any explicit indications on suicide. Nevertheless,
Muslim commentators refer to Koran 4.29 "Do not kill yourselves (*anfusakum*),
surely God is merciful to you" understood as a prohibition of suicide or as a prohibition to kill other Muslims; both interpretations can be accepted at the same
time.[84] Koran 3.145 appears complementary to this, where it is recalled that the
time of death is in the hands of the Creator only: "Nor can a soul die except by

---

[81] Adib S.M., Kawas S.H., et al., End-of-Life Issues as Perceived by Lebanese Judges, *Developing World Bioethics*, 2003, 3 (1), 10–26.

[82] Ibid., 23–24.

[83] Euthanasia was taken to mean "the deliberate administration of an overdose of medication to an ill patient at their request with a primary intent to end their life"; assisted suicide means "prescribing a medication (e.g. narcotics), or counselling an ill patient to be able to use an overdose to end their life"; see Ahmed A.M., Kheir M.M., et al., Attitudes Towards Euthanasia and Assisted Suicide among Sudanese Doctors, *Eastern Mediterranean Health Journal*, 2001, 7 (3), 551–555.

[84] Rosenthal F., On Suicide in Islam, *Journal of the American Oriental Society*, 1946, 66, 239–259.

God's leave, the term being fixed as by writing". Both Abu Ja'far al Tabari (d. 923) and Fakhr al-Din al-Razi (d. 1209) (in *"al-Tafsir al-Kabir"*, *The Great Commentary*) prefer to interpret 4.29 as a prohibition against killing another Muslim. In the same work, *"al-Tafsir"*, al-Razi specifies that at times it may be of use to interpret the Koranic passage referring to suicide especially because "we see many Muslims killing themselves for such reason we have listed [i.e. the difficulties of life]. So in this case, the prohibition is beneficial".[85]

On the contrary, three "sayings" of the Prophet are more direct. "In old times there was a man with an ailment that taxed his endurance. He cut his wrist with a knife and bled to death. God was displeased and said "My subject hastened his end: I deny him Paradise".[86] In a second *hadith* (in Muslim and Bukhari) he who commits suicide by stabbing himself will continue to stab himself in Hell (for eternity).[87] A third *hadith* (in Bukhari) recalls the case of a warrior who had distinguished himself in battle but, wounded several times, lost heart and killed himself. The Prophet declared that, due to this action, the warrior had annulled all the merits he had conquered on the battle field for God and was condemned to Hell.

Therefore, the Prophet disapproved of suicide, even refusing – in one case – prayers for the dead man. This case was that of a *hadith* (in Muslim) which says: "(The dead body) of a person who had killed himself with a broad-headed arrow was brought before the Apostle of Allah (may peace be upon him), but he did not offer prayers for him."[88] Whether those who have committed suicide are entitled to prayers is a historically debated subject by jurists. Nawawi (in *Sahih Muslim bi-sharh al-Nawawi*), in commenting upon the *hadith* recalled that some "scholars" did not pray for those who had committed suicide (e.g. Abd al-Aziz and al-Awza'i) whilst the majority prayed (e.g. Malik, Abu Hanifa, al-Shafi'i, etc.).[89] The different schools however, have maintained different opinions and the practice has changed in time. In the end the most charitable opinion has prevailed.

In the cases of unrequited or illicit love in literature, suicide could be justified.[90] Moreover, popular Hellenistic philosophy influenced the idea that death is

---

[85] Brockopp J.E., The "Good Death" in Islamic Theology and Law, in Brockopp J.E. (ed.), *Islamic Ethics of Life*, Colombia (South Carolina), University of South Carolina Press, 2003, 177–193.

[86] *Islamic Code of Medical Ethics*, Chapter 8.

[87] Bukhari, *Sahih*, *Medicine*, Volume 7, Book 71, Number 670 in www.usc.edu/dept/MSA/: "Narrated Abu Huraira: The Prophet said, "Whoever purposely throws himself from a mountain and kills himself, will be in the (Hell) Fire falling down into it and abiding therein perpetually forever; and whoever drinks poison and kills himself with it, he will be carrying his poison in his hand and drinking it in the (Hell) Fire wherein he will abide eternally forever; and whoever kills himself with an iron weapon, will be carrying that weapon in his hand and stabbing his abdomen with it in the (Hell) Fire wherein he will abide eternally forever."

[88] Muslim, *Sahih*, *The Book of Prayers*, Book 4, Number 2133, in www.usc.edu/dept/MSA/fundamentals/hadithsunnah/muslim/smtintro.html.

[89] Brockopp, art. cit., 185.

[90] Rosenthal, art. cit., 247 and 255.

preferable with respect to a dishonourable or intolerable life. For example, Abu Hayyan al-Tawhidi (d. 1023) indicated episodes in which human life has value if accompanied by virtues otherwise it is the equivalent of a non-life and therefore a despicable life can be avoided by recourse to suicide.[91]

Attempted suicide and suicide do not appear to be punishable in this life. The majority of jurists also refuse the blood price. If necessary, God will provide for punishment after death.

However, suicide appears unacceptable as a way out in the face of personal failure or pain. It would take on the appearance of an act of cowardice or an inadmissible flight from reality also because patience and strength of mind will be rewarded after death. The evaluation of suicide as an act of cowardice seems to be prevalent (in the *ahadith* as well) with respect to highlighting the desperate situations which may provoke it. Yusuf al-Qaradawi, in his famous work, *Al-Halal wa-l-Haram fi-l-Islam* (The lawful and the prohibited in Islam) uses a language which expresses the combative attitude of Islam on the subject: the Muslim must be resolute in facing up to any difficulty; the believer has been created to fight not to sit around lazily. His faith and character do not allow him to flee from the "battlefield" of life also thanks to the means at his disposal: unshakeable faith and moral steadfastness.

The difference between suicide and martyrdom is often difficult to identify but for Muslims this difference is important because the martyr goes to Paradise whilst the person who has committed suicide risks Hell. The exaltation of martyrdom is based on Koran 4.74 "Let those fight in the cause of God Who sell the life of this world for the Hereafter. To him who fighteth in the cause of God, whether he is slain or gets victory – soon shall We give him a reward of great (value)" and 47.4–6.[92]

To understand the positions in the debate of martyrdom in Muslim countries, there are two preliminary remarks to be made, the first more general, the second with reference to the Palestinian question. The first refers to Islam as a "complete" religion, i.e. a Religion–State system (*Din wa Dawla*) in which the two spheres are, at least in theory (but in fact at the origins of Islam), inextricably merged. This fusion is clear in the statements made by religious figures which we show below. The political dimension of Islam imposes defending with all means possible the Community of the faithful (the *Umma*), that is to say, defending the "true religion", the lands where Muslims live, the Holy Places, etc.; to this end Islam requires a total mobilisation, i.e. social, moral, financial, military and so on.

---

[91] Ibid., 248–251.

[92] 47, 4–6: "4 Therefore, when ye meet the Unbelievers (in fight), smite at their necks; at length, when ye have thoroughly subdued them, bind a bond firmly (on them): thereafter (is the time for) either generosity or ransom: until the war lays dawn its burdens. Thus (are ye commanded): but if it had been God's Will, He could certainly have exacted retribution from them (Himself); but (He lets you fight) in order to test you, some with others. But those who are slain in the way of God, He will never let their deeds be lost. 5. Soon will He guide them and improve their condition, 6. and admit them to the Garden which He has announced for them."

The second preliminary remark to be made, referring to the Palestinian situation, can be identified in a passage in a recent communiqué signed by 28 personalities of Al-Azhar and is fundamentally agreed with by all Muslims: "The Zionist entity [al-Kiyan as-Sihyuni, i.e. Israel] is a racist and colonialist entity, aiming for military settlement, made up of usurpers brought to Palestine. They have usurped the land, they have killed and expelled the inhabitants, they have destroyed homes and sacred places such as churches and mosques. For these reasons they can only be seen as aggressors and usurpers, and holding them innocent is an error and a lie . . . . The correct distinction must be between . . . aggressors and victims. All those who usurp land, violate dignity, desecrate holy places are considered 'fighters' regardless of the clothes they wear."[93]

Concerning the suicide–martyrdom relationship, the aforementioned communiqué specifies: "The attempt to confuse 'martyrdom' with 'suicide' is wrong, because the person who commits suicide is desperate due to his own life, whilst martyrdom is a heroic act carried out by a person who sacrifices his soul on the straight path of God to defend himself, his country, the community, dignity, honour, religion and holy places."[94]

The two preliminary remarks above can explain why desiring death to defend the true religion or dying in a religiously intolerable situation is required by the Shari'a as an act aimed at defending the community of believers (Umma). However, how this protection is concretely achieved is not always clear nor are the Doctors of the Law in agreement. For example, a fatwa dated 21 April 2001 by the Grand mufti of the Kingdom of Saudi Arabia, sheikh Abdulaziz al-sheikh in the Saudi newspaper Al-Sharq al-Awsat, reopened the debate on Palestinian suicide "terrorists" stating[95] that the jihad represents one of the best actions in Islam, but expresses doubts on suicide actions amidst the enemies, namely if they are approved by the Shari'a. Fighting to strike the enemy is required but without infringing Muslim law. The Al-Azhar sheikh, M.S. Tantawi, has added: the actions undertaken by the Palestinians represent legitimate defence and a type of martyrdom as long as their purpose is to kill soldiers, but not women and children.

In December 2001, after an attack by Hamas which caused the death of 25 civilians, M.S. Tantawi explained[96] that the Shari'a "rejects all attempts on human life. In the name of Shari'a we condemn all attacks on civilians, whether they were carried out by a state [Israel] or by other groups [Palestinian organisations]".

One of the most authoritative Muslim jurists, Yusuf al-Qaradawi, answered on the Qatari satellite channel, Al-Jazeera, arousing great consensus amongst viewers: the Palestinian acts are acts of martyrdom; Tantawi's opinion should

---

[93] Gli Ulama di Al-Azhar. Le operazioni di martirio sono il più alto grado di jihad, "Al-Sha'b" (Egypt), in www.aljazira.it.

[94] Ibid.

[95] Abou El-Magd N., The Politics of Fatwa, Al-Ahram Weekly, 3–9 May 2001, no. 532.

[96] Saad R., Weapons of the Weak, Al-Ahram Weekly, 13–19 December 2001, no. 564.

not apply to the Palestinians as they are victims, not aggressors. Furthermore, the suicide attacks are the only weapon left for the Palestinians. The real terrorist is the Jewish State which came into being by "butchering and displacing them [Palestinians], to settle Jews from different part of the world". Qaradawi refuses the idea that Hamas operations kill innocent civilians because the Israelis themselves say they are "a nation in arms", all Israeli men and women are soldiers and, therefore, occupying troops.[97]

At the end of 2001, N.F. Wassel, the Grand *mufti* of Egypt, made this statement: "the martyrdom operations that Palestinians have been carrying out are aimed at putting an end to injustice, defending Islam's holy sites and preserving the dignity of the Arab nation. Thus, they are legitimate . . . . Jihad and the defence of the homeland and Islam's holy sites are a commitment that every Muslim honours more than ever in Palestine".[98]

In April 2002, the Al-Azhar *sheikh*, M.S. Tantawi, changed his mind. On a website associated with Al-Azhar he stated that the martyrdom operations (suicide) and killing civilians are lawful acts that should be carried out more frequently. Martyrdom against any Israeli is legitimate, according to the *Shari'a*, until Palestine is won back and the cruel Israeli occupation withdrawn.[99] On the same site, Ahmad al-Tayyeb, the new Grand *mufti* of Egypt after Wassel, reconfirmed that the solution to Israeli terror lies in an increase in the number of martyrs' attacks against the enemies of Allah and these attacks should be supported by Muslim governments and Muslim peoples.

In a *fatwa*, a South African *mufti*, Ebrahim Desai refuses suicide as a solution to any type of personal suffering. However, when it is the whole community that suffers, it is lawful to fight the *Jihad* at the risk of one's life. The Prophet himself, according to Desai, praises the person whose army has been defeated but courageously confronts the enemy and is killed. The action is praised as the sacrifice is made in the interest of the Muslim Community. The case of the Palestinian suicide bombs is similar. The juridical justification of this martyrdom is based on the principle of the lesser evil: a suicide bomb is an evil but this evil is opposed to a greater evil, Israeli occupation.[100]

There are other jurists with opposing positions. For example, *sheikh* Abdullah al-Sabil, imam of Al-Haram al-Sharif Mosque in Mecca, totally condemns attacks on the People of the Book (i.e. Muslims, Christians and Jews): "Those who try to harm unarmed civilians do not understand *Shari'a*." A more decisive condemnation is expressed by a *fatwa*[101] of the Saudi *sheikh* Ibn al-Uthaymeen

---

[97] Ibid.
[98] Ibid.
[99] www.lailatalqadr.com/stories/n040401.shtml, 4 April 2002 in MEMRI (The Middle East Media Research Institute), *Un autorevole ecclesiastico governativo egiziano invoca "attacchi da martirio che terrorizzino i cuori dei nemici di Allah"*, Servizi Speciali, no. 363, 7 April 2002.
[100] Fatwas on Suicide Bombs, in *Lexington Area Muslim Network*, 29 November 2000.
[101] *Fatwas on Suicide Bombs*, art.cit.

according to whom anyone who drives a car with explosives against the enemy, knowing that he will die, is to be condemned as suicide. Even if the suicide victim presumes, out of ignorance, that the act is good and appreciated by God, this act remains condemnable and only God may forgive it. The *sheikh* underlined that the main reason for these actions is revenge and hatred against the enemy, independently of the evaluation that the act is *halal* (lawful) or *haram* (prohibited) according to the *Shari'a*.

Regarding the American intervention in Iraq, on 6 April 2003 in the Egyptian newspaper Al-Ahram, Tantawi specified that all those who blow themselves up against the American occupant in Iraq are martyrs of Islam, adding "all those healthy of mind must oppose the invader until they get their rights back". In this case, he added, the *jihad* is of different kinds: the use of the body, money, word, or everything that a person can do or say so that the truth wins is *jihad*. The Academy for Islamic Research of Al-Azhar pronounced itself in similar terms.[102]

As far as the suicide attacks against the Twin Towers of September 11th, 2001 are concerned, Tantawi defined them as acts of terror against innocent people, an ignoble and odious action that gives every Muslim state the right to defend itself and react, once the real responsibilities have been identified.[103] In his opinion, the countries that give shelter to these terrorists must be punished. Tantawi, moreover, clearly distinguished between the suicide attacks by Palestinians and those of 11th September as "there is a very big difference between these terrorists and those who defend their land. We are in solidarity with the Palestinian people because they are right. As for terrorism, we denounce and combat it because it is a flagrant act of injustice against the human race". A similar distinction was made by Qaradawi when he explicitly condemned the attacks of 11th September. According to him, the kamikaze attacks in Israel have the aim of liberating an occupied land by means of their own bodies, whereas "the attacks on the US had as their aim terrorising others and the means of those who carried out the acts were the bodies of others".[104]

Lastly, the Grand *mufti* of the Kingdom of Saudi Arabia, *sheikh* Abdulaziz al-sheikh, in a *fatwa* issued a few days after 11th September,[105] stated that "these matters that have taken place in the USA and whatever else is of their nature of plane hijackings and taking people hostage or killing innocent people, without a just cause, this is nothing but a manifestation of injustice, oppression and tyranny, which the Islamic *Shari'a* does not sanction or accept, rather it is expressly

[102] Ajur A.W., *Appello al mondo civilizzato e a tutte le forze amanti della pace*, 10 March 2003, in www.aljazira.it. See also the clarifications in Shahine G., Debating Jihad, *Al-Ahram Weekly*, 27 March to 2 April 2003, no. 631.

[103] Halawi J., A forbidden Alliance?, *Al-Ahram Weekly*, 20–26 September 2001, no. 552.

[104] Saad, art. cit.

[105] Shaikh (Al) A., *Fatwa from Grand mufti of Saudi Arabia on USA Terrorism*, in www.sunnahonline.com.

forbidden and it is amongst the greatest of sins". The exception "without a just cause" should be noted.

DEATH

Whilst at times it is tolerable to ignore the religious customs of a general patient, it can be unforgivable to ignore them in a critical clinical condition and close to death due to the risk of accentuating the patient's mental and physical suffering. Sometimes it is possible to adapt the hospital room to the religious culture of terminally ill Muslim patients in a non-Muslim environment. The dying patient's head should be turned towards Mecca, modifying the position of the bed at whose side the family and close friends gather. The terminally ill patient is exhorted to pronounce the *Shahada* or the "profession of faith" (one of the pillars of Islam)[106] and, if this is impossible, he must be made to repeat it slowly. If this is also not possible, the invocation of the name of God is sufficient. An imam, although not obliged, may be present at the death of a Muslim and officiates at the burial, the prayer is whispered in the ear of the deceased. Shrouds for the deceased should not be elaborate. According to the *Shari'a*, infidels may touch the corpse only with gloves and it should be the family that washes and lays out the body. The deceased must be buried as soon as possible whilst cremation of the corpse is traditionally prohibited. The modesty of a dead Muslim woman must be respected as if she were alive.

The following guidelines are to be followed for the corpse of a Muslim[107]: (a) Close the eyes, mouth and, when buried, turn the face towards Mecca. (b) Lay out the body. (c) Place the hands on the chest with the right hand on top of the left hand. (d) The body must be washed after death by a Muslim of the same sex. (e) Cover the body with a sheet.

In Tunisia, when children die, the parents almost always refuse a post-mortem out of fear that the dead child will suffer, even when it is explained to them that other children could benefit from the results obtained from dissection. Generally, Doctors prefer to inform the father rather than the mother of the death of the child.

In the Koran and in the *Sunna* there is no specific definition of the end of life, which is left to scientific study, human experience and inevitably reflects the medical knowledge available at the historical period of the various authors.[108] When asked by the Iranian Ministry of Health what the criterion of death (cardiorespiratory or brain death) is according to the *Shari'a*, the *ayatollah* Tabrizi

---

[106] The *Shahada* is the profession of faith which is performed with the legal act of pronouncing in front of witnesses the sentence "La ilaha il Allah, Muhammad-ur-Rasool-Allah", i.e. "None has the right to be worshipped but Allah, and Muhammad is the Messenger of Allah"; this declaration is sufficient to include the person pronouncing it in the community.

[107] Lawrence and Rozmus, art. cit., 231.

[108] Recommendations, in Mazkur (Al) K., Saif (Al) A., et al. (eds.), *Human Life. Its Inception and End as viewed by Islam*, op. cit., 628.

replied that it is essential to refer to a specialist to decide whether the first or second criterion corresponds to the definition of death adding, however, that it consists of the separation of the soul from the body.

Similarly, addressing Doctors of the Ain Shams Medical College in February 2000, the *sheikh* of Al-Azhar, M.S. Tantawi, recalled that death is separation from life and that only Doctors can ascertain it. If the doctor deems that the patient is dead when the heart beats because of a machine whilst the brain is dead, this is the competence of the doctor and there is no error in asking the relatives permission to switch off the machine.[109]

However, for theologians, death remains a direct divine act; in fact, it is not defined as the termination of cerebral activity or the stop of cardiac activity but as the separation of the soul from the body.

The Koran says nothing on the nature of the soul or on its localisation in the body. Amongst the consequences, the acceptance of the brain death criteria – which represent the compulsory passage to ethically perform an explanation from a corpse – remains an object of discussion.[110]

It must be noted that all the criteria of brain death are increasingly accepted in the medical field, whilst the objections are common amongst jurists (there are authoritative oppositions to the brain death criteria but also advocates of the brain stem death criteria); these contrasts influence the position of certain countries which still do not have a specific law (e.g. Egypt) and contribute to the general difficulty in finding organs from corpses. In the last place, the criteria referring to "cortical death" do not appear to arouse any interest.

POST-MORTEMS

The expectation in the Resurrection of bodies on the Day of Judgement has led to many prescriptions regarding the corpse: these include the duty of burying it as soon as possible (respecting its external appearance and dignity) and the prohibition of mutilation and cremation. If a part of a corpse that has already been buried is found elsewhere, it should be washed, the ritual prayers are said and it is buried in the same grave as the corpse without exhumation being necessary. Regarding burial times, it is important to recall that generally in Muslim countries very few hours pass between death and burial.

Muslim law presents two potentially contrasting principles on autopsies. The first states that "mutilation is prohibited in Islam" with the purpose of abolishing the pre-Islamic custom of mutilating the bodies of enemies killed in tribal struggles. The second principle states that the needs of the living have priority over those of the deceased; this principle is based – in turn – on that of public good (*maslaha*) and necessity. As man is a noble creature in the eyes of God,

---

[109] Brockopp, art. cit., 178–179.
[110] See section "The Debate on the Criteria of Death" in Chapter 7.

autopsies and transplants would appear – at least on first sight – contrary to the respect due to man. Every operation on the human body without therapeutic purposes is equivalent to a desecration of the body; this is the case of the post-mortem which is useful for the living but not for the deceased. However, in cases of necessity, just as it is possible to overcome specific prohibitions on food (Koran 2.173), similarly jurists tend to allow post-mortems if they are useful for saving human lives.

The jurists in favour are currently aware of the religious obligation of respecting the corpse as if it were the body of a living person in observance of the *hadith* that says "Breaking a dead man's bone is like breaking it when he is alive" (*hadith* collected by Abu Dawud[111] and Ibn Maja). In the past, the *Shi'ites* al-Hilli (d. 1277) and al-Tusi (d. 1068) expressed themselves in the same way. According to al-Tusi, the status of a deceased person is similar to that of a foetus and for both the same *diyya* (blood price) is paid when they are damaged; for any other lesion on the corpse the "blood price" corresponds to that to be paid to a living person in a similar case.[112] The *Malikite* Ibn Al-Haj (d. 1336) expressed himself in the same terms: any improper act on a living Muslim must not be caused after death, with the exception of what is allowed by the *Shari'a*.

Amongst the situations that can historically represent the anticipations and legitimisations of a post-mortem, the *Shari'a* allowed removing from the stomach of a corpse a sum of money swallowed when alive and belonging to another person; this was in order to return to the owner his due and to avoid objections to the legitimate heirs of the deceased (Ibn Qudama, d. 1223). Furthermore, for *Malikites* and *Shafi'ites* it was lawful to open the womb of a woman who had died during pregnancy if the foetus was believed to be still alive.[113] On the contrary, the *Hanbalites*, who doubted that the foetus could still be alive, seemed contrary to the intervention on the dead woman.[114]

Any uncertainty on establishing the death of an individual had been analysed by the classic jurists. Al-Shafi'i (d. 820) recommended waiting for two to three days before burial as a state of unconsciousness or a shock could affect those presumed dead by drowning, storms, war, attack by fierce animals, etc. Ibn Qudama limited postponement of burial to the appearance of the characteristic signs of death. Ibn Hajar al-Haythami (d. 1567), a *Shafi'ite*, allowed postponement until the arrival of camphor (Arabic *kafur*) to wash the deceased.[115]

Another problem debated by jurists concerns the transfer of the corpse from one place to another[116] (as inevitably is the case to perform a post-mortem). The

---

[111] Abu Dawud, *Sunan, Funerals*, Book 20, Number 3201, in www.usc.edu/dept/MSA/fundamentals/hadithsunnah/abudawud/satintro.html.

[112] Rispler-Chaim V., *Islamic Medical Ethics in the Twentieth Century*, Leiden, Brill, 1993, 76.

[113] Ibid., 81–82.

[114] Ibid., 76–77.

[115] Ibid., 79.

[116] Ibid., 80.

Shari'a prefers the deceased (independently of the cause) to be buried where he died or in the nearest cemetery. Amongst the historical examples, Prophet Muhammad had those who fell at the Battle of Uhud (625 AD) buried near the battlefield instead of in the cemetery in nearby Medina. The conquerors of Damascus were buried where they died and not all in the same place. 'A'isha, the Prophet's wife, criticised the transfer of her brother, Abu Bakr, who died in Abyssinia, to Mecca, as this caused a deterioration of the body. Ahmad b. Hanbal (d. 855) accepted a corpse being moved if there were valid reasons, an opinion that can be used to legitimise the movements connected with a post-mortem. Amongst the reasons indicated in some contemporary *fatwas*,[117] we discover the need to avoid the corpse being submerged by a flood or to make relatives' visits more frequent, or if a cemetery has not been used for more than a century and the location is destined for the building of a mosque; in this case there is recourse to the principle of the benefit of the community as being greater than that of the individual. Amongst the *Shi'ites, ayatollah* Khomeini deemed it lawful to exhume a corpse to extract a baby that was still alive in the womb of a dead mother or if there is the fear that a fierce animal could devour the body of the deceased or floods can occur or enemies take possession of it in war.[118]

At present post-mortems are becoming widespread in Muslim hospitals for scientific and legal purposes (in particular in the case of murders, suicides and accidents). Precisely due to the respect accompanying the corpse, there is the habit, after dissection, to bring together the parts of the body, stitch them back together and wrap the corpse in the shroud. When the faithful ask about the lawfulness of this in the light of the precepts of the *Shari'a*, the Doctors of the law in favour of autopsies answer that, despite the violation of the body and the delay in burial, the information that can be obtained satisfies the public good (e.g. in the event of murders) therefore the autopsy is justified.

Still with the *Shi'ites, ayatollah* Khomeini prohibited the dissection of the corpse of Muslims whilst he deemed lawful that of non-Muslims.[119] The Grand *ayatollah* of Iraq, al-Sistani, deems it prohibited to dissect the corpse of a Muslim for medical, investigative reasons, etc.; conversely the dissection of the corpse of an infidel is lawful.[120] According to other *Shi'ite* juridical responses,[121] it is prohibited to dissect the corpse of a Muslim but if it is done, pecuniary compensation must be paid (blood price or *diyya*) (decree 778). It is lawful to dissect the corpse of a non-Muslim or a person of uncertain faith (779). If the life of a Muslim depends on the dissection of the corpse of another Muslim when it is not possible to use a non-Muslim, then it is lawful to dissect the Muslim corpse (780).

---

[117] Ibid.

[118] *Ayatollah* Khomeiny, *Principes politiques, philosophiques, sociaux et religieux*, Paris, Éditions Libres-Hallier, 1979, 150.

[119] Ibid., 160.

[120] Sistani (Al) USAH, *Al-Fatawa al-Muyassarah*, Beirut, 2002, 416.

[121] Bostani (al) A.A. (ed.), *Le Guide du Musulman*, Publication du Séminaire Islamique, 1991, 268.

On several occasions the Grand *muftis* of the Arab Republic of Egypt have expressed *fatwas* (from 1959 to 1974) according to which, in the case of deceased whose relatives are unknown, the corpse can be dissected for legal and scientific purposes and for transplants; if the relatives are known, their prior authorisation is essential. In an opinion of the Committee of the *Fatwas* of Al-Azhar (1982) a post-mortem is accepted in the case of an unidentified corpse (e.g. victim of a road accident). This licence regarding an "unidentified corpse" may be due to the absence of relatives who could report the damage undergone by the corpse.[122] In the case of deceased Muslims, the consent of the next of kin should follow this order of priority[123]: (1) father, son, mother; (2) brothers and sisters, wife, grandfather, grandson; (3) cousins of paternal and maternal uncles. Ignoring the objections of the family remains lawful for the priority benefit of the community and for medical–legal reasons.

The first *fatwa* in the Arab world devoted to post-mortems is believed to date from 1910 and is that of the Egyptian *sheikh* Rashid Rida (d. 1935) entitled "Postmortem examinations and the postponement of burial" which appeared in *Al-Manar*.[124] Rida agreed with the need to delay burial to allow the doctor to confirm death, avoiding hurried burials of apparent deaths. Rida added that, when a non-Muslim government performs this practice on a Muslim, this should not be seen as an anti-Muslim act therefore, as a consequence, Muslims are not obliged to flee to the Ottoman Empire to avoid the delay of burial. Contrasting voices can also be found on this subject such as those expressed, more recently, by the *sheikh* of Al-Azhar (1973–1978), Abd al-Halim Mahmud (d. 1978), who considered the delay of burial always to be a sin, except to perform ablutions and lay out the corpse.

A second authoritative *fatwa* on the subject was that of H.M. Makhluf, Grand *mufti* of Egypt in 1945, published in 1952, expressing the position of the *Sunnite* juridical school on post-mortems[125]: a doctor is competent only when he has a thorough knowledge of the inside and the outside of the human body which is possible only through dissection. The *Shari'a* allows it as it is useful for science and justice. Medicine in the past was primitive, being limited to external symptoms, which today is inadmissible. Makhluf quotes a previous *fatwa* by *sheikh* Yusuf al-Dajawi which legitimised post-mortems, in cases of need, with recourse to *maslaha* (public good) and the observation that the advantages are greater than the damage. Dajawi compared the violation of the corpse with the permission granted by the S*hari'a* to remove money from the stomach of the deceased person who, in his lifetime, had incurred a debt and swallowed the money by

---

[122] Rispler-Chaim V., The Ethics of Postmortem Examinations in Contemporary Islam, *Journal of Medical Ethics*, 1993, 19, 164–168.
[123] Dada M.A., Islam and the Autopsy, *Journal of the Islamic Medical Association of South Africa*, 1998, Vol. 4, no. 3.
[124] Rispler-Chaim, op. cit., 77–78.
[125] Ibid., 73–74.

mistake. Lastly, Makhluf, to the objection that the post-mortem did not exist before his century, replied that medicine is currently based on science and technology whilst Muslim medicine cannot remain behind Western medicine (a very heartfelt topic).[126]

The reference to the West may also lead to an opposite position as shown by the piece by an anonymous writer in the Muslim bimonthly *Al-Nur* in 1997 in which post-mortems were attacked as a crude, useless and dishonourable practice that mutilates the human body and a practice that is part of an anti-Muslim conspiracy by infidels.[127] Contrary to all violations of the sanctity of the body remains *sheikh* Abd al-Fattah, according to whom any lesion caused on the deceased is a sin (*Al-Ahram*, 15 April 1983) although, according to the principle of necessity, it can be performed on an anonymous person who has died in a road accident.

Despite the elements contrary to post-mortems – i.e. the delay in burial, the movement of the corpse and the violation of the integrity of the body – the *muftis* generally propend for the practice but under specific conditions. An emblematic opinion was offered by the Committee of *Fatwa* of Al-Azhar in January 1982 in which it was stated that a post-mortem can be performed if it is of benefit for students to gain knowledge, for justice to prevail and to keep infectious diseases under control; in these cases the benefits exceed the damage but the practice must be limited to the situations of necessity.

Regarding the penal context, in the *Shari'a* system, monetary compensation and *lex talionis* are closely associated. If the murderer confesses immediately to the crime before a judge and pays the blood price (if the heirs accept), in this case it is not necessary to perform a post-mortem on the corpse of the murder victim[128]; in the event that the culprit confesses but without paying the "blood price" or if the relatives of the deceased refuse it, then the judge can order the post-mortem. According to Koran 2.282 "The witnesses should not refuse when they are called on (for evidence)", therefore the doctor cannot refuse his evidence on the results of the post-mortem.

The opinions generally in favour of post-mortems are not disregarded by the State, together with the criteria of their limitations. An example is given by the law on transplants of the Arab Republic of Syria of 23 August 1972 according to which a post-mortem can be performed: (a) if the Doctors deem it socially useful but only if the deceased when alive was not against post-mortems or if relatives up to the third degree do not oppose it; (b) the relatives' objections can be overcome if the post-mortem is performed for scientific reasons or to prevent an epidemic. In Egypt the examination of a corpse can be

---

[126] Rispler - Chaim, art. cit., 167.

[127] Dada, art. cit.

[128] Hussain W., Post-Mortem Examination: The Quranic View, *Hamdard Islamicus*, 1992, XV (3), 85–92. A post-mortem also seems to be avoided when the assassin is killed immediately or captured after the crime. See also Dada M.A., art. cit.

performed when there is the suspicion that the death was caused by foul play and not in other cases. In Saudi Arabia a post-mortem is not usually performed except in cases of forensic interest and limitedly to the parts necessary for the post-mortem as manipulation of the corpse remains prohibited.[129] Ministerial Resolution no. 288 of 1990, in art. 22 in reference to post-mortems, states that the doctor cannot provide a death certificate without having ascertained the cause of death. The doctor has the faculty of not issuing the certificate if he suspects that death is the consequence of a crime and in this case he must advise the competent authorities. In the event of suspected crime or poisoning (22.2.L.), the doctor is obliged to inform the police immediately and indicate the wounds found on the body in a report to be forwarded to the forensic expert as well; the latter will perform a general examination of the corpse. In the event that he were to deem a post-mortem necessary to identify the cause of death, he must obtain permission from the authorities and collect all valid evidence (e.g. clothes, bullets, etc.). In neighbouring Qatar, Law no. 8 of 2003 allows post-mortems for penal, pathological-diagnostic and teaching purposes[130]. The approval of relatives is not essential in the first two cases. A post-mortem for diagnostic-pathological and teaching purposes requires the authorisation of the *Shari'a* court. A post-mortem carried out to study the human body (i.e. for teaching purposes) must be approved in writing by the person concerned before death or with the consent of the heirs; this type of post-mortem is also lawful on unidentified corpses or where there are no known heirs or relatives. Art. 7 specifies that male Doctors cannot carry out post-mortems on female corpses except for teaching purposes and when no female doctor is available.

To conclude, regarding the ritual rules for burial, reference to those indicated by the model of the "Last Will and Testament" by the Connecticut Council of Masajid mentioned above may be of use, in the part on "Burial Arrangements", in which the Muslim in the USA requests[131]:

1. My body be buried according to the rules of *Sunnite* Islam.
2. "Under no circumstances may my body be voluntarily turned over for an autopsy, or embalming or for organ donation."
3. My corpse be prepared for burial by *Sunnite* Muslims.
4. No non-Muslim rite or custom be performed at any phase of the funeral and burial. No images, crescents, decorations, crosses, flags, symbols or music of any kind may be present at any time.
5. "My body may not be transported over any unreasonable distance from the locality of death unless necessitated by the circumstances or consensus of my Muslim family members."

---

[129] Sohaibani M.O., Autopsy and Medicine in Saudi Arabia, *Annals of Saudi Medicine*, 1993, 13 (3), 213–214.

[130] Bhootra B.L., Forensic Pathology Services and Autopsy Law of Qatar, *Journal of Clinical Forensic Medicine*, 2006, 13, 15–20.

[131] Connecticut Council of Masajid, op. cit., 3.

6. The grave must be dug according to Muslim custom and "it should face in the direction of the Qiblah (towards the Ka'abaa at Makkah, Saudi Arabia)".
7. My corpse be buried without a coffin or other encasements which separate the shroud from the surrounding earth.
8. The grave be covered only by earth without engravings or symbols, at the most a simple stone.
9. The burial must take place as soon as possible, preferably before sunset on the same day as death or the following day; "under no circumstances should the burial be unduly delayed".

CONCLUSION

The protection, especially psychological, of the seriously or terminally ill patient, even at the price of hiding the truth, seems to prevail in the Muslim world. Some social characteristics may reinforce this protective approach. For example, the differences between the sexes in which the husband or father or a male relative can claim the tradition role of guardians, protectors or representatives of the woman; or the social importance of the family that can become a shield defending the patient in different ways. This protective dimension is accompanied by a doctor–patient relationship which very often remains paternalistic, independently of what is stated by codes and regulations.

The continuous social transformations brought about by modernisation tend to increase the opportunities for information and the freedom of the individual. At the same time, various phenomena including the invasive nature of contemporary medicine and its capacity to prolong a very serious illness (which previously was fatal) for a very long period of time in conditions which at times are pitiful have stimulated, in some countries, the so-called "medical anti-paternalism" aimed at a radical protection of the patient's rights.

These transformations challenge the traditional power of the doctor and require, in Islamic areas as well, a new balance in the doctor–patient relationship which must be inserted into a wider set of social transformations. Classical Muslim medical tradition can provide multiple models of reference to be adapted to the new circumstances.

The Living Will can represent an attempt to impose the will of the individual in controlling the last phases of his life. This situation is little known to traditional cultures and, in the more radical and individualistic forms, does not appear applicable to the Muslim world.

FEMALE GENITAL MUTILATION IN SPECIFIC MUSLIM AREAS

SOME HISTORICAL-JURIDICAL ELEMENTS   305
THE OPINIONS IN FAVOUR OF GENITAL MUTILATION   309
THE OPINIONS AGAINST GENITAL MUTILATION   311
THE DEBATE IN SOME COUNTRIES   316

Female genital mutilation (FGM) is currently present, to a varying degree and in different ways, in at least 40 countries, the majority of which are in Africa (amongst populations of every faith: Muslim, Christian, animist, etc.) as well as in some Middle Eastern and Asian countries (including Muslim countries such as Bahrain, Yemen, Pakistan, Bangladesh, Indonesia and Malaysia) and also Peru, Brazil, Mexico, etc. International organisations claim that there are more than 140 million circumcised women in the world, but this number continues to increase due to the high rate of demographic growth in the countries where it is most practised.

In Muslim countries (e.g. Egypt, Somalia and Sudan), these practices arouse particular interest for two main reasons:
1. The possible juridical–religious justifications for the custom play a significant role combined with the weight of local traditions in favour of it. In non-Muslim areas, on the other hand, local tradition represents the legitimisation for genital mutilation.
2. Due to the serious physical injuries and psychological consequences it produces for women.

SOME HISTORICAL-JURIDICAL ELEMENTS

The expression "female circumcision" still frequently used by the mass media appears inappropriate for the most radical techniques, as it would suggest a female variant of male circumcision. As the latter is generally inoffensive and hygienically beneficial (although divergences in opinion exist in this regard), these characteristics risk being transferred to the female practice. The reality is different in that the extension of these mutilations and their serious psycho-physical consequences (including death) for women make the term of "female genital mutilation" (FGM) more correct and is now adopted internationally to condemn all variants of the operation.

The *Shari'a* plays a fundamental role in guiding human actions, classifying them into five categories: actions can be compulsory (*fard* or *wagib*), recommended (*mandub, mustahabb*), free (*ja'iz, mubah*), reprehensible (*makruh*) and prohibited (*haram, mahzur*). The same act can simultaneously be classified in different ways depending on the context in which the action is carried out.

Dariusch Atighetchi, Islamic Bioethics: Problems and Perspectives.
© Springer Science+Business Media B.V. 2009

In the medical context, three types of FGM are currently distinguished:
1. Ablation of the prepuce of the clitoris and, at times, the tip of the clitoris; for Muslims, this is in general the *sunna* practice, i.e. complying with the Tradition of the Prophet or, according to a different opinion, the custom in His time
2. Ablation of the clitoris (clitoridectomy or excision) and, sometimes, of part of the labia minora and/or labia majora
3. Ablation of the clitoris, labia minora and labia majora and subsequent stitching of the two edges of the vulva, leaving a tiny passage for urine and menstrual blood (infibulation, Sudanese or Pharaonic circumcision)

This triple division often appears theoretical in consideration of the variety of techniques from one population to another or from region to region, according to the age of the circumcised female, her social class, the instruments available, etc.

According to evidence from poetry and the "sayings" (*ahadith*) of Prophet Muhammad, female circumcision (the term most frequently used in classical Arabic for the *sunna* version is *khafd* or *khifad*, i.e. "reduction", which can also indicate clitoridectomy) was a widespread custom in the Arabian peninsula in the pre-Islamic period[1] and in some of the populations who had been won over to the new faith in the first centuries after the death of Muhammad (e.g. present-day Egypt, Sudan and Somalia). Despite conversion, many of these people have kept this archaic custom to the present day and which, in the meantime, has taken on religious values.

As far as the information present in the Sources of the *Shari'a* is concerned, the Koran completely ignores female circumcision. Conversely, there exist indications in some "sayings" or "stories" (*ahadith*) of Prophet Muhammad in collections considered less authoritative or unauthentic. The most important *hadith*, taken from the *Sunna* by Abu Dawud (d. 888), reports the piece of advice given by Muhammad to Umm ʿAtiyya, a "cutter of clitorises" (*muqatti'at al-buzur*): "to be moderate when performing the operation of circumcision on women and to cut off only a small portion of the prepuce of the clitoris, for that is better fitted to preserve femininity and more welcome to masculinity"; then "circumcision is a *sunna* for men and only a *makruma* for girls. Lightly touch and do not wear out (*ashmi wa là tanhiki*). The face will become more beautiful and the husband will be delighted" (Ghazali),[2] where *sunna* refers to an act complying with the example of the Prophet whilst *makruma* refers to a meritorious or noble but not compulsory action.

In consideration of the little attention given to circumcision in the Sacred Sources, Muslim law generally shows little interest in male and female circumcision.

---

[1] There are reports in Saudi Arabia today of the survival of female circumcision amongst some Bedouin of the western provinces.

[2] See, respectively, Giladi A., Normative Islam versus Local Tradition: Some Observations on Female Circumcision with Special Reference to Egypt, *Arabica*, 1997, XLIV (2), 254–267. Bouhdiba A., *La sexualité en Islam*, Paris, Presses Universitaires de France, 1986, 216.

This has not prevented the two practices from having opposite fates. Whilst male circumcision has become a distinctive sign of the Muslim man, FGM remains overall absent, except in specific regions where it is currently the subject of fervent debate.

Despite the absence of FGM in the Koran, there is no shortage of significant positions taken by figures from the past on some aspects of FGM as shown by the case of the Caliph Omar (634–644) when he made a man, who had by mistake cut off part of the penis of a boy during circumcision, pay compensation. The same Caliph requested that a woman practising female circumcision (*khafd*) pay monetary compensation (*daman*) for having erroneously provoked the death of a girl.[3]

The initiators and masters of the four canonical juridical currents of *Sunnite* Islam made pronouncements on the practice, also in reference to specific consequences of a ritual nature.[4] On the level of general evaluation, for Abu Hanifa and Malik (who gave rise to the *Hanafite* and *Malikite* schools) both male and female circumcision are praiseworthy but not compulsory; for Ibn Hanbal (initiator of the *Hanbalite* school) circumcision is compulsory for males but only recommended for women; for Shafi'i (initiator of the *Shafi'ite* school) both types of circumcision are compulsory.[5]

The favourable inclination towards female circumcision finds authoritative supporters in the Muslim Middle Ages. The *Hanbalite* theologian Ibn Taymiyya (d. 1328) judged the uncircumcised woman inevitably a victim of libido, which caused the spread of adultery and prostitution amongst the Tartars and the Franks, contrary to the situation amongst Muslims.[6] In the opinion of this jurist, circumcision (*khitan*) consists of cutting the upper part of the skin which is like a cock's crest.

Ibn Qayyim (d. 1350) took for granted (in the debate between jurists) the agreement on the definition of the practice as *mustahabb* (recommended), whilst he recognised that divergences essentially concerned the decision whether it was obligatory or not.[7] Qayyim added that when a man is not circumcised, the fact is visible and consequently more shameful. Nevertheless, female circumcision remains "a sign of devotion" (*alam ala l-ubudiyya*).

---

[3] Rispler-Chaim V., *Islamic Medical Ethics in the Twentieth Century*, Leiden, Brill, 1993, 67.

[4] For example, references to circumcised male and female genitals appear in discussions on the rules of ritual purification to be performed after sexual intercourse without complete penetration. See Giladi, art. cit., 262 and note 56.

[5] Atighetchi D., Il contesto islamico: problemi etico-giuridici ed il dibattito in Egitto, in Mazzetti M. (ed.), *Senza le ali. Le mutilazioni genitali femminili*, Milano, Franco Angeli, 2000, 41–53. Masry (El) Y., *Le drame sexuel de la femme dans l'Orient arabe*, Paris, Laffont, 1962, 45–46. Anees M.A., Genital Mutilations: Moral or Misogynous? *The Islamic Quarterly*, XXXIII (2), 1989, 101–117.

[6] Michot Y., Un célibataire endurci et sa maman: Ibn Taymiyya (m. 728/1328) et les femmes, *Acta Orientalia Belgica*, 2001, XV, 165–190.

[7] Giladi, art. cit., 263 and 265.

Muhammad b. Rushd, *cadì* in Cordoba and bearing the same name as his more famous grandson Averroes (d. 1198), deemed it compulsory for believers to have their slaves and male sons circumcised whilst he considered the practice *makruma* (recommended) for their daughters as it was an act of purification.[8]

Amongst the opinions of other classic jurists,[9] Ibn al-Jallab (*Malikite*, d. 988) considered circumcision as *sunna* for men and women; for al-Nazawi (*Ibadite*, d. 1162) circumcision is compulsory for men; Ibn Qudama (*Hanbalite*, d. 1223) considers the male practice compulsory and female circumcision meritorious (not compulsory); according to al-Nawawi (*Shafi'ite*) circumcision is compulsory for men and women; al-Musili (*Hanafite*, d. 1284) deemed circumcision a *sunna* practice for men and meritorious for women; according to Ibn Juzay (*Malikite*, d. 1340) male circumcision is a *sunna* practice and not obligatory; al-Mardawi (*Hanbalite*, d. 1480) deemed it compulsory only for men; according to al-Bahuti (*Hanbalite*, d. 1641) the practice is compulsory for men and women.

Amongst the *Shi'ites*, a tradition going back to Ali Ibn Abi Talib, cousin of the Prophet and the first *Shi'ite imam*, considers male circumcision compulsory whereas female circumcision is not; al-Tusi (d. 1067) considered male circumcision compulsory but female circumcision only meritorious; al-Hilli (d. 1277) judged male circumcision compulsory and female circumcision recommended (*mustahabb*); according to al-Amili (d. 1559), it is compulsory for men and recommended for women. The *Shi'ite* Ibn Babawayh (10th century) stated: "there is no harm (*la ba'sa*) if a woman is uncircumcised whereas for a man circumcision is unavoidable (*la budda minhu*)".[10]

According to these opinions, the *Sunni* currents (although differentiated internally) would appear to be prevalently orientated towards three main opinions in reference to female circumcision: (1) a compulsory (*wajib*) practice; (2) a good practice following the example of the Prophet but without being compulsory; (3) a noble or meritorious or free act which confers dignity on the woman.

The theories according to which FGM is a meritorious practice are historically prevalent but are limited to tolerating the *sunna* (lighter) form of mutilation.

There remains one fundamental consideration: the "sayings" of the Prophet do not specify exactly what female circumcision consists of and the descriptions of the ways of performing it, provided by authoritative Muslim jurists over the course of time, generally remain vague. For example, the jurist Shafi'i described it as the ablation of a small portion of skin in the upper area of the genitals.[11] Another *Shafi'ite* jurist, al-Mawardi (d. 1058), held that *sunna* circumcision consisted of cutting the

---

[8] Kister M.J., And He was Born Circumcised. Some Notes on Circumcision in Hadith, *Oriens*, 1994, 34, 10–30.

[9] Berkey J.P., Circumcision Circumscribed: Female Excision and Cultural Accommodation in the Medieval Near East, *International Journal of Middle East Studies*, 1996, 28, 19–38. Abu-Sahlieh S.A.A., *Circoncision Masculine. Circoncision Féminine*, Paris, L'Harmattan, 2001, 168–169.

[10] Giladi, art. cit., 263, note 60.

[11] Anees, art. cit., 112.

epidermis without performing a total ablation whilst Ibn Qayyim stated "If the circumciser extirpates the whole clitoris the woman's passion will weaken and her husband will respect her less."[12] Today, the Egyptian *mufti* al-Sukkari accepts an operation performed by a male or female Muslim surgeon in the presence of the mother or tutor, using more modern instruments to reduce pain. For Jad al-Haq, former *mufti* of the Arab Republic of Egypt in the early 1980s and former *sheikh* of Al-Azhar (who died in 1996), the operation consists of cutting the skin present above the urinary orifice without exaggerating and without extirpating it. There is no shortage of supporters of a symbolic operation consisting of a small cut and a few drops of blood (e.g. in Guinea) in harmony with the words of the Prophet "lightly touch and do not wear out".

However, these are vague descriptions. This fact has contributed to the survival of the practice according to very different pre-Islamic methods although Muslim law formally accepts only the *sunna* version, which is why jurists condemn both excision and infibulation. Behind this facade in Egypt, ablation of the clitoris and of the labia minora continues whilst infibulation or Pharaonic circumcision prevails in Sudan, Somalia, Djibouti, Mali and other countries.

Popular language itself shows the connection between circumcision and religion through the use of the term *sunna*, which indicates both the removal of the prepuce of the clitoris and a practice in compliance with the example of Muhammad. Moreover, although classical Arabic uses the term *khifad* ("reduction") to designate FGM, in colloquial Arabic the word *tahara* is preferred which refers to the state of ritual purity required by anyone about to pray: uncircumcised genitals are impure.

THE OPINIONS IN FAVOUR OF GENITAL MUTILATION

Historically, the first important jurists in favour of FGM based their opinion on two main arguments: the need to curtail female desire and an aesthetic motivation. Today, tolerance or the acceptance of some form of FGM is often based on medical and rational motivations.

The "guaranteed" benefits brought about by FGM appear to be fundamentally three in number, which are also agreed with by religious figures from different religious backgrounds (e.g. Copts or animists in the same geographical areas):
1. It keeps women from carrying out prohibited acts, protecting their morality and honour. This is the most frequently mentioned benefit.
2. Hygiene and elimination of unpleasant odours are ensured by cutting out the clitoris and the labia minora.
3. The operations protect women from the earliest age against nervousness, preserving their sexual sensitivity. The reduction of the female sexual instinct

---

[12] Giladi, art. cit., 264.

may be evaluated positively even in relation to the decrease of the instinct in the husband due to his growing old.[13]

The positions in favour of the custom expressed by highly authoritative bodies and doctors of the Law refer to these three fundamental reasons.

On 28 May 1949, the Egyptian Committee of the *Fatwa* judged refusal of female circumcision a non-sin. On 23 June 1951, the same Committee considered it recommended as it allowed controlling female "nature" and also considered the opinions of doctors on the damage caused by the practice irrelevant, as medical theories on illnesses and their relative therapies change with the passing of time; therefore it is impossible to challenge the words of the Prophet.[14]

In 1981 a *fatwa* by the same Committee (headed by Jad al-Haq) presented female circumcision as an obstacle to perversion and a check on the sexual desires of women, as these desires are stronger today due to free circulation (promiscuity) of the sexes. Similar to what was expressed in 1951, Jad al-Haq opposes doctors who are against the practice as their science changes continuously, unlike the teachings of Muhammad.[15]

It is to be underlined how in the *fatwas* of 1951 and 1981 the value of the *ahadith* is judged superior to the opinion of doctors.

The well-known *sheikh* Yusuf al-Qaradawi prefers to leave the choice to the parents although opting – personally – for recourse to mutilation as it contributes to the morality of the woman. *Sheikh* al-Sha'rawi made a similar pronouncement: "The Prophet has forbidden exaggerated excision and has recommended a very light ablation."[16]

Sheikh Al-Sukkari belongs to the ranks of the more resolute supporters of the operation.[17] In order to obviate the challenged value of the supporting *ahadith*, he had recourse to custom (*urf* and *adat*, subsidiary sources of the *Shari'a*). In his opinion, FGM has never been openly prohibited by the *Shari'a* and has become a rule of conduct as it has been practised for a very long time. In fact, as what is not prohibited is allowed, it follows that it is better to apply conduct that has never been prohibited (FGM) rather than abandon it. Muhammad himself did not make any pronouncements on the harmfulness of the practice, therefore it is illicit to prohibit it; if it were harmful, the Prophet would have forbidden it himself. The injuries and the complications that affect circumcised women today are believed to be derived from a procedure that does not respect the indications given by the *sunna*; vice versa, their respect in past centuries would have prevented the

---

[13] Another consequence of a social order is connected with the possible absence of libido or orgasm in the circumcised woman, namely the use of drugs by husbands to offer performances that can satisfy their mutilated wives. Reports of this were frequent, starting with the famous text by the Egyptian El-Masry Y., *Le drame sexual de la femme dans l'Orient Arabe*, 1962. In this regard, the Cairo magazine *Al-Tahrir* stated on 20 August 1957 "if you want to fight drugs, ban excision".

[14] Abu-Sahlieh S.A., To Mutilate in the Name of Jehovah or Allah: Legitimization of Male and Female Circumcision, *Medicine and Law*, 1994, 13, 575–622.

[15] Rispler-Chaim, op. cit., 90; Abu-Sahlieh, art. cit., 586 and 594.

[16] Chams D., Excision - Retour à la case départ, *Al-Ahram Hebdo*, 9–15 July 1997.

[17] Abu-Sahlieh, art. cit., 582 and 594–595.

negative consequences reported today. Lastly, arguing with those faithful who are against female circumcision, he deems that the real cause of a similar critical attitude lies in the harmful Western influence that is succeeding in imposing its materialistic and secularised point of view on all matters on Muslims.

According to the opinion of the former *sheikh* of Al-Azhar, Jad al-Haq, because of the predominant promiscuity of the sexes girls must be excised, otherwise they are subject to multiple sexual stimuli which lead them to vice and perdition in a depraved society.

The position of several experts on the issue is uncertain. Some consider the practice inseparable from the criterion of beneficiality, according to which God does not impose anything harmful on man, therefore the operation should be carried out if deemed of use; in this case, it would be up to the doctors to make the final decision after having examined each girl (in this way the jurists are freed of all responsibility).

The attitude of the well-known *mufti* Mahmud Shaltut appears complex, according to whom in the sources of the *Shari'a* there are no justifications for male or female circumcision. Islamic law allows certain harmful acts to be carried out when the benefits that derive from them are greater than the disadvantages. Male circumcision is useful in that the elimination of the prepuce represents a hygienic measure that is effective in preventing various diseases (including cancer). On the contrary, female circumcision is devoid of any health legitimisation; therefore it does not appear, from the juridical point of view, to be either compulsory or optional. It can however belong to *makruma* (meritorious) acts for the husband who is not used to feeling the protuberance of the female genital organ (aesthetic argument) whilst for the woman this would be the equivalent of a simple act of bodily hygiene.[18]

THE OPINIONS AGAINST GENITAL MUTILATION

Opponents of FGM refer to four main juridical–religious arguments, although it is difficult – in the case of religious authorities – they define it as explicitly prohibited in any situation; in this they show their discomfort in refusing a tradition that is deeply rooted in the collective mentality and equivocally interwoven with religious values.

1. The practice derives from the local pre-Islamic tradition (e.g. *sheikh* Abbas, Islamic Institute of the Mosque of Paris[19]) and has been inherited by some Muslim countries.
2. The "sayings" of the Prophet in favour of circumcision are not "authentic".

---

[18] Naggar (Al) A., Position de l'Islam vis à vis de la circoncision chez les filles, 4. Abu-Sahlieh, art. cit., 590 and 597.

[19] In the opinion of Abbas, whilst male circumcision (not compulsory) has a hygienic and aesthetic purpose, there is no Islamic religious text of value in support of female excision, as shown by the total absence of the practice in the majority of Muslim countries. When some individuals mutilate, this is due to the survival of customs practised before the conversion of these people to Islam. See Abu-Sahlieh, art. cit., 582.

3. It is illicit to manipulate what has been created in the light of some passages of the Koran (e.g. 4.119 and 32.7). This is the position of Selim al-Awwa, Professor of Muslim Law at the Zagazig University of Cairo, according to whom the *Shari'a* invites Muslims not to manipulate creation. This interpretation allows the violation of physical integrity only for a higher therapeutic purpose (e.g. operations). It has to be noted, however, that those who practise female circumcision themselves deem it a beneficial operation, both morally and physically, when performed according to the *sunna* criteria.
4. It is impossible to think that God orders woman to mutilate a healthy organ that is the fruit of His creation.

The traditional position of Muslim law has been recalled by *sheikh* Abdurahman al-Naggar, one of the *ulama* of Al-Azhar: an action is compulsory only if prescribed by a verse of the Koran or by an authentic "saying" of the Prophet or by the consent (*igma'*) of the community (*umma*) of the faithful. As none of these is the case for FGM, it follows that it is not compulsory. Similarly, in 1946, the *mufti* of Sudan *sheikh* Ahmed al-Taher stated again that terms such as "embellishment", "preferable" and "recommendable" do not imply obligation[20] [but nor do they imply prohibition].

Amongst the most combative opponents of FGM, we can mention the Sudanese *sheikh* H.A. Abu-Sabib for whom (1984) the *ahadith* in favour are unreliable and the Sacred Sources do not impose suffering for an operation that science itself proves to be harmful.[21] Indeed, in his opinion, the duty to protect the integrity of the human being is rendered explicit in the Koran 2.195 "and make not your own hands contribute to (your) destruction" and in the *hadith* "Who harms a believer, harms me and who harms me, harms God."[22]

A fifth argument of great importance put forward against FGM (but not immediately belonging to the juridical–religious objections from which it is distinguished) concerns the serious psychological–health consequences that can occur and this harm is often underestimated by Muslim jurists. It can be classified in at least two groups[23]:
1. Short-term harm caused by mutilation: haemorrhage consequent to injuries of the blood vessels; shock from loss of blood and pain as the operation is performed without anaesthesia; infections (e.g. tetanus) are very frequent; lacerations of other organs (bladder, urethra, vaginal walls, etc.) due to mistakes by the practitioner or resistance by the victim; retention of urine occurs in all small girls in the first few days after the operation due to the pain and

---

[20] Dorkenoo E. and Elworthy S., *Mutilazioni Genitali Femminili: Proposte per un cambiamento*, Roma, AIDOS, 1995, 1–40.
[21] Abu-Sahlieh, art. cit., 598 and 583–584.
[22] Abu-Sahlieh, art. cit., 584.
[23] Rushwan H., Female Circumcision, an Ethical Concern, in *Proceedings of the First International Conference on Bioethics in Human Reproduction Research in the Muslim World*, Cairo, IICPSR, 1992, 65–66.

swelling of the tissues and, in turn, the incapacity to urinate increases suffering and the possibility of infection.

2. Amongst the long-term complications are the hardening of the scars with problems during sexual intercourse and labour; infibulation can cause enormous cysts which have to be removed surgically to avoid infection; the orifices which are too narrow cause the accumulation of urine and menstrual blood with the possible formation of "stones" in the vagina and consequent fistulae leading to incontinence of urine and faeces; chronic infections of the genital organs as the vagina – in the case of infibulation – becomes a semi-closed organ etc. Other possible consequences can affect the psycho-sexual sphere: frigidity, absence of orgasm, depression, absence of libido, psychosis, etc. up to obstetrical complications in infibulated women, who may lose their babies. The earlier a woman is mutilated, the greater the damage she suffers.

The possibility for an excised woman to feel pleasure is compromised in proportion to how radical the ablation is[24]; the surgeon may limit complications avoiding her further suffering but he cannot return her capacity to feel pleasure if the ablation has been radical. In any case, clitoridectomy reduces sensibility but does not make sexual desire disappear as this is a "psychological attribute."

The context in which the excision takes place, i.e. in the village in the presence of relatives and neighbours, may accentuate the feeling of humiliation and oppression in the young girl; an environment that approves the role of the woman as a second-class citizen who has to be mutilated as otherwise she is incapable of controlling her sexual charge and controlling her behaviour.

In the opinion of the Egyptian doctor Nawal al-Saadawi (one of the best known opponents of the practice), the excessive importance given to virginity in Arab societies is the main factor that contributes to the diffusion of female circumcision despite a growing trend, limited to high urban classes, to consider it a harmful custom.[25] Excision represents a "chastity belt" that offers, in exchange, the guarantee of finding a husband.[26] It is men themselves who prefer excised women, fearing that non-excised women are sexually so demanding as to consume their husbands' energy. The presence of mutilating practices in societies and amongst individuals for whom the woman's virginity is an indispensable prerequisite for marriage and where extramarital relations are often very harshly punished is no coincidence.

---

[24] Soad, aged 40, saw her marriage fail as the excision she had undergone as a child made sexual intercourse repugnant to her and therefore she obtained no pleasure. Her husband began to take drugs to boost his sexual performances. It should be pointed out that he was the one who had wanted an excised wife as a guarantee of virginity. Account in Darwich D. and Khalifa D., La blessure de la chasteté, *Al-Ahram Hebdo*, 4–10 February 1998, 28.

[25] Saadawi (El) N., *The Hidden Face of Eve*, London, Zed Press, 1980, 33.

[26] An Egyptian dressmaker wanted to have her 12-year-old daughter, suffering from kidney disease, operated, even at the cost of her life: "I prefer to see my daughter die rather that see her unmarried", account in Darwich and Khalifa, art. cit., 28.

Contrary to common belief, women are often not the passive victims of genital mutilation. The mutilating practice must be assessed in the light of the psychological, cultural, social and religious context in which the values shared by family tradition and local culture are deeply rooted and the woman is often an authoritative representative of these values. In other words, the woman is frequently an active and convinced advocate of traditional practices that are considered necessary instruments to convey positive social values (chastity, purity, cleanliness, etc.). The majority of Egyptian women seem in favour of FGM, and in particular rural women consider it with pride as a necessity.[27] A study of 1978 on 290 women in Mogadishu (Somalia)[28] circumcised at an early age (88% infibulated or excised at the age of 7 on average) has shown this consensus. The motives given to justify the operation undergone were: religious motives (69.5%),[29] remaining virgins to be able to marry (20%), traditional motives (9.5%) and hygienic motives (1%). The same women unanimously declared they wanted to have any future daughters of their own circumcised according to the following practices: *sunna* practice (10%), infibulation (50%), clitoridectomy (40%). To date almost all Somali women appear to have undergone some form of genital mutilation.

The words of Jomo Kenyatta, former President of Kenya and leader of the anti-colonialist struggle after the Second World War, reveal the meanings attributed to the practice in his own tribe, the Kikuyu[30]; these meanings can be found amongst numerous African ethnic groups of Muslim faith (Kenyatta was not, in fact, a Muslim). The President of Kenya, as a strenuous defender of tribal traditions against every attempt at acculturation, declared (in 1938): "Excision and infibulation unite us tightly; they prove our fecundity" and adds "clitoridectomy – as indeed circumcision among Jews – is a bodily mutilation, viewed somehow as the condition sine qua non for receiving a complete religious and moral education".[31] Overcoming the ordeal of pain in the rite of excision (an essential passage in many ethnic groups) represented the viaticum for the "second birth" that allowed the "family spirit" to take possession of the young woman who was entering adulthood.

Where FGM is practised, there is a situation that does not come under the logic of medical ethics. Indeed, the mutilation of a healthy organ becomes the condition

---

[27] Qabil D., Le développement à sens unique, *Al-Ahram Hebdo*, 12–18 June 1996, 5.

[28] Dirie M.A. and Lindmark G., Female Circumcision in Somalia and Women's Motives, *Acta Obstetricia Gynecologica Scandinavica*, 1991 (70), 581–585.

[29] This is the ritual formula that accompanies an infibulation in a Muslim context in Djibouti: The traditional operator says a short prayer: Allah is the greatest and Muhammad is His Prophet. May Allah keep away all evils. She then spreads some offerings on the floor to Allah (split maize or, in the urban areas, eggs). Then the old woman takes her razor and cuts the clitoris. This is followed by infibulation: the operator, with her blade, takes her razor from top to bottom of the small lips and scrapes the flesh from inside the large lip. This nymphectomy and scraping are repeated on the other side of the vulva. In Dorkenoo and Elworthy, op. cit., 6.

[30] Saurel R., L'Enterrée vive: Charcutage et safari au pied du Mont Kenia, *Les Temps Modernes*, 1979, 34 (392), 1352–1372.

[31] Abu Sahlieh, op. cit., 315. Kopelman L.M., Female Circumcision/Genital Mutilation and Ethical Relativism, *Second Opinion*, 1994, 20 (2), 55–71.

for the establishment of a normal social, psychological and spiritual life; in other words, it is preferred to violate the physical dimension for the purpose of protecting the psychological and spiritual dimension of the person (possibility of socialising, getting married, having children, etc.).

Returning to the Muslim world, one of the principles of bioethics and Islamic medical ethics requires the protection of the integrity of the whole person in their physical, psychological and spiritual dimensions. The amputation of an organ or of any of its parts is licit if it is the only way to save the rest of the organ or the person's life, save a different provision of the *Shari'a*. In addition, any invasive operation by medical personnel must be regulated according to the principle of the lesser evil for the purpose of the respect, within the limits of the possible, of the integrity of the individual. In the light of these criteria, it can be stated that, in Muslim areas where FGM is widespread and tolerated by Muslim law, the main obstacle to its perpetuation can be identified precisely in Islamic medical ethics.

However, independent of the cultural context in which FGM takes place, the operation is not considered by the protagonists as mutilation but, on the contrary, as an improvement of the conditions of genitals which are otherwise dirty, disgusting and dangerous, the mutilation of which is to be considered desirable for every woman. Pubic hair is often shaved for the purpose of producing a smooth and "clean" body (in Egypt an uncircumcised girl can be called *nigsa*, i.e. unclean). Very often the community exercises a strong psychological pressure on the girl so that she becomes convinced that her genitals and clitoris are "dirty" or "dangerous" therefore she will feel psychologically gratified by being mutilated and becoming the same as others. The majority of excised women appear to ignore the function of the clitoris, reducing it to an organ devoid of importance. Calling a woman "uncircumcised" is an offence that can affect her children in the future.[32] Failure to be excised or infibulated can prevent the girl from finding a husband, having children and being integrated into social life; the woman is marginalised, scorned and humiliated and excluded from the purposes of her existence: marriage and motherhood. It is precisely the fear of not being able to marry off their daughters that induces mothers to have them mutilated. In Sudan, for example, infibulated women, i.e. the majority,[33] enjoy greater social prestige, whilst the matrimonial dowry from

---

[32] In the Muslim Middle Ages, the expression "son of an uncircumcised woman" (*ibn al-bazra*) could be used as an insult. Precisely to avoid the repetition of this, Halid b. Abdallah al Qasri (d. 743–744), Umayyad Governor of Mecca and then in Iraq, forced his Christian mother to be circumcised. See Giladi, art. cit., 263.

[33] Sudanese circumcision or infibulation deserves some mention as it is one of the most violent techniques. The practice – developed in many ways – mainly consists of the excision of the clitoris, the labia minora and labia majora whilst the orifice may be sewn up with strips of sheep's intestine or thorns or other materials leaving a tiny space for the flow of urine and menstrual blood. The stitching is cut and opened up on marriage to allow penetration by the husband during intercourse. The stitching is also opened up during labour to allow the foetus to be born but the opening is subsequently closed up again (re-infibulation).

the husband is more generous than that given to a non-circumcised woman (virginity becomes a source of income).

Many women, furthermore, psychologically tend to deny even the possibility that the operation can cause serious psychological and health problems. The practice is often carried out in secret in the presence of women only, far from indiscreet eyes (including those of their husbands). There exists an opposite trend that desires to openly show the appreciation and pride once mutilation has taken place, with celebrations and so on.[34] In rural areas, it is often a profession that is handed down from mother to daughter, representing a useful economic resource which contributes to perpetuating the custom.

THE DEBATE IN SOME COUNTRIES

In the Muslim countries where FGM is more deeply rooted, a lively debate is in progress between its supporters and its opponents. We will dwell, in particular, on the Egyptian case which represents a model of the difficulties and problems that face those who would like to see FGM eliminated.

Decree no. 74 of 1959 of the Egyptian Ministry for Health (the result of a heated internal debate) did not prohibit FGM.[35] The text laid down that: (1) doctors were prohibited from performing female circumcision; however, on request, partial circumcision only may be performed; (2) female circumcision is also prohibited in the clinics run by the Ministry for Health; (3) authorised midwives are not allowed to perform any surgical operation, including FGM. In other words, the custom in the *sunna* version was allowed which, due to the reduced possibility of medical intervention, tended inevitably to be performed by unscrupulous "backstreet" practitioners. The critics of the 1959 decree stated that this led to more radical operations such as clitoridectomy being performed, as in fact was the case.

More recently, in September 1994, on the occasion of the International Cairo Conference on "Population and Development", the Egyptian Minister for Health, Dr. Maher Mahran, became the spokesman for a bill to prohibit FGM. Women's organisations supported this bill, enjoying at the same time the support of S. Tantawi, then the Grand *mufti* of the Arab Republic of Egypt, who deemed

---

A more radical closure may be performed on divorcees who "regain" their "virginity" until they remarry. It is superfluous to recall the serious clinical and psychological complications following these customs, which are also performed without the use of anaesthetics. After the operation, the girl's legs are tied together keeping them straight for 10–40 days for the cicatrisation to begin.

[34] See Ezzat D., A Savage Surgery, *The Middle East*, 1994, 1, 35–37.

[35] A commission created on 24 June 1959 by the Ministry for Health gave its opinion shortly before the decree in a similar way. On that occasion, excision practised by a doctor was exclusively accepted, using the prohibition of the religious leaders to practise total excision of the external genital organs as it was harmful both before and after marriage and extraneous to the rituals of Islam. See Abu-Sahlieh, art. cit., 602 and note 173.

weak, defective and unreliable the "sayings" of the Prophet tolerating FGM.[36] In a document, signed with other Egyptian personalities (including the Minister for Health) on 9 October 1994, Tantawi denied the lawfulness of the practice but, at the same time, "tactically" delegated all decisions to doctors, for the purpose of avoiding a conflict with the supporters of the practice. In addition, the document maintains that the laws currently in force, which prohibit non-doctors from performing surgical operations, are sufficient to contrast those who practise circumcision illegally.

Maher Mahran's bill was blocked by strong opposition from certain Parliamentary circles inclined to follow the indications proposed in October 1994, immediately after the Cairo Conference, once again by a *fatwa* by Jad al-Haq, *sheikh* of Al-Azhar, who presented circumcision as being good for women, stating "if a district stops, by common agreement, practising [male and female] circumcision, the Head of State declares war on it because circumcision is part of the rites of Islam . . . . This means that male and female circumcision are compulsory".[37]

After a few initial uncertainties, the next Minister for Health, Ali Abd al-Fattah, issued a decree on 19 October 1994 that allowed mutilation exclusively in public hospitals, but not in private ones, in order to guarantee medical control.[38] According to these measures, both male and female circumcision were to be performed exclusively by doctors in hospitals. In the latter case, a commission including a doctor and a religious expert were to explain to parents the risks connected with mutilation in order to dissuade them after which, if the parents did not change their minds, the doctors were to operate respecting health precautions. However, it appears that the "commission" to dissuade parents from the operation was absent from the hospitals.[39] In other words, after 40 years, FGM was re-introduced into hospitals.

As the practice involves the majority of Egyptian girls, those supporting the new decree maintained that medical supervision of the operation represented the only realistic approach to try and limit the phenomenon. According to this approach, it is impossible and counterproductive to prohibit it explicitly (e.g. Mahmoud Korayem, Professor of Obstetrics and Gynaecology at Ain Shams University) as it would encourage illegal practices whilst, on the contrary, the presence of a doctor would limit the damage.[40]

Conversely, those contrary to the decree of the Minister for Health stated that the government position was too soft. It represents a legitimisation of the operation precisely by the doctors for whom it was more difficult to operate

---

[36] Ezzat D., Female Circumcision contested again, *Al-Ahram Weekly*, 20–26 April 1995.

[37] Abu-Sahlieh, op. cit., 415.

[38] Abu-Sahlieh, op. cit., 413–414; Howeidy A., Divided over Female Circumcision, *Al-Ahram Weekly*, 3–9 November 1994.

[39] See Qabil D., Le développement à sens unique, *Al-Ahram Hebdo*, 12–18 June 1996, 5.

[40] Howeidy, art. cit.

according to the measures of the law of 1959. However, the non-governmental organisations (NGOs) advised against a direct confrontation on the subject with the major Muslim and Copt theologians (who were also prevalently in favour of the practice) as the attention devoted to mutilation by the mass media (especially Western) could induce many men of religion to opt for the most extreme positions in harmony with those of Jad al-Haq; for them the attack on a deeply rooted tradition in a Muslim society is the equivalent of attacking Islam. In fact, the widespread and constant historical presence of circumcision in certain regions associated with its tolerance by jurists is interpreted as the existence of the fundamental consent of the community (*igma'*) such as to justify female circumcision; this is the theory of Sukkari. One example was given by the Cairo magazine *Al-Itissam* when in December 1980 it protested against the undue interest by the WHO in the practice, inviting the jurists of Al-Azhar to be wary of ideas from abroad in order to fight them and prove their falsity, protecting Muslim customs.

The position of doctors has never appeared clear. It is true that many of them refuse to operate, considering that mutilation is prohibited by religion, also referring to arts. 241 and 242 of the Egyptian Penal Code, which punishes all voluntary physical injury. At the same time, it is well known that circumcision is often performed for payment by doctors and medical students instead of the classic "matrons". With great perspicacity, N. Saadawi recalls that the medical profession is practised in the respect of the moral and social values characteristic of a certain society; therefore, it becomes inevitable that the profession is used to defend and perpetuate these values. This means that many doctors share (perhaps implicitly) the conviction which is deeply rooted in local culture according to which removal of part of the female genital organs reduces sexual desire in favour of female virginity, the honour of the family and, in short, to the advantage of the whole of the Muslim community.

A demonstration of this thesis can be found in the conclusive document of the First International Conference on Bioethics in Human Reproduction in the Muslim world, which was held at the University of Al-Azhar in 1991 with the participation of many authoritative doctors and jurists. The mention in the conclusive guidelines of FGM presents an unclear attitude as it states that "some types of extensive female circumcision are harmful to health and progeny and may cause psychological and marital problems"; regarding these techniques, the document invites acting in order to eliminate them.[41] This appears to mean that the practice is not condemned in all its variants, not even in all the most invasive and extensive ones.

Another document in this direction is that signed on 25 October 1994 by 15 authorities including the Chairman of the Union of Egyptian Doctors, the Secretary of the Board of the Union, the Chairman of the Committee of Ethics of the

---

[41] Serour G.I. (ed.), *Ethical Guidelines for Human Reproduction Research in the Muslim World*, Cairo, IICPSR, 1992, 23.

Union and other university professors. The text[42] recalls that "female circumcision is part of the rites of Islam and Muslim law prohibits total ablation" (art. 10); the promulgation of a law prohibiting female circumcision is refused (art. 2); it is recalled that in correctly performing the operation, the "surgical, professional and religious" laws must be respected with the aim of "limiting oneself to levelling the protruding part without exaggerating", or touching the labia and the clitoris only with moderation (art. 1 c). Article 10 expresses hope for the application of Ministerial Decree no. 74 of 14 June 1959.

In respecting the Ministerial Decree of 19 October 1994, the circumcision operation by Egyptian doctors opened up a serious ethical problem. The operation respected the requests of the parents, applying the decree, but infringed the classic principle of medical ethics whereby the patient must not be harmed except for a licit and superior therapeutic reason; the dilemma was aggravated by the intentions of the parents who, in the face of the possible refusal by doctors to operate (but also following operations performed by doctors which are not considered regular by the parents), may turn to unscrupulous private practitioners subject to no control but only with the desire to earn money.

During the Cairo Conference of 1994, the Arab Republic of Egypt also signed the international agreements against any form of discrimination towards women and to fight torture; the more or less "tactical" tolerance shown regarding FGM could be interpreted as a betrayal of this solemn commitment.

Following the internal and international protests after the decree of 19 October 1994, which allowed FGM in public hospitals, on 17 October 1995 another decree by the Minister al-Fattah prohibited female circumcision in public hospitals.

On 8 July 1996,[43] the Minister for Health Dr. Ismail Sallam issued Decree no. 261 which prohibited all forms of FGM in all hospitals or clinics, whether private or public, except for pathological clinical cases; the infringement of the decree by non-physician was prosecutable under the law.[44] This was the first official measure against excision after the government decree of 1959.

Critics of the decree of the Ministry for Health (8 July 1996) have stated that the ruling does not possess the value of a law and inevitably encourages the illegal and uncontrolled development of the practice. A law of the Republic still appeared a far-off dream for many. Confirmation came from Amina al-Guindi, Secretary General of the National Council for Maternity and Infancy, according to whom all the attempts to include an article prohibiting FGM in the Bill on the rights of childhood were aborted due to the opposition of the majority of members of Parliament.[45]

---

[42] Abu-Sahlieh, op. cit., 414–415.

[43] I refer here to Atighetchi, *Il contesto islamico: problemi etico-giuridici e il dibattito in Egitto*, art. cit., 50–52.

[44] *International Digest of Health Legislation*, 1997, 48 (3/4), 318; Ezzat D., Government Ban on FGM, *Al-Ahram Weekly*, 25–31 July 1996.

[45] Ezzat, *Government ban on FGM*, art. cit.

Hoda Gad, lawyer for the NGOs, made the bitter comment that a similar law makes sense only if accompanied by an awareness-raising campaign for the public also involving husbands (often in favour of mutilation) whilst, because of how the situation was evolving, "we are sure of the development of a black market of excision" which had become well developed since 1959.[46]

The decree should not create any illusions on an immediate efficacy, especially due to the extremely widespread belief in Egyptian society at every social and cultural level according to which FGM contributes to controlling female sexuality preventing premarital sexual intercourse and promiscuity and thus becoming an instrument of equilibrium and social morality and enjoying the often unconditioned favour of women themselves.[47]

In June 1997, the Council of State had recommended the annulment of the decree of the Ministry for Health of July 1996. According to the Council, as the rite was not yet prohibited by law, circumcision was licit, without being compulsory, whilst the choice has to be left to the family.

A group of citizens led by *sheikh* Youssef al-Badri and Mounir Fawzi, Professor of Gynaecology at Ayn Shams University, petitioned the Administrative Court of Cairo against Ministerial Decree 261 of July 1996. The three reasons for the appeal clearly appear to be religious[48]: (1) Ministerial Decree no. 261 infringes art. 2 of the Egyptian Constitution according to which the *Shari'a* is the main source of legislation; (2) the decree ignores the consent of the doctors of the Law on female circumcision as a legitimate practice as it is compulsory or recommended by the "sayings" of the Prophet; (3) the Government does not have the power to modify a passage from the Koran or a compulsory or recommended *hadith*.

Following the petition, on 24 June 1997 the Administrative Court again authorised the practice of excision in Egyptian hospitals, inverting the previous decree

---

[46] Youssef Y., Les craintes d'un marché noir, *Al-Ahram Hebdo*, 28 August–3 September 1996. A nurse at a private hospital in Asyut reports that, despite the new decree, the practice is carried out under the eyes of the owner of the hospital without him having the courage to protest "otherwise there would be an uprising!". A citizen of Oum Gomaa deems the anti-excision decree as bizarre whilst doctors and midwives continue to operate without increasing the fees. Another citizen of Aswan said: "This is our custom and habit, we won't change them because those in power want us to. And there's no control by the State to be worried about." Amnesty International reported that Amira Mahmoud Mohammad (aged 4) and Warda Hussain al-Sayyid (aged 3) had bled to death in October 1996 after the FGM performed at home by a state doctor in Armant, a small town in Upper Egypt, see Amnesty International–Report-Mde 12 November 1997, *Egypt: Women Targeted by Association*.

[47] Several women made lapidarian statements when interviewed by Egyptian newspapers in reference to the decree prohibiting FGM in hospitals. Amongst them, the nurse Oum Gamal said: "We cannot apply this ban. We have to control the feelings of the woman and guarantee her chastity." Oum Essam, office worker, commented: "God be praised that I had my granddaughter operated on before this decree" replying to Tantawi that religion has always been there so there is no reason to challenge excision which has always been accepted. See Qabil D., Excision. Une décision en attendant l'application, *Al-Ahram Hebdo*, 31 July–6 August 1996. A male shopkeeper said that girls always have to be circumcised, otherwise they will be born like men and will not be able to find husbands.

[48] See Dupret B., Sexual Morality at the Egyptian Bar: Female Circumcision, Sex Change Operations, and Motive for Suing, *Islamic Law and Society*, 2001, 9 (1), 42–69.

of prohibition of the practice by the Minister of Health, Ismail Sallam. According to the Court, it was prohibited to forbid a practice that was religiously lawful but not compulsory. In addition, as the practice of circumcision was not yet illegal, the Minister for Health did not have the authority to prohibit it, as only the Parliament had this authority. In other words, the Minister was alleged to have overstepped his prerogatives as it was an issue to be solved legislatively and not through a ministerial decision[49] (until then no law prohibiting FGM had existed).

This decision can be considered as a temporary victory of a heterogeneous coalition between "religious fundamentalists" such as *sheikh* Youssef al-Badri (who celebrated saying "God be praised, we have won and we will apply Islam again"[50] or "This practice is an integral part of Islam. We cannot reject it just because some Western people say it is barbaric. We have to have all Muslim women circumcised"[51]) and "reactionary" doctors such as the gynaecologist Mounir Fawzi. The aim of the coalition was to legalise FGM in all public and private hospitals. Whilst Badri considers excision recommended by a *hadith* but not compulsory, Fawzi defines excision a "hygienic" practice with the benefit of limiting the sexual desire of women to a reasonable level. For both, circumcision, if adequately practised, is a very simple type of operation which leaves the clitoris fundamentally intact whilst the measure prohibiting doctors from operating would only encourage more radical techniques to the detriment of women.[52]

In the meantime, the weekly magazine *Al-Mussawar*, dedicating a dossier to the issue of FGM, revealed that the Grand *mufti* of the Egyptian Republic, Nasr Farid Wassel, had joined the long list of authorities in favour of circumcision together with *sheikh* al-Sha'rawi who specified literally: "The Prophet prohibited exaggerated excision and recommended a very light ablation, maintaining the honour of the woman. And this is because she must often be solicited by her husband and not vice versa. She must not let herself by carried away by lust."[53]

Certain of the final victory, *sheikh* al-Badri revealed his next step: to use the Administrative Court ruling against the Ministry of Education to have any mention of the negative impact of FGM on female health eliminated from textbooks, replacing them with more "correct" teachings on the "positive" aspects of the operation.

The reactions of the opponents of FGM were obviously bitter and disillusioned and they planned to appeal to the Supreme Administrative Court whilst new deaths of mutilated girls were reported in the press.[54] Amongst the reactions, we can quote that of Mohamad Abdel-Moneim, editorialist of *Al-Ahram*, who in

---

[49] Chams D., Excision – Retour à la case départ, *Al-Ahram Hebdo*, 9–15 July 1997.

[50] La Torture de l'excision, *Le Monde*, 26 June 1997, 17.

[51] Ezzat D., Court upholds Genital Mutilation, *Al-Ahram Weekly*, 26 June–2 July 1997.

[52] Lancaster J., Egyptian Court Overturns Decree that Banned Female Circumcision, *The Washington Post*, 25 June 1997, A 26.

[53] Chams, *Excision – Retour à la case départ*, art. cit.

[54] Like the case of 13-years-old Amal, who died during excision due to a mistake by the anaesthetist in the clinic of Dr. Mohamad Abdel-Wahab in Manchiyet Nasser. In the working-class neighbourhood, Al-Chaféi, the barber, continues to mutilate little girls every day, preferably when they are 6-years-old: orphans and blind girls are operated on free of charge.

reference to mutilation declared "We cannot get rid of it. But apparently, the affair goes far beyond ritual mutilation. It is not simply a question of excising girls but excising brains"[55] in reference to the responsibility of adult circumcisers.

The most recent turning point of the issue took place on 28 December 1997 when the Supreme Administrative Court (against the decisions of which appeal cannot be made) confirmed the decree of the Ministry for Health of July 1996, prohibiting all types of FGM as it is a practice contrary both to the principles of the *Shari'a* and positive Egyptian law. The Court has made the distinction devised by the Supreme Constitutional Court between[56]: (a) principles of the *Shari'a*, the origins and interpretations of which are fixed and absolute, and (b) principles open to interpretation. The legislature may intervene only on the latter, with recourse to the power of interpretation. The ruling also states: "Female circumcision does not represent a right to be exercised by anyone over anyone else, as it is neither a duty (*fard*) nor an obligation (*wajib*) according to its ruling (*hukm*) under Islamic law. In contrast, the majority of the medical profession considers . . . it a detrimental act to women, inflicting substantial damage which cannot be permitted except for the sake of medical treatment. And according to Islamic jurisprudence law . . . there is the firm principle: *lâ darar wa lâ dirâr* [no damage and no infliction of damage]."[57] In addition, the measure prohibits the practice even when there is the consent of the girl or her parents; moreover, the "mutilating" doctor is liable to imprisonment and disqualification from the medical profession. Barbers and midwives risk two-years imprisonment. This is to date the most important ruling issued in Egypt on the question and may represent a real turning point until a State law is passed that prohibits all forms of FGM without compromise. The weak point of the prohibition lies in the possibility of the doctor performing a circumcision when he deems it "necessary". This clause has been criticised as a concession to the trends in favour of FGM. Amongst the confirmations, the gynaecologist Mounir Fawzi has openly declared that he would continue to perform circumcisions having recourse precisely to the necessity for medical reasons.

In June 1999, the Egyptian Gynaecological Society intervened on this latter aspect to specify that any "medical necessity" is in fact almost non-existent.[58] In fact, congenital deformation or hypertrophy of the clitoris (which some believe inevitable if circumcision is not performed) is extremely rare and can be cured without clitoridectomy. The supporters of FGM replied that the practice has a cultural and religious value that cannot be determined or challenged by a State court of law or by a government.[59] *Sheikh* Tantawi subsequently stated

[55] Chams, *Excision – Retour à la case départ*, art. cit.

[56] Dupret, art. cit., 55–56.

[57] Bälz K., Human Rights, the Rule of Law and the Construction of Tradition. The Egyptian Supreme Administrative Court and Female Circumcision (Appeal no 5257/43, 28 December. 1997), *Égypte/Monde Arabe*, 1998, 34, 141–153.

[58] Digges A., Circumcisors Circumvent the Law, *Cairo Times*, 6–19 August 1999, 2 (12).

[59] Chelala C., Egypt takes Decisive Stance against Female Genital Mutilation, *The Lancet*, 10 January 1998, 351, 120.

once again the absence of a direct and clear relationship between FGM and the *Shari'a*. The organisations involved in the fight against FGM consider the ruling of the Supreme Administrative Court without simple illusions: it is an important symbolic success but the real victory passes only through the work of conviction and personal dissuasion of women in their recourse to mutilating practices, and this challenge will take a very long time.

Despite the unexpected defeat, *sheikh* al-Badri commented on the ruling of December 1997 with sarcasm: "The judge is a man and can make mistakes" adding "This case is not only mine but belongs to all Muslims. I will continue to fight with legal means for the defence of Islam."

In June 1998, Mohamed al-Chahawi, Deputy in the People's Assembly, presented a bill to put an end to excision, a practice that continues to be rampant. According to the bill (which has aroused considerable interest),[60] each girl's medical booklet will have to show whether she has undergone excision or not. In the former case, the parents and the operator would be punishable by sanctions varying from a fine ranging between Egyptian pounds 1,000 and 5,000 and 6-months imprisonment. Each girl ought to be examined each year at the beginning of the school year until she is 15, in order to avoid any infringement. The medical booklet should be presented on the day of marriage before the signature of the contract in order to allow the future husband the right to accept or refuse an excised woman.

Although the majority of circumcisions are performed in private homes, the death of a young girl on 18 July 1999 confirmed once again[61] the complacent role of doctors. In this case it was a surgeon (the director of a private clinic), who operated thanks to the leeway granted by the measure of December 1997 (the circumcision for medical reasons).

Encouraging signals come from the world of university students according to a survey carried out through interviews of 1,020 students at Cairo universities.[62] The wide sample included an equal number of medical students and students in other subjects as well as an equal number of males and females. Out of all the

---

[60] Ismail G., Le livret medical contre l'excision, *Al-Ahram Hebdo*, 3–9 June 1998.

[61] A mother went to the director of a private hospital in Salaam City to have her 11-year-old daughter, Mona Abdel Hafees, and her two nieces circumcised. The director-surgeon accepted operating them the same evening for Egyptian pounds 80 each. The director and a second surgeon began the operation on the first niece without an anaesthetic, but following the girl's screams of pain she was given an injection afterwards. The second niece was given an anaesthetic before the circumcision. In the case of Mona, the director said that the girl had not reacted to the first anaesthetic; therefore, she was injected with a second dose. An hour later she was transferred to another hospital in Cairo. After the threats of the mother to destroy the waiting room of the hospital if she was not told the truth, the director-surgeon admitted Mona's death. The Ministry ordered the closure of the hospital of Salaam City for infringement of the prohibition of circumcision whilst the public prosecutors gave bail to the two surgeons responsible for the girl's death pending the conclusion of the investigation. See Digges, art. cit.

[62] Allam M.F.A., Irala-Estevez J., et al., Students' Knowledge of and Attitudes about Female Circumcision in Egypt, *The New England Journal of Medicine*, 1999, 341 (20), 1552–1553. (Letter).

students, 734 (72%) said they were in favour of the abolition of FGM (the percentage of women and medical students was higher). The majority of them also said they were sure that the practice does not have religious grounds and that circumcision is not useful to preserve chastity or to protect hygiene.

A more recent survey on a larger sample appears to confirm a limited decline in female circumcision in Egypt alongside an increased direct involvement of doctors and paramedical staff in operations instead of traditional practitioners.[63]

In Sudan,[64] three main types of FGM are practised: infibulation (Sudanese or Pharaonic circumcision), the oldest and most widespread in the country; circumcision (*sunna*), the least invasive; an intermediate form[65] between Pharaonic and *sunna*. The age range of the girls is between five and nine. Anaesthetics and the use of antibiotics are fairly frequent in the cities. In Sudan the fight to abolish FGM dates back to the early decades of the 20th century, involving British and Sudanese administrators, men of religion, politicians, etc.; however, the results have been disappointing as 80% of women are sexually mutilated and the tradition continues. Nevertheless, limited to groups of cultivated women in an urban milieu, there is a shift from Pharaonic circumcision to the *sunna* and intermediate versions. Even more than in Egypt, the practice of circumcision in Sudan appears directly linked with the literacy level of the subjects involved.[66] The reasons that explain the resistance of FGM to any modification include the classic beliefs we have already seen: circumcision decreases the female sexual desire; a circumcised woman is "clean"; men want to marry infibulated women as the narrow orifice procures greater pleasure; Islam requires it despite the continuous denials by religious leaders (e.g. *sheikh* Abu Sabib) although their prohibition appears to be limited to the forms of extensive excision and infibulation. Similarly, in 1946, the *mufti* of Sudan *sheikh* Ahmed al-Taher stated that terms such as "preferable" and "recommendable" do not constitute an obligation to mutilate.

In 1943, the Governor-General of Sudan created a Health Committee to study the phenomenon which materialised in a bilingual publication stating (with the support of the men of religion) that FGM should be abolished due to its harmfulness.[67] In 1946 (under a colonial government), legislation prohibited Pharaonic

[63] Gibaly (El) O., Ibrahim B., et al., The Decline of Female Circumcision in Egypt: Evidence and Interpretation, *Social Science and Medicine*, 2002, 54, 205–220.

[64] Nagar S.E.H., Pitamber S., et al., *Synopsis of the Female Circumcision Research Findings*, Babiker Badri Scientific Association for Women Studies, Sudan, 1994, 1–44, in particular 8–12. (Photocopy).

[65] This is a method that spread as a consequence of the 1946 Law which prohibited Pharaonic circumcision and by which the aim was to increase the *sunna* version (the least harmful). However, due to the practical difficulty of abandoning the Pharaonic version, this "intermediate" version consisted of the ablation of the labia minora and part of the labia majora. The subsequent stitching can vary in the space left open.

[66] See amongst others Herieka E. and Dhar J., Female Genital Mutilation in the Sudan: Survey of the Attitude of Khartoum University Students Towards this Practice, *Sexually Transmitted Infections*, 2003, 79, 220–223.

[67] Dorkenoo and Elworthy, op. cit., 27.

circumcision, prescribing prison and fines for those found guilty. Some interpreted this act as interference by the British on local practices and values; the protests raised by the first arrest under the application of the law made it practically inapplicable. Another intrinsic weakness in the prohibition lay precisely in prohibiting only Pharaonic circumcision, which allowed performing other practices (e.g. infibulation). Article 284 (A) of the Penal Code of 1974 authorises FGM in the *sunna* version. In 1977, the 5th Congress of the Society of Sudanese Gynaecologists requested the abolition of all forms of FGM. In 1978, the Faculty of Medicine together with the Ministry for Health and the WHO started a research project from which it appeared that 90% of the women studied had been circumcised. In 1979 at a seminar of the WHO on the subject, Awatif Osman, Director of the Nurses' College of Khartoum, recalled how in previous years *sunna* circumcision was taught in nursing schools in order to impose the least pain with antiseptic methods. Since 1979, schools no longer teach the technique whilst obstetricians are advised not to practise it.[68] In 1988, the Ministry for Internal Affairs decreed the creation of the National Committee for Eradication of Female Circumcision. Further initiatives include programmes educating on FGM by the Psychiatric Health Association and the National Population Committee. All the programmes and the initiatives against FGM, independently of who launches them, are minimised or neutralised by the military and economic crisis that has swept across Sudan and several other African states (e.g. Somalia). The Sudanese Penal Code of 1991 does not explicitly prohibit FGM.

The situation in Indonesia, the country with the largest Muslim population in the world (more than 200 million inhabitants, 90% of whom are Muslims), is not well known and not very clear. One of the most recent studies[69] shows that amongst Muslim communities, female circumcision is widespread and perceived as a traditional custom but also as a religious duty. The prevailing juridical school of thought in the country is the *Shafi'ite* one, which traditionally considers circumcision a compulsory practice (*wajib*) both for males and for females. Some studies consider that female circumcision entered Indonesia with the arrival of the Muslims as these practices are absent in the areas of non-Muslim ethnic groups. The growth of Islamic fundamentalism appears to be accompanied by an intensification of widely varied techniques of FGM due to the religious legitimisation.

Until 2003, the authorities made no move to officially hinder the practice. About 92% of the families taking part in the survey said they were in favour of circumcising their daughters; more than two-thirds of the mothers consider the practice beneficial whilst only 5% considered it harmful. The study shows that distribution of female circumcision in the various areas, by the age of 18, oscillates from 86% to 100% of girls. According to the four categories of FGM drawn up by the

---

[68] Dorkenoo and Elworthy, 28.

[69] Budiharsana M., Amaliah L., et al., *Female Circumcision in Indonesia*, Jakarta, Population Council Jakarta – USAID, September 2003.

WHO in 1998,[70] the practices of female circumcision most widespread in Indonesia belong to the fourth category which groups together practices which vary greatly from one to another: these range from bloodless and symbolic techniques to actual excision.

Amongst the 2,215 cases studied, 68% of the operations were performed by traditional providers whilst the remaining 32% were performed by health care providers (HCP), especially midwives.[71] In fact, medicalisation of the practice is extending (including in hospitals) by medical practitioners (mainly midwives) based on the false idea that their operations are safer (from the health point of view) than those of the traditional practitioners. Reality would show that the opposite is true as midwives and HCPs in the majority of cases practise some form of excision and incision of the clitoris using scissors; on the contrary, the traditional providers prefer to use penknives in more symbolic actions of scraping, rubbing, pricking or piercing on the external genital organs.[72]

Lastly, FGM is prohibited in many states with a strong Muslim component such as Burkina Faso, the Central African Republic, Djibouti, Ghana, Guinea, Togo and the Ivory Coast. Senegal (a nation with a large Muslim majority, like Burkina Faso, Guinea and Djibouti) has joined this list and, in 1999, approved a law prohibiting FGM, thanks to the financial support provided by UNICEF to local NGOs. The law lays down heavy fines and six-months imprisonment for infringements.

---

[70] The first three categories are: type 1: excision of the prepuce, with or without partial excision of the clitoris; type 2: excision of the clitoris with total or partial excision of the labia minora; type 3: partial or total excision of the external genitals and sewing up of the vaginal opening (infibulation). Budiharsana – Amaliah, op. cit., 3 and 6.

[71] Budiharsana – Amaliah, op. cit., 25.

[72] Budiharsana – Amaliah, op. cit., 30, 40, 42.

THE KORAN AND MODERN SCIENCE

INTRODUCTION 327
SCIENTIFIC EXEGESIS OF THE KORAN 332
MODERATE CONCORDISM 345
THE OPPONENTS OF SCIENTIFIC EXEGESIS 347
CONCLUSION 349

INTRODUCTION

The fundamentals of Islam can be summarised as the faith in one, single and transcendent God and in His absolute freedom and omnipotence. Islam can also be defined as "Koran-centric" due to the unique role played by the Koran, which is the direct and literal Word of God and the first source of Muslim Law, in the mental and social universe of Muslims. Muslim theocentrism has greatly influenced the characteristics and aims of science which have been elaborated over the centuries by Muslim scholars.

At the same time, the interest of the Muslim Holy texts (Koran and *Sunna*) in science is shown by the very word of God as at least 750 verses of the Koran (about one-eighth of the total) encourage the study of nature with references of a cosmological, physical, biological, etc. order; an example of this is Koran 3.190 "Behold! In the creation of the heavens and the earth, and the alternation of night and day, – there are indeed Signs for men of understanding" (or 50.7; 88.18–21; 29.20; 51.20–21). Famous "sayings" (*ahadith*) of Prophet Muhammad express similar thoughts, for whom increasing knowledge is compulsory for Muslims who are invited to seek it "even as far away as in China".

According to a very common Islamic interpretation, science aims to demonstrate the unity and the coherence of everything that exists in order to guide man to venerate the creator of this unity, namely God.[1] The unity of creation, which is continuously evoked, is accompanied by the emphasis on its intrinsic harmony, avoiding radical dualisms (of a philosophical and mystical origin), ranging from that of the soul-body of man, to that between the microcosm and macrocosm and so on. The unification of all the sciences forms the path to the discovery of universal harmony and approaching the Creator. This unifying vision of reality is then translated into a system of thought where theology, philosophy and science try to meld flexibly, avoiding relentless conflicts between the different branches of knowledge. The consequence is the refusal of the idea of a real opposition between science and religion as a "typically Western" contrast

---

[1] Nasr S.H., *Scienza e civiltà nell'Islàm*, Milano, Feltrinelli, 1977, 17–18.

Dariusch Atighetchi, Islamic Bioethics: Problems and Perspectives.
© Springer Science+Business Media B.V. 2009

and which is extraneous to the totalising and all-inclusive character of Islam. In this regard, it is often recalled how Islam has never known a case such as that of Galileo Galilei as any persecutions, excommunications or similar came from doctrinal rather than scientific differences. Spiritual and temporal are two faces of the same coin and the concept of immanent uniqueness (*tawhid*) is at the heart of Muslim epistemology; "while religion and science are two different epistemic categories in the Western mind", for Muslims these are "a continuum complementing each other".[2]

According to a 1982 study by Mümtaz A. Kazi – later President of the Islamic Academy of Sciences (Amman, Jordan) and Coordinator General of the Organisation of the Islamic Conference (OIC) Standing Committee on Scientific and Technological Cooperation (COMSTECH) – the glorious scientific period from the 8th to the 12th century was mainly the result of the religious impulse for the study of nature and empirical research. Muslim scientists and philosophers accepted the principles of the faith and their research, pervaded by moral zeal, was carried out for the benefit of humanity and for spiritual ends.[3] Science and religion operated in symbiosis. Everything changed with the renaissance of science in Europe. For internal social reasons, the religious-secular separation brought about "value-neutrality of science" with consequent moral relativism and ethical anarchy. The conflict between Church and Science "has led Western science to arrogate to itself an exclusive prerogative to apprehend truth and reality, thus reducing man's capability to develop an integrated appreciation of reality and react to it rationally and holistically".[4] The great scientific and technological progress of the West has produced a "value-neutral" society without morals.

The encouragement to observe nature and increase knowledge is not an end in itself. In the first place, recourse to rationality and the senses would lead "spontaneously" to appreciating the wonders of creation and of the Creator. Scientific research becomes a basis to accept divine existence.[5] In addition, rational activity with its conclusions is legitimised only if subordinate to the Revelation and finalised to a greater knowledge of God. It is no coincidence, in the analysis of the ayatollah Amuli, that the Revelation and demonstrative reasoning are complementary; what is discovered by reason is supported and confirmed by the Revelation, whilst what is inaccessible to human reason has been revealed to us. The Revelation stimulates and guides human reason in its inferences but its

---

[2] Anees M.A., *Islam and Scientific Fundamentalism. Progress of Faith, Retreat of Reason?*, in www.ummah.org.uk/bicnews/Articles/science.htm.

[3] Kazi M.A., Islamization of Modern Science and Technology, in *Islam: Source and Purpose of Knowledge*, Proceedings and Selected Papers of Second Conference on Islamization of Knowledge, 1982, International Institute of Islamic Thought, Herndon, Virginia, 177–186.

[4] Kazi, art. cit., 180.

[5] Nadvi M.S., Rise and Fall of Muslims in the Realm of Science (I), *The Muslim World League Journal*, 1999, 26 (11).

teachings are not exhausted in the "horizontal dimension" of phenomena but, on the contrary, they are inserted into a transcendental context referring them to the origin of all phenomena (i.e. to God) and their ultimate end (God again).[6] As summarised by Ibn Hazm (994–1064) faith and intellect participate in the enterprise of approaching God although they start off from different angles.

Some thinkers (such as the Egyptian "radical" Wahid ad-Din Han) appear less open to the possibility of reason, since the Koran establishes two rules: (1) man must believe in the existence of supernatural realities but cannot reach the truth through direct observation as suggested by Koran 2.2–3: "This is the Book; in it is guidance sure, without doubt, to those who fear God; who believe in the Unseen." (2) Man must recognise his own incapacity to discover the laws of existence.[7]

This set of conceptions is more comprehensible when we consider that in general Muslim culture and science prefer to refuse the reduction of natural phenomena to events that are not connected to the superior orders of reality. Evidence in this sense could be attributed to the interest shown by the Arab-Muslim civilisation in the study of mathematics: this passion was aroused by its natural function of acting as a bridge between the multiplicity of what is created and divine unity. Absolute divine omnipotence is translated into the refusal of the "causae secundae" which was agreed with by the most widespread theological current, the ash'arite one,[8] according to which every relationship of cause-effect in nature is the result of the action of a single motor, namely God. The very existence of immutable natural laws limiting divine freedom can be challenged.

The quest for a causal explanation rarely attempted, and in any case unsuccessfully, to find satisfaction outside the faith[9] as took place, on the other hand, in medieval Christianity; in other words, attempts at rationalisation aroused little interest also due to the fear of establishing a degree of independence of natural phenomena from their Principle or embracing the Infinite within some finite system[10] excluding God.

The figure of the Muslim scientist also appears to traditionally take on connotations that make him considerably different from his Western counterpart. According to a Muslim interpretation, in the West, the guiding intellectual role has been embodied by different figures (the Benedictine monk, the scholastic doctor, the lay scientist, etc.); on the contrary, in the Muslim world, the figure of the sage (*hakim*) remained unchanged for a long time, embracing religious and natural sciences in different periods (incarnated in

---

[6] Jawadi Amuli A., Divine Revelation, Human Reason and Science, *Al-Tawhid*, 1984, 1 (2).

[7] Wahid Ad-Din Han, Les critères de la science moderne attestant que la législation islamique est une nécessité d'ordre universel, in Roussillon A., Science Moderne, Islam et Stratégies de Légitimation, *Peuples Méditerranéens*, 1982, 21.

[8] Baffioni C., *Storia della filosofia islamica*, Milano, Mondadori, 1991, 90–92.

[9] Nasr, op. cit., 22.

[10] Ibid.

the encyclopaedic unity of knowledge represented by figures such as Avicenna, Nasir al-Din Tusi, Razi, etc.).

There was also the intention in Muslim science to contribute to spiritual perfection. For many centuries, this approach characterised the scientific leadership of the Muslim world (approximately 8th–14th centuries), thanks also to the favourable action of unrepeatable historical and social factors. In particular, the greatest contribution to the growth of science is summarised by the historian George Sarton according to whom the main, and least obvious, conquest of the Middle Ages was the emergence of the experimental spirit and this was due in the first place to the Muslims from the 12th century onwards.[11] In turn, the experimental spirit may derive from the empirical and pragmatic approach that are very much present in the Koran and in the *Sunna* as fundamentals of Muslim law. The authoritative Pakistani *sheikh* M. Shihabuddin Nadvi emphatically recalled that only Muslims "sowed the seeds of modern applied sciences through empirical and experimental processes spread over centuries"[12] taking as their starting point the stimuli present in the Koran.

The decline and decadence of Arab-Muslim civilisation from approximately the 14th century was caused by countless factors, of which the main one, according to the Pakistani A. Salam (Nobel Prize for Physics in 1979), is of an internal nature and can be attributed to the exhaustion of scientific initiative.[13] This was accompanied by discouraging innovation (*ijtihad*) in all fields in favour of Sufism, which advocated detachment from the world; the emergence amongst jurists of a more rigid "orthodoxy" and the progressive exclusion of natural sciences from the curricula of the theological schools (*madrasa*) with the exception of astronomy and mathematics. There was an imposition of thought that was more closed to the critical spirit, accompanied by an atrophy of science[14] whilst

---

[11] Salam A., *Religione e scienza con particolare riferimento alla scienza nell'islamismo*, in Galbiati D., Eligio Padre, et al. (eds.), *Scienza ed Etica alle Soglie del Terzo Millennio*, Società Italiana di Fisica, Conference Papers, Varenna 28–30 September 1992, Vol. 36, 1993, 57–95, 61–62; See also Qadir C.A., *Philosophy and Science in the Islamic World*, London, Croom Helm, 1988, 104–121.

[12] Nadvi, art. cit., 36.

[13] External factors can include the severe destructions caused by the Mongols and which culminated in the fire of Baghdad in the 13th century; in the end of the Caliphate of Cordoba; the fall of the Emirate of Granada (1492); the shift of the trade routes due to the discovery of the Americas and the Cape route; the loss of control over economic exchange and the growth of the Spanish and Portuguese empires. All these events increased the isolation of the Muslim world.

[14] Salam, art. cit., 62–63, note 2, Appendix IV. Included in the concrete examples of the negative attitude Towards innovations taken by "official" thought – after the golden age – Salam highlights the opposition to printing (in Egypt until the arrival of Napoleon in 1798, in Turkey printing was accepted for secular books from 1839 and from 1874 for the Koran due to religious opposition) and the demolition in 1580 of the last astronomic observatory of Islam (built in Constantinople in 1355) due to a religious faction. Religious opposition to scientific thought can also be exemplified in the response to the attempt of the Ottoman Sultan Selim III to modernise the army in 1799 with studies of algebra, ballistics and mechanics under the guidance of European teachers; the project failed due to the contempt by jurists and theologians for the new technologies (Salam, 63 and 85).

in the West, thanks to the Renaissance, it was reawakening, with Islam playing an important role thanks to the Arabic translations of the works of the Greek world and the works of Muslim doctors and scientists.

Despite the rapid progress made in the past few decades, the general state of technical-scientific inferiority and the social and economic crisis of the Muslim world, apart from some exceptions (according to Abdus Salam the organisational difference with respect to the West remains "abyssal"), is the reason for the considerable malaise of the communities following the last and perfect Revelation (the Koran) and the last and perfect Divine Law (*Shari'a*); this discomfort is embodied by the interrogative raised by an Egyptian intellectual: "How is it possible that the world has become Hell for believers and Paradise for infidels?".[15] Munawar A. Anees, an Islamic bioethics expert and a member of the Royal Academy of Jordan, openly states that "a crisis of knowledge of immense proportions overwhelms contemporary Muslim civilisation" and this crisis "has thrown the *Ummah* into an abyss. No exotic claims about alien intervention can absolve Muslims of their intellectual docility".[16] Although Muslim communities are highly differentiated, "there is not a single Muslim country today that meets the criteria of modern political and social governance, religious liberty, economic evolution, gender equality, cultural prosperity and human dignity. Muslims continue to live under dictators, autocrats, kings and authoritarian rulers in grossly oppressive conditions".[17] Lastly, these communities are suffocated "by the loss of pluralism and progressive thought" which were distinctive characteristics of their glorious past.

The "dazzling" scientific and technological predominance of the West leads many developing nations to feel pride in imitating the models of the way of life of developed nations. This also concerns many Muslims. *Sheikh* M.S. Nadvi denounces this phenomenon as a form of "mental apostasy".[18] As Western ideologies are a vehicle of atheism and materialism, Islam has to return to a competitive scientific level to fight this invasion as the only language that modern nations understand is that of science.[19] This has to be the *Jihad* of modern times. Only technical and scientific knowledge can be imported from the West, not their ideologies and philosophies. As the Middle Ages showed, Muslim science is the only one that can unite the spiritual and material dimensions (the latter is currently a Western monopoly). Only in this way can Muslims regain "their lost glory and pride".

The meteoric expansion of Islam from the 7th century (from present-day Mauritania to Central Asia) and the subsequent golden age (8th–14th century) are still often summarily perceived by the collective imagination as the signs of a divine blessing which subsequently failed to appear. It is not surprising that

[15] Branca P., *Introduzione all'Islàm*, Cinisello Balsamo, Paoline, 1995, 272.

[16] Anees, art. cit.

[17] Anees M.A., From Knowledge to Nihilism: Redeeming Humility, *Journal of Islamic Philosophy*, 2005, 1, 5–10.

[18] Nadvi M.S., Rise and Fall of Muslims in the Realm of Science (II), *The Muslim World League Journal*, 1999, 26 (12).

[19] Ibid., 47.

today a good number of Muslims deem it sufficient – uncritically and ahistorically – to return to the mythical and successful origins of Islam (reintroducing the *Shari'a*) to win back the greatness of the golden age; this approach is very dear to a large number of "fundamentalists" according to whom Islam is the solution to all problems. Nor is the Nobel Prize-winner Abdus Salam exempt from the logic of the "supremacy of Islam" when he wonders "Can we go back along the course of history to win back the supremacy of scientific progress in the world?" The close link between scientific-technological power and political redemption is underlined by many scientists and scholars. The words of *sheikh* M.S. Nadvi are very clear: thanks to science, "Muslim society should attain the power so that it can obtain political and military supremacy"[20] or "the Divine Religion attains political and material supremacy in the world".[21]

The general approach to scientific speculation in the Muslim world of the 12th century often overlooked the complex epistemological and methodological questions that have had and still have a fundamental importance in modern Western scientific thought, representing the indispensable introduction for the successes obtained. What has mainly been lacking in the history of science in the Muslim world – compared to Western development – is precisely an in-depth methodological reflection on the operative methods of scientific research, including an epistemological awareness of the autonomy of the fundamentals of knowledge. This analysis agrees with the theory of Roussillon, according to whom the attempt to Islamise science is perhaps made by reinforcing those "epistemological obstacles" against which contemporary scientific thought was constructed.

SCIENTIFIC EXEGESIS OF THE KORAN

I refer here to a distinction which I have made elsewhere, according to which there are at least three different levels or degrees of scientific exegesis of the Koran (radical, moderate or ultra-moderate concordism and, in the last place, anti-concordism).[22] In the Muslim world, there is a common current advocating scientific "Concordism" (a Western term) between the Koran and modern science which has its origins in particular in the verses 6.38 "Nothing have we omitted from the Book" and 16.89 "and We have sent down to thee the Book explaining all things, a Guide, a Mercy and Glad Tidings to Muslims". According to the "concordist" interpretation of these verses, several parts of the Koran anticipate, directly, indirectly or by allusion, modern scientific information which was, however, impossible to have in the times of the Prophet (7th century). These

---

[20] Ibid., 45.

[21] Nadvi, *Rise and Fall of Muslims in the Realm of Science (I)*, art. cit., 38.

[22] See Atighetchi D. Islam, Corano e scienza moderna, *Il Mulino*, 1997, 5, 969–978; L'esegesi scientifica del Corano, in Atighetchi D., *Islam, Musulmani e Bioetica*, Roma, Armando Editore, 2002, 231–251.

verses are believed to represent evidence of the scientific miracle of the Koran. Various attitudes can be distinguished regarding the "concordist" interpretation. The most radical identify information present in contemporary science in the eternal Word of God. Radicalism may even consider the Holy Text as a criterion to determine the validity of a scientific hypothesis. Consequently, the path towards progress appears anticipated and outlined by the Holy Text as "a book of science that will never be superseded by any human scientific discovery" and capable of providing answers to problems that man is not yet capable of formulating (B. Torki). According to the Egyptian Wahid ad-Din Han, the most serious shortcoming of Muslims today concerns the inability to identify scientific discoveries as proof of what is established by the Koran.

The second current (but the boundaries are often blurred) can include the moderate and the ultra-moderate "concordists". In the opinion of the moderates, it is possible that God mentioned some scientific laws and natural phenomena for the purpose of strengthening the faith of believers whilst the Koran remains fundamentally a religious text (and not also a scientific text as for the radicals) therefore it is possible to see a contrast between the eternal contents of the Word of God and the scientific discoveries relative to the times and cultural contexts; this latter contrast is highlighted by the opponents of concordism.

"Concordism" developed in the 19th century (a symbolic date is when the French landed in Egypt in 1798) as one of the answers to the shocking contact with Western technological superiority. Moreover – and this is the most important aspect – it corresponds to an inevitable conceptual temptation in the relationship between religion and science in the Muslim world. This temptation is rooted in the eternal, totalising and all-inclusive value of the word of God and His Law (the *Shari'a*) which must be contended with. The idea that contemporary science can be found – to varying degrees – in the Koran is today agreed with by a fair number of researchers and scientists who, fearful that Western science outclasses the true religion, show the miraculous inimitability of the Koran and the capacity of Islam to adapt to every new situation. Indeed, precisely because it is based on the Last Revelation, "Islam is the only ultra-modern religion."[23] If for modernists it is necessary to "modernise Islam", for strict "concordists", it is better to "Islamise modernity". This trend may even represent a sort of "return to the origins" if all the discoveries and developments of Western science are believed to have taken place thanks to the assumption of the principles on knowledge contained in the Koran and in the *Sunna* and which were re-formulated during the Muslim Middle Ages.

Amongst the historical supporters of "concordist" interpretation, the greatest Muslim theologian (called the "proof of Islam") Ghazali (d. 1111) is often quoted and who, in his work "Revivification of the religious sciences" stated: Everything that the human spirit understands with difficulty, everything that

---

[23] Wajid H., Post-Mortem Examination: The Quranic View, *Hamdard Islamicus*, 1992, XV (3), 85–92.

is the object of diverging theories or considerations, all this is examined by the Koran that speaks of them by signs or allusions.[24] An identical theory is shown in another of his works, *The Jewels of the Koran*: the fundamental principles of all the sciences . . . are contained in the Koran and all indeed are drawn from the boundless sea of the knowledge of God the Highest; All the science of the ancients and of the modern peoples converge in the Koran; Reflect on the Koran and investigate its marvels for the purpose of finding all the complex of knowledge of the ancients and of modern peoples, and the synthesis of their principles. Meditation on the Koran is of use in going from synthesis to minute specification, which represents an ocean the limits of which cannot be seen.[25] Ghazali probably meant to say that "the principles" of all sciences are based on the Koran (which can be defined as a deep ocean embracing all the sciences), instead of saying that all the sciences and the specific knowledge of the individual sciences are contained in the Koran.

The problem lies in the very character of the Koran which is "the" Holy Book par excellence, an omni-comprehensive and inimitable miracle in the hands of the faithful. It is difficult to accept, in these terms, that it does not deal at all with important scientific information, which would decrease its insuperable value. From this point of view, "radical concordism" intends to connect the divine word with a precise scientific correspondence[26] exposing, however, the Koran to multiple risks and contradictions that are rooted in continuous scientific discoveries and laws.

There are numerous theologians and jurists who, since the 19th century, could be included, for various reasons, amongst the "concordists". Amongst these can be mentioned al-Kawakibi (d. 1902), the advocate of "Islamic reformism", according to whom the majority of Western discoveries have been, to some extent, pre-announced in the Koran. Mustafa Sadiq al-Rafi'i, an Egyptian scholar, in his book *The inimitability of the Koran*, maintains that the Koran contains all the sciences with their principles. Mohammad al-Iskandarani (19th century), a doctor and author of *The revelation of the luminous secrets of the Koran on the celestial and terrestrial bodies, the animals, the plants and mineral substances. Sheikh* Muhammad Bakhit said "The Koran is the source of all the sciences and human civilization" and, lastly, *sheikh* Tantawi Jawhari (d. 1940)[27] regretted not living long enough to identify the technical and scientific discoveries contained in the Holy Text.[28]

---

[24] Jomier J. and Caspar P., L'Exégèse scientifique du Coran d'après le Cheikh Amin Al-Khouli, *MIDEO*, 1957, 4, 269–280.

[25] Ghazali (Al), *Le perle del Corano*, Milano, Rizzoli, 2000, 123 and 125.

[26] This is the theory fully agreed with by Naguib M., *The Scientific Religion*, Cairo, Nadwat Ansar el Koraan (n.d.).

[27] See Jomier J., Le Cheikh Tantawi Jawhari et son commentaire du Coran, *MIDEO*, 1958, 5, 114–174, in particular 147–154.

[28] Regarding the relationship between Islam and modernisation and for information on some of the figures mentioned, see Jomier and Caspar, art. cit., 273–274 and Branca P., *Voci dell'Islam moderno*, Genova, Marietti, 1991.

The "scientific anticipations" present in the Holy Text in forms which are allusive to greater or lesser degrees are the result of subjective deductions by interpreters and refer to astronomy, flora, fauna, human reproduction, physical phenomena, phenomena of society (e.g. AIDS), human history and geography.[29] Any event deemed important for the fate of humanity can be found by its fortunate "discoverer" in the word of God. For example, when the two Soviet sputniks were launched into space in 1957, a group of authoritative "scholars" met in Cairo to evaluate the event from the Muslim point of view. There emerged the two classic trends already mentioned in the relationship between the Koran and science. The former (e.g. upheld by Mohammad al-Banna) defended the presence of scientifically accurate indications in the Koran, including the possibility of inter-planetary travel in verse 55.33 "If it be ye can pass beyond the zones of the heavens and the earth, pass ye!"[30]

At times "radical concordism" is not limited to finding indications of the most modern scientific discoveries[31] in the Koran, on the contrary it can aim at consolidating an alternative vision of the world (including from the scientific point of view) in harmony with the totalising character of Islam. The Algerian

---

[29] On some of these subjects, see for example: Rostron I.P., An Islamic Perspective on Quantum Mechanics, *The Islamic Quarterly*, 2002, XLVI (1), 79–94. On p. 89, the author states: "My thesis is that these Islamic principles [the principles of the faith], which are at the heart of Islam are also at the heart of my explanation of Quantum theory"; Jibril B.Y., Towards a Faith-Based Teaching of Chemistry, *The Muslim World League Journal*, 2001, April, 36–40 and 2001, May, 25–28; Nadvi M.S., Plants Confirm Unity of God, Prophethood and Resurrection Day, *The Muslim World League Journal*, 1997, 24 (11), 30–34; Saleem H.M., Origin of Life (With Reference to the Holy Qur'an and the Modern Sciences), *Islam and the Modern Age*, 1991, XXII (2), 137–148; Mahroof M.M.M., Islamization of Mathematics, *The Muslim World League Journal*, 1997, 25 (6), 31–35.

[30] In the opinion of al-Banna, Koran 34.3 refers to the discovery of the atom and sub-atomic particles "by Him Who knows the unseen, from whom is not hidden the least little atom [ar. *dharra*] in the heavens or on earth: nor is there anything less than that, or greater, but is in the Record Perspicuous [i.e. the Koran]"; 81.6 is believed to allude to the hydrogen bomb through the expression "When the oceans boil over with a swell"; in 17.1 reference to the night ascension by Muhammad to the Ultimate Temple (Jerusalem) is believed to anticipate the advent of aviation; Koran 22.65 "He withholds the sky (rain) from falling on the earth except by His leave" is connected to the universal gravity of Newton. See Jomier and Casper, art. cit., 269–270. Other scientists and men of religion are convinced that the Koran 17.44 contains a reference to the Brownian movement, i.e. the perpetual movement of molecules in water in praise of God. 13.2 is alleged to mention universal gravity "He has subjected the sun and the moon (to his Law)! Each one runs (its course) for a term appointed".

[31] In 75.3–4 the Koran underlines the infinite capacity of God to restore in the Resurrection every detail of the human body: "Does man think that We cannot assemble his bones? Nay, We are able to put together in perfect order the very tips of his fingers." This passage is believed to show the characteristic of different fingerprints for every individual which was known to the Koran whilst the British discovered this only in 1884. See Syed I.B., The Human Cloning: An Islamic Perspective, *The Muslim World League Journal*, 1997, 25 (2), 43–48.

physicist B. Torki fits into a similar context, in an explicitly stated "concordist" work that meets the aim of confronting the materialistic aggression of the West where nature replaces God, man is assimilated with animals (Darwin) and reduced to sexuality (Freud) or economy (Marx). This aggression aims to destroy the precepts and values of Islam, to separate Islam from society, the State and science; at the same time the aggression aims at instigating women to rebel against men, children against their parents, etc.[32]

Maintaining this alternative vision, the "Commission on Scientific Signs in Quran and Sunna" (established on the decision of the Supreme International Council of Mosques in Mecca in the mid-1980s) includes amongst its objectives that of giving the physical sciences a Muslim character, by conceiving a methodology to discover, study and investigate the scientific signs present in the Holy Texts.[33] The Saudi *sheikh* Abdul Majeed Zindani, Secretary of the aforementioned Commission as well as member of the Supreme International Council of Mosques, summarised the position of "radical concordism" by stating textually: "we are at the dawn of an even more important era, that of the agreement between science and religion and this can only take place between the true science and religion of Islam that Allah has protected from all falsification and every alteration"; "Muslims can direct the path of the scientific movement, give science its real place and make it proof of faith in Allah."[34] According to Zindani, the importance of the scientific inimitability of the Koran and of the *Sunna* lies in the fact that divine science is totalising and without errors whilst human sciences are limited, susceptible to grow and exposed to error[35]; there cannot exist any contradiction between the Revelation and experimental sciences. If there is a contrast, scientific theory has to be rejected because the Koran is the Revelation. If there is harmony, the Koran proves the validity of the scientific theory.[36]

---

[32] Torki B., *L'Islam religion de la science*, Edition Chihab (n.d.), 11–13. In the opinion of the Algerian physicist, amongst the countless scientific affirmations present in the Koran, the temporal relativity discovered by Einstein was already suggested in the verse 22.47 "Verily a Day in the sight of thy Lord is like a thousand years of your reckoning" (Torki 105–108). The Soviet hypothesis of Pontecorvo-Smordinski of a recompacting of the whole universe in a single point with the precipitation of matter and a subsequent Big Bang can be found in 29.19 "See they not how God originates creation, then repeats it: truly that is easy for God" (Torki 119). The relationship between the human species with other animal species is challenged (together with evolutionism) as scientifically weak or inconsistent also because the Holy Book in 2.30 maintains the extra-terrestrial origin of man "thy Lord said to the angels 'I will create a vicegerent on earth' (Torki 181). The Koran 13.15 or 17.44 anticipate the forthcoming discovery of the existence in the cosmos of other creatures as well as angels and spirits (183–184).

[33] Commission on Scientific Signs in Quran and Sunnah, *The Muslim World League Journal*, 2000, 28 (2), 14–15.

[34] Il Miracolo scientifico del Corano, in *Il Musulmano*, UCOII monthly publication, 4 May 1994, 1–7, 5–6 and *Il Musulmano*, 2 March 1994, 1–18.

[35] Mabruk A., Sauver l'humanité de la science, *Al-Ahram*, 1 October 1985 in Botiveau B. and Jacquemond R. (élaboré par), *L'Inimitabilité coranique en matière médicale*, CEDEJ, Vol. 20, 1985, 113.

[36] Ibid.

The same relationship is dealt with by Sharaf al-Qudat, who has a more moderate position. In his opinion, the authentic Holy Texts cannot contradict scientific truth; this derives from the prodigious character of Islam "unlike the other religions of our time". Regarding dubious *ahadith*, Qudat suggests examining them in the light of scientific truth: if they do not comply, they must be rejected because they are not authentic.[37]

The speech by Hosni Mubarak (*Al-Ahram*, 25 September 1985), at the time (and still) President of the Arab Republic of Egypt, appeared similar to Zindani's approach, at the international congress held at the University of Al-Azhar on 24–26 September 1985 on "The Inimitability of the Koran in the medical field", under the patronage of the highest civil, religious and scientific authorities of Egypt and with the participation of representatives from 28 Arab-Muslim countries. Mubarak, after having recalled the principal scientific miracles presented by the Word of God concluded: "We Muslims must understand that the discoveries made by science are not at all proof of the veracity of the Koran. On the contrary, it is the Koran, in what it states that represents the intangible proof, whether by conviction in the faith of Muslims or scientifically for eternity."[38]

On the sidelines of the Congress, *sheikh* Mahdi Abd al-Hamid Mustafa recalled that during the selection of the more than 200 scientific studies presented, attention was paid to their agreement with the scientific data of the Koran adding: "the Koran is never in contrast with scientific truth; if there is a difference between the Text and theory, the latter is necessarily wrong or insufficient and therefore it must be revised".[39] In these expressions the Koran no longer represents subsequent confirmation of what science discovers independently, but

[37] Abu Sahlieh A.A., Le Statut juridique du foetus chez les musulmans hier et aujourd'hui, in AA.VV., *Éthique, Religion, Droit et Reproduction*, Paris, GREF, 1998, 63–80.

[38] "Et nous, musulmans, devons comprendre que les découvertes auxquelles la science parvient ne sont en rien une preuve de la véracité du Coran. Au contraire, c'est le Coran, dans ce qu'il a rapporté, qui constitue la preuve intangible, que ce soit par conviction dans la foi des musulmans ou scientifiquement pour l'éternité . . ." in Mubarak H., *Inimitabilité manifeste*, in Botiveau and Jacquemond (art. cit.), 110. According to the Egyptian President, the first Muslims were guided by the Koran, "a divine constitution free of error" therefore the inimitability of the Koran and of the *Sunna* in the medical field was considered taken for granted. For example, when in the Koran 39.6 God says "He makes you, in the wombs of your mothers, in stages, one after another, in three veils of darkness," this is a miraculous revelation dating back 14 centuries based on which the embryo has 3 membranes, invisible to the naked eye, which protect it from the light; modern science has only recently discovered these membranes. Science has discovered the existence of sperm cells with a head, neck and tail similar to a leech, whilst the Koran revealed their presence many centuries before the discovery of the microscope. These beings fertilize the ovule giving rise to the "clot of congealed blood" (Koran 96.2). The development of the embryo finds in the Koran 23.12–14 an anticipation of what has been discovered only today. The uniqueness of the fingerprints of each individual was already known to the Koran 75.3–4 "Does man think that We cannot assemble his bones? Nay, We are able to put together in perfect order the very tips of his singers." Lastly, several verses proclaim the use of medicines such as camphor and ginger or the value of natural breastfeeding or dental care, etc.

[39] Botiveau and Jacquemond, art. cit., 121.

vice versa it is the Koran that tells the truth whilst scientific demonstration must subsequently adjust itself. The criterion of truth risks being placed upstream and outside scientific reasoning.

On the conclusion of the Congress, Hafiz Mahmud in *Al-Gumhuriyya* of 29 September 1985 said: "It is in the way indicated to us by the Koran when it reveals to us the secrets of life that we find the straightest path to submit science, art . . . the whole of civilisation . . . to the certainty of humanity . . . . Indeed, it is pure 'temporal' science that wanted to produce the atomic bomb, the hydrogen bomb and chemical arms that reduce the human dimension of civilisation to nothing. But if, in scientific research and experimentation, we accept being guided by the inimitability of the Koran, we offer the world a new domain in which we can emanate science, knowledge."[40] In this article, mention is made of the contrast between "pure temporal science" and science based on the inimitability of the Koran. Roger Garaudy, the French philosopher and a convert to Islam, stated "Islam came to submit science and technology to God and to save humanity from destruction."[41]

Precisely due to the importance of the medical information present in the Koran, Dr. Mamduh Gabr, President of the Egyptian Medical Association and Chairman of the Congress of Cairo in 1985, requested during this congress that similar scientific meetings on "concordism" be repeated periodically for the benefit of medical teaching where the importance of religion in the training of doctors must be highlighted.

The conclusive document of the Congress, addressing universities in Muslim countries, recommended theses in medicine and scientific subjects on the relationship between the Koran, the *Sunna* and modern science. The recommendation was also made to the Ministry for Education to revise the scientific methods that oppose or are not very compatible with the truths of the Koran, as would be the case with biology.[42]

Dr. Muhammad Hasan al-Hifnawi, Vice President of the Egyptian Medical Association and Secretary-General of the Congress, in an interview with *Al-Musawwar* of 4 October 1985 stated that there does not exist any clear scientific truth that is opposed to a text of the Koran and, in the event that there is disagreement, this must provoke an effort for independent reasoning (*ijtihad*). For these reasons, before the continuous therapeutic and diagnostic progress of contemporary medicine, "the doctor wishes to be ensured that everything he does is in compliance with the *Shari'a*".[43]

According to one of the most prominent Egyptian *sheikh*s, Mutawalli al-Sha'rawi, there are many observations in the Koran on the laws of the universe.

---

[40] Mahmud H., Le Guide, *Al-Gumhuriyya*, in Botiveau and Jacquemond, art. cit., 108.

[41] Mabruk, art. cit., 112.

[42] Botiveau and Jacquemond, art. cit., 118.

[43] Marmuch L., Niveaux de lecture, *Al Musawwar*, 4 Octobre 1985, in Botiveau and Jacquemond, art. cit., 115.

These observations were incomprehensible to men in the period when the Koran was revealed[44]: the roundness of the earth, the existence of an atmosphere surrounding it, embryology, the earth rotating on its own axis, the relativity of time, etc. The Prophet did not dwell on these subjects, leaving the possibility of understanding them to the capacity of each generation according to the scientific knowledge available. And so, said Sha'rawi, giving concrete examples, the Koran allows us to discover scientific truths adapted to the cultural level of different periods.

From 10 to 12 April 2000, the "Sixth International Congress on Scientific Signs (Miracles) in the Koran and in the Sunna"[45] was held in Beirut. The event was organised by the Muslim World League (MWL) and by the Lebanese *Dar-al-Fatwa*. The opening was chaired by the Prime Minister of Lebanon, Dr. Salim al-Hoss, before more than 1,350 delegates, with the participation of the Secretary-General of the MWL, A.S. al-Obeid, the Grand *mufti* of Lebanon, *sheikh* Muhammad Rasheed Qabbani, as well as Ministers of the Lebanese government, Members of Parliament and intellectuals. In his opening speech, the Secretary of the MWL, al-Obeid, invited believers to increase "meditation on all aspects of science" in order to reorganise the relationship between man and other creatures but all with the aim of strengthening the relationship between man and God, which is the only relationship that allows us not to deviate from the right path or to get lost in the mass of information. In addition to his invitation to protect the eco-system, which finds its balance only in a closer relationship with God, al-Obeid said that the benefits brought about by modernity have their roots in different sources, but this must not lead to forgetting the Creator, which would lead to irresponsible actions. According to Obeid, modern technological developments recall the need to recover the moral values and beliefs in the context of an authority based on the faith. To this end, the Secretary-General of the MWL recalled the usefulness of congresses and conferences such as that of Beirut with the aim of introducing "mankind, with the help of the scholars in science, academicians, researchers and intellectuals, to the various Signs, which the modern science goes on discovering each passing day".[46] In addition, similar meetings

---

[44] Sha'rawi (Al) M.M., *Le Miracle du Coran et en quoi il diffère?*, in www.islamophile.org

[45] MWL Commission on Scientific Signs in Quran & Sunnah, *The Muslim World League Journal*, 2000, 28 (2), 7–12. Two main topics were examined during the Congress: (1) the medical signs in the Koran and the *Sunnah*, (2) the signs in the Cosmos. In the medical field, the Saudi Arabian M.A. Albar gave his opinion on contagious diseases highlighting how the decline of morals coincides with the spread of venereal pathologies. The Egyptian doctor Abul Majd reasserted the efficacy of prayers in fighting diabetes as these contribute to reducing the rate of diabetic foot. The Tunisian doctor Zuhar Rabeh Qarami drew attention to the advice of the Prophet Muhammad in practising ablutions (*wudu*) to quell anger; in addition, they are also effective in activating the circulation of the blood and the organs involved. Prayers calm and relax the nerves. Sitting down and prostrating oneself thanks to prayer helps the circulation of the blood and reinforces the bones. Dr. Ahmad al-Qadhi (a doctor who works in the USA) has provided examples of how Islamic medicine is useful in improving the condition of cancer patients at a very advanced stage.

[46] Ibid., 9.

are of benefit to inform non-Muslims of the scientific miracles of the Koran and of the *Sunna* as well as providing the opportunity to contact them, through Western scientific channels: "These contacts and interactions have often resulted in the conversion of some of them to Islam after they have been convinced of the scientific Signs mentioned in the Koran and Sunna."

In his turn, the Grand *mufti* of Lebanon expressed himself according to a strictly "concordist" point of view as the Koran is a guide for all nations independently of space and time. Its marvels are inexhaustible; all the information it contains concerning the universe is the object of science that man can use to improve his life. Whenever new facts and discoveries come to light, they lead us back inevitably to the existence of God "Who has created man from clay" [an explicitly "creationist" position]; there exist many verses in the Koran which speak of the secrets of life, creation, human beings "and other worlds in the heavens and the earth" and all this confirms the celestial origin of the Koran.

As shown by the words of Abdullah al-Obeid, the "concordist" perspective is not limited to the Muslim world, but aspires to take on the rest of the world to make proselytes and spread Islam. The Organising Committee of the Sixth International Congress on Scientific Signs in the Koran pronounced itself in this sense, adopting – inter alia – the following conclusive recommendations:
(a) To publish the research and studies presented at the Congress in Arabic and in other foreign languages.
(b) To solicit universities in the Arab world to make known the developments emerging from studies under the auspices of the Commission on Scientific Signs in the Koran.
(c) To put the subjects, the studies and the research of the Commission on the Internet to diffuse their beneficial effects.

On this last point, it is worthwhile remembering that currently many of the Muslim sites concerning medicine and bioethics in Islam contain a considerable number of pages dedicated to the medical-scientific miracle of the Koran and of the *Sunna*.

An allusion to AIDS can be identified in Koran 7.33 "The things that my Lord hath indeed forbidden are: shameful deeds, whether open or secret" which is associated with a "saying" of Muhammad (referred by Ibn Maja) according to whom when a monstrous sin is common in a community, it will be struck by the plague or other unknown diseases. Zindani apologetically connects these predictions with the European and American situation where homosexuality, adultery and prostitution are tolerated, causing the spread of these pathologies.

In conclusion to the book *La Bible, le Coran et la Science* Maurice Bucaille, a French surgeon, deems having demonstrated the perfect harmony existing between the revelation of the Koran and modern science whilst nothing similar can be found in the Old and New Testaments.[47] With these theories, Bucaille can be included amongst the concordists, independently of whether he is Muslim or Catholic. Referring to human reproduction, the Koran shows several notions in

---

[47] Bucaille M., *La Bible, le Coran et la Science*, Paris, Seghers, 1976, 254.

agreement with the latest discoveries of embryology. For example, *sura* 16.4 "He has created man from a sperm-drop" uses the term *nutfa* to indicate that a very small amount of liquid is necessary to procreate.[48] In 32.8 "and made his progeny from a quintessence of the nature of a fluid despised" the term quintessence-sap is "*sulatat*", i.e. the best part extracted from a thing, which would indicate the possibility that one sperm cell from tens of millions penetrates the egg fertilising it. The seminal liquid is indicated by various expressions including "mingled sperm" (Koran 76.2 "Verily We created man from a drop of mingled sperm"); only today science has demonstrated that sperm is made up of different secretions from different glands. Koran 96.2 "ton Seigneur, celui . . . qui forma l'homme de quelque chose qui s'accroche"[49] (English: "[God] created man, out of a (mere) clot of congealed blood [Ar. *alaqa*]"). The French "qui s'accroche" [Ar. alaqa] can be translated by "something hanging from the womb" or "something which clings". The term *alaqa* would appear to reveal a surprising modernity and is opposed controversially by Bucaille to the translation "clot of blood" or "adherence" frequently used in the West in order to diminish or deny the scientific value of the Koran. Consequently, even the authoritative translation by Yusuf Ali is disputed: "[God] created man, out of a (mere) clot of congealed blood [Ar. *alaqa*]."

Again concerning the references of the Koran to the embryo, the Code of Ethics (2001–2002) of the Pakistan Medical and Dental Council states, in paragraph 7.0: "The Qur'an itself has a surprising amount of an accurate detail regarding human embryological development, which informs discourse on the ethical and legal status of the embryo and fetus before birth."[50]

In situating the discoveries of physics, "radical" concordism often shares a "fixed" position, according to which the statements of science are true or false, *tertium non datur*, a position associated with a conception of scientific knowledge understood as not being modifiable according to a point of view that is superseded by contemporary epistemological viewpoints. Moreover, this rigidly binary pattern appears the most suitable to relate to the eternity of the divine word. It is worth observing that, methodologically, the "concordist" mechanism functions after something has been discovered or invented, i.e. subsequently identifying this information in the Koran possible. On the contrary, in order to increase scientific knowledge, it would be necessary to take the opposite path: to take an idea, hypothesis or theory from a Koranic passage that can explain specific phenomena, hypothesis to be ascertained afterwards in practice through an adequate scientific procedure. Unfortunately, apart from explicit invitations to proceed in this direction, there is no trace as yet of any significant results having been obtained following such an application.

The relationship of "concordism" with the "evolutionist" theory is very controversial. *Sheikh* Tantawi Jawhari seemed to maintain that the law of survival of the fittest was already present in Koran 13.17–18, whilst the prevalent opinion

---

[48] Ibid., 201–204.

[49] Ibid., 204–205.

[50] The Pakistan Medical and Dental Council, *Code of Ethics*, 2001, in wwww.pmdc.org.pk/ethics.htm

states that the "evolutionist" theory is alleged to be one of the multiple secularised visions of human nature produced by Western culture. The Koran indicates the existence of three categories of living creatures on earth: animals, men and *jinn*. All three types of creatures descend from inorganic substances. Koran 21.30 maintains the theory (currently dominant in science) of the origin of terrestrial life from water "We made from water every living thing" with the exclusion of *jinn* (spirits created from fire without smoke) and men. Regarding the latter, Koran 15.26 specifies "We created man from sounding clay, from mud moulded into shape" (see above the words of Rasheed Qabbani, Grand *mufti* of Lebanon). The Sacred Text does not approve of the hypothesis of the simultaneous evolution of man in different geographical areas as Adam, the first man, was created by God: "O mankind! Reverence your Guardian-Lord, who created you from a single Person, created, of like nature, his mate, and from them twain scattered (like seeds) countless men and women" (Koran 4.1). In the light of these verses, several Muslim scholars refuse to consider human intelligence, the capacity to read and write (gifts of God according to Koran 96.4 and 55.4) as the results of some "natural selection", also because no theory is believed to exist that is capable of explaining in a convincing way the evolution of man from the ape in each individual stage.

From the statistical point of view, the neo-Darwinian theory of the origin of life from a casual combination of nucleic acids, enzymes and proteins is not very plausible just as the theory of the Big Bang is not very probable; both raise more doubts than answers, whilst the answers are present in the creationism of the Koran.[51] The "notion of gradual evolution, mutation or transition is both blasphemous and unfounded" according to Muslim tradition. Physically and spiritually, man has always been the same; his physical and spiritual constitution was conceived and created in its entirety once and for all by God. Amongst the incongruities of the evolutionist theory we can find[52]: (a) natural selection brings out the most from the least, i.e. man from inferior species. (b) Natural selection, in Darwinian terms, is a casual and spontaneous process, without intelligence. The complex structure of the world and of the universe becomes the effect of a gigantic lottery. These are unacceptable ideas for Islam as disorder, non-intelligence and non-conscience cannot generate order, intelligence and conscience except through the action of a creative intelligence. In everything that exists, God expresses a project and a purpose, similarly to the fact that man has been created the "Vicar of God" and not uselessly. (c) Natural selection as a casual process without intelligence and will tends towards a model of "extreme determinism" therefore man himself is not characterised by a freedom of action and thought but is determined by chance. This is why the scientists have to abandon all preconceptions and prejudices against the validity of the Revelation on natural phenomena and, on the contrary, understand that it guides them to the truth. In rejecting evolutionism, Torki explicitly refers to the extra-terrestrial origin of man "I will create a vicegerent on earth" (Koran 2.30). Indeed, Koran 13.15 or 17.44 preannounce the forthcoming discovery of new creatures in the cosmos.

---

[51] Mohamed Y.A., Fitrah and the Darwinian View of Man, *The Muslim World League Journal*, 1994, 22 (3), 43–46, 44. See also Awad M.N., The Scientific Criticism of the Theory of Evolution. The Emergence of Life on Earth, *Rissalat al-Jihad*, 1987, 56, 52–58.

[52] Mohamed, art. cit., 45–46.

A Malaysian Muslim scholar, commenting upon Muslim positions opposing the theory of evolution, recalls that "Muslim creationists, like their Christian counterparts, fail to offer the world of modern science a more plausible explanation of the origins of biological life based on natural cause to replace Darwinian theory of evolution."[53]

The concordist attempt inevitably involves various Koranic precepts such as the prohibition of drinking wine (e.g. Koran 5.90), eating pork (Koran 16.115; 6.145, etc.) or ritual slaughter, the benefits and validity of which tend to be reinforced with the help of contemporary science and discoveries.

In consideration of the disasters produced by alcoholism, Muslim authors highlight the correspondence between alcohol and the dangers for the health (the liver, the stomach and the mind) and even death, whilst these dangers can be avoided thanks to the divine prohibition. It is interesting to note how any discoveries or scientific indications of an opposite sign – contravening the divine prohibition and a long apologetic tradition in support of the precept – tend to be overlooked. In the West the problem of the analysis of the psycho-physical consequences of alcohol consumption has followed – in the more recent developments – a more elaborate route.

To the question: Is alcohol always bad for a person? the answer is increasingly: at present, it has in fact been ascertained that a moderate and regular consumption of wine or beer is beneficial for the health, reducing cardiovascular mortality. The benefit, however, is limited exclusively to the "moderate quantity", disappearing in those who drink large quantities or who are chronic drinkers and in subjects with pathologies for whom the consumption of alcohol is not recommended. One or two glasses of wine a day (or a bottle of beer) during meals protect the heart by decreasing the risk of thrombus due to the reduction of the amount of fibrinogen on the blood and the increased release of tissue-type plasminogen activator (T-PA).[54]

---

[53] Loo S.P., Islam, Science and Science Education: Conflict or Concord?, *Studies in Science Education*, 2001, 36, 45–78, 58.

[54] Here is a summary of the benefits of a moderate and regular consumption of wine: "Increase in HDL-cholesterol, inhibition of thrombin-induced platelet aggregation, raised fibrinolytic activity of the serum and possibly an increase in insulin sensitivity. The polyphenol compounds found in wine have been shown to have antioxidant properties that inhibit the oxidation and aggregation of LDL-cholesterol. These compounds inhibit not only thrombin-induced platelet aggregation and show anti-proliferative effects on the growth of smooth-muscle cells and tumour-cell lines in vitro", cfr. Scholl J., Doctor, is wine good for my heart?, *The Lancet*, 7 August 1999, 354, 514 (Letter). Amongst the numerous contributions confirming, to varying degrees, this information, the following can be mentioned: Hendriks H.F.J., Veenstra J., et al., Effect of Moderate Dose of Alcohol with Evening Meal on Fibrinolytic Factors, *British Medical Journal*, 16 April 1994, 308, 1003–1006; Thun M.J., Peto R., et al., Alcohol Consumption and Mortality among Middle-aged and Elderly US Adults, *The New England Journal of Medicine*, 1997, 337, 1705–1714; Renaud S.C., Guéguen R., et al., Alcohol Consumption and Mortality in Middle-aged Men from Eastern France, *Epidemiology*, 1998, 9, 184–188; Gorinstein S., Zemser M., et al., Moderate Beer Consumption and Positive Biochemical Changes in Patients with Coronary Atherosclerosis, *Journal of Internal Medicine*, 1997, 242, 219–224; Tjønneland A., Grønbœk M., et al., Wine Intake and Diet in a Random Sample of 48763 Danish Men and Women, *The American Journal of Clinical Nutrition*, 1999, 69, 49–54.

On the subject, it is curious to note how the two greatest doctors in the history of medieval medicine, namely Razi (*c.*865–925) and Avicenna (980–1037), not only were oriented in a different direction with respect to the measures of the *Shari'a* and even appear, on the subject, anticipators of some modern scientific trends. In the *Kitab al-Murshid* (*The Guide*), Razi textually states: "For the preservation of the health and the improvement of the digestion, the utility of wine is certain, if it is given the place it should have, and if by the quantity, quality and time when it is taken, it is in compliance with the rules of the art of medicine. It fertilises the body, expels all residues, forcing them out of the body, intensifies the innate warmth". Razi continues: "When wine is abused and it is always used to become totally inebriated, the result is great damage."[55] Similarly, Avicenna said: "Wine taken in small quantities is useful; in large quantities it is dangerous", however it should not be taken every day nor on an empty stomach or after light or acid food. Do not become inebriated, at the most once a month.[56]

With the usual realism, juridical reflection has regulated the most differing situations so that, for example, drugs containing alcohol for disinfectant purposes or to dissolve a drug are licit. According to the juridical doctrine, recourse to wine is licit if it is the only means available to avoid, for example, suffocating, based on the principle of the necessity to save a human life. A *fatwa* by the Egyptian *sheikh* Mahmud Shaltut published in 1966 in *Al-Fatawa* specified[57] that the Muslim prohibition on drinking wine was not focused so much on the drink itself as on the effects of its consumption such as the loss of lucidity which destroys the dignity of man, the alteration of good relations between individuals or even leading to homicidal behaviour, etc. These are alleged to be the spiritual and social reasons which are rooted in Koran 5.90–91. Shaltut refers to the official data in France of 1956 on the damage and deaths produced by alcoholism, concluding that God has pity on man by prohibiting alcohol. By analogy, this prohibition is extended to other drinks, food, powders, etc., which produce similar effects; these include drugs such as hashish, opium, cocaine, etc. Although in the absence of a literal text that specifies the prohibition, the latter agrees with the spirit of the *Shari'a* and with the general principle of preventing damage and avoiding any instrument of corruption.

Attempts at scientific justification also occur to legitimise the prohibition of the consumption of pork whilst in current Western hospital diets, it is frequently appreciated precisely for its highly nourishing properties and its digestibility, for example as a post-operation meal.

Obviously, the value of a divine precept does not require, logically speaking, any justification by human science (the support of which is nevertheless welcome). Indeed, the precept has an intrinsic value independent of its scientific

---

[55] "Enjoyment can be had from inebriation when it is not a systematic habit but is practised only once or twice a month". Razi, *Medico nomade del deserto*, Como, Red Edizioni, 1983, 61–62.

[56] Avicenne, *Poème de la medicine*, Paris, Les Belles Lettres, 1956, 66.

[57] Rispler-Chaim V., *Islamic Medical Ethics in the 20th Century*, Leiden, Brill, 1993, 115–117.

demonstrability as it can represent only an ordeal – which is anything but arduous – God requests from the faithful.

Every statement of the Koran and of the *Sunna*, forming the object of any medical-scientific interest, lends itself easily to a "concordist" interpretation and approach.

MODERATE CONCORDISM

The critics of "concordism" have a more lively sense of history regarding scientific transformations in the face of which they deem it a priority to protect the eternity of the "religious" contents of the Koran, putting "in parentheses" or suspending the search for concordances between the words of God and sciences due to the intrinsic relativity and transience of these new things and until further verifications are made. Nevertheless, the concordist temptation remains strong on individual subjects and discoveries even in scholars who are not open to "radical concordism".

The critics of radical scientific exegesis of the word of God include, in the first place, the "moderate concordists" who do not completely refuse the existence of scientific references.

In his analysis of the virtues and defects of scientific exegesis of the Koran[58] *sheikh* al-Khouli (1885–1966) asked: Is it lawful to refer to the Koran for information on medicine, astronomy, geometry, chemistry, etc. when these are continuously modified? This attitude is more harmful than useful. The references to nature and creation in the Koran strike the feelings and imagination of men but there are no references to precise scientific data. However, if the advocates of the total agreement between the Koran and science insist, it can tactically be accepted that "no Koranic text that is well understood clashes with scientific truth, the study of which has shown that it is part of the laws of the universe . . . . This is sufficient to ensure the Koran's adaptation to life, agreement with science and immunity to criticism".[59] In other words, al-Khouli is still midway between a radical refusal of concordism and its tactical acceptance.

M.A. Kazi defines as "ignorant" those scholars who look for scientific details in the Koran, reducing it to a scientific text. It is a "code of ethics and morals". At the same time, Kazi states that the Koran indicates on several occasions fundamental laws of nature and speaks of scientific facts, but it is not a scientific text. Nevertheless, he continues: "One can draw inspiration from this reservoir for the study of all knowledge, that is, what is known today and what will be discovered in the future."[60] Indeed, in writing books on various specialised branches of science, Muslim authors "must provide additional proof by quoting

---

[58] Jomier and Caspar, art. cit., 278–279.
[59] "Aucun texte coranique bien compris ne heurte de verité scientifique dont l'étude aurait montré qu'elle fait partie des lois de l'univers . . . . Cela suffit pour assurer au Coran l'adaptation à la vie, l'accord avec la science et l'immunité face à la critique" in Jomier and Caspar, art. cit., 278.
[60] Kazi, art. cit., 183.

a relevant reference from the Qur'an and the *Sunnah* if available. This will not only impress the enquiring Muslim mind but strengthen its faith and belief".[61]

According to Yusuf al-Qaradawi, in *Islam, the Future Civilization*, the Koran contributes to forming a scientific mind in man, i.e. a man who believes in scientific and empirical evidence. The future will confirm the many pieces of scientific knowledge in the Koran in a more or less veiled form. According to the Iranian physicist M. Gholshani, the Koran represents the guiding Text for human development containing everything necessary for the faith and human behaviour. As it is not a scientific encyclopaedia, it is incorrect to adapt it to changing scientific theories. Nevertheless, it contains references to natural phenomena which help to bring people closer to God; indeed, scientific progress makes the topicality of certain Koranic passages more comprehensible. According to *sheikh* Mustafa Maraghei,[62] former Rector of the University of Al-Azhar, the Koran is not a scientific manual although it contains general principles which are of use on the spiritual and material level for man; he prefers to avoid making forced scientific interpretations of the word of God except when the passage lends itself explicitly. Another Egyptian scholar, M. Abu Zohra, recalled that the Koran has as its principal aim that of guiding man in the religious sphere, although strong references are made to reflecting on the natural phenomena (which the Old and New Testaments do not do) as a first step to increase faith. The moderates may include the Nobel Prize winner Abdus Salam who maintains that science is the maximum expression of human rationality; science and religion must remain clearly separate whilst remaining complementary and not in opposition. Salam seemed reluctant to use the expression "Islamic science" and, in a paper in 1984, made a comparison with science in China, Japan and India: "These societies are not seduced by slogans of 'Japanese' or 'Chinese' or 'Indian' science. They do not feel that the acquiring of science and technology will destroy their cultural traditions; they do not insult their traditions by believing that these are so fragile."[63] Abdus Salam maintains that there is no contradiction between Islam and modern science, a statement of principle which however does not lead him to commit himself to specific concordances between the Koran and science.

Words that were almost identical to those of A. Salam were pronounced, moreover, by John Paul II in the Vatican (9 May 1983): "the Church is convinced that there cannot be a real contradiction between science and faith". However, in a West involved in a process of permanent scientific revolution, no representative Catholic authority analyses the Gospels in search of precise scientific "concordances" which, in any case, would have little impact on the faithful, now accustomed to the relativity imposed by continuous scientific discoveries and information. On the contrary, in the Muslim world a scission between the divine

---

[61] Ibid., 185–186.
[62] Golshani M., The Scientific Dimension of the Quran, *Al-Tawhid*, 1408, 5 (1), September–November 1987, 27–37.
[63] Loo, art. cit., 55–57 and in particular 56.

level and nature is not easily accepted, therefore a concordist interpretation of the Koran aims to reassert this connection as a sign of the truth of Islam. Indeed, it can be said that today Islam is the only historic monotheism where a significant number of faithful try to identify concordances between contemporary scientific discoveries and the Holy Texts (Koran and *Sunna*). In this regard, there are an increasing number of books and articles that agree with this perspective, as well as inter-Muslim and international congresses and conferences on these topics.

THE OPPONENTS OF SCIENTIFIC EXEGESIS

The radical refusal of scientific exegesis takes its cue from different justifications in individual authors. In fact, many perplexities date back to past centuries, as shown by the Andalusian writer Abu Ishaq al-Shatibi (d. 1388) – in *The Harmonies* (*Al-Mowafaqat*) – for whom the Companions of the Prophet and the wise men close to him in time did not speak of the Koran as a container of science[64]; Koran 6.38 "Nothing have we omitted from the Book" is believed to refer to duties and acts of worship and not to any scientific contents.

For a strong opponent of "concordism", this current comes from a basic illusion as contemporary science is successful in the physical sphere which is far simpler than the field of laws of the human soul and the invisible realities dealt with by the Koran; the eternal Text does not require superfluous scientific proof.[65] Amongst the various theories of the "anti-concordism", we discover that the Koran cannot represent the guiding text for humanity when commented upon according to increasingly rapid transformations which alter the text and concepts according to the needs of the commentators and current scientific theories with the risk of containing contradictory theories[66]; the presence of the greatest new information and technical-scientific discoveries would reduce it to a science and technology manual that can be constantly updated until religion is made into a sort of trade; it is not correct to interpret the words of the Koran in a way that was impossible for the contemporaries of the Prophet; if the Koran contains all human knowledge, the intellect becomes lazy and freedom deprived of meaning.

Furthermore, some critics underline that the continuous attempts to find concordances between science (medical and otherwise) and the Koran inevitably lead to a weakening of the sacred character of the Word of God, representing

---

[64] Jomier and Caspar, art. cit., 274–276.

[65] Hussein K., Le commentaire "scientifique" du Coran: une innovation absurde, *MIDEO*, 16, 1983, 293–300.

[66] For example, the recent discovery of a fossil micro-organism (probably bacteria) on a fragment of rock from Mars and collected in the Antarctic could represent a scientific confirmation of the extra-terrestrial origin of life (and of man) based on the Koran 2.30 ("I will create a vicegerent on earth") which, however, contradicts the theory of the origin of terrestrial life from water which can in turn be found in the Koran 21.30 "We made from water every living thing."

a plot against religion by Zionism and imperialism, aiming to spread encourage the division of the Muslim faithful.[67]

The theories of other authors against scientific concordism clearly show the wealth of perspectives in this position.

The Egyptian Sayyid Qutb (1906–1966), leader of the Muslim Brotherhood, refused any concordance between the Koran and modern science and refused a literal interpretation of the Holy Text.[68] According to Qutb, the agreement between the Koran and modern science is the product of the Western rationalist logic, which is very far from the Islamic way of thinking. More precisely, in the book *Fi-Zilal al-Quran* (In the Shadow of the Quran), he underlined the harmony existing between the book of the universe and the Koran: science and faith are closely united. The harmony of the cosmos confirms the Koranic principle of uninterrupted creation (Koran 19–20, 29; 36, 78–81, etc.). Nevertheless, Qutb clearly distinguishes the religious dimension from the scientific one, stating that it is not lawful to confirm the Koran with science nor is it lawful to look for scientific information in the Koran. This does not mean, in Qutb's words, that science ought to be separated from faith; on the contrary, he criticises the atheist temptation of pure science. As human reason changes according to times and places, reason itself needs Islam to be coherent and well guided.

In the end, Qutb accepted technology and industrialisation, which are characteristic aspects of modernity, but refused secularism and the Westernisation that went with it.[69]

M.A. Anees remarks how there exists considerable confusion in the Muslim world in the relationship between science and religion. In this regard, Anees refuses the position of those who "cling to the alleged value neutrality of knowledge".[70] At the same time, he refuses the Koran as a handbook of science and technology. In his opinion, to Islamise knowledge, it is not sufficient to add the word Islam to make Islamisation real in a general context in which the idea of "Islamic science" remains "poorly articulated and epistemologically weak".[71] In any case, according to Anees, all the positions that have just been mentioned betray the inimitable relationship between faith, spirituality and rationality which is characteristic of Islamic science. The present state of Islamic science can be redeemed only by increasing free inquiry and developing "positive interfaces between science and religion, toward a greater harmony in knowledge".[72]

---

[67] Ukacha A., Opportunisme, *Al-Ahram*, 25 August 1985, in Botiveau and Jacquemond, art. cit., 124–125.

[68] Carré O., Eléments de la 'aqidah de Sayyid Qutb dans "Fi Zilal al-Quran", *Studia Islamica*, 2000, 91, 165–197.

[69] Mumtaz Ali M., Contemporary Islamic Thought on Modernization: Reflections on Reconceptualization, *Hamdard Islamicus*, 1998, 21 (4), 19–32.

[70] Anees, *Islam and Scientific Fundamentalism. Progress of Faith, Retreat of Reason?*, art. cit.

[71] Anees, *From Knowledge to Nihilism: Redeeming Humility*, art. cit., 7.

[72] Ibid., 8.

An important jurist, the Egyptian *sheikh* Mahmud Shaltut, said that God did not reveal the Koran to instruct man on scientific theories and technological practices as the eternal would be dragged down to the level of scientific theories and hypotheses.[73]

CONCLUSION

Concordism, namely the belief in a scientific correspondence between the Koran and modern science, is a temptation that is continually present in Muslim culture. As long as it is a cultural, ideological, religious phenomenon which does not have an impact on the scientific progress of research, its contribution remains sterile. This is because concordism is limited to identifying in the Holy Texts specific discoveries or events *after* their discovery on a scientific level. The approach would be scientifically valid (with unimaginable consequences on the ideological and religious levels) when the opposite procedure were true, i.e. a scientific hypothesis is formulated from the Koran to explain specific phenomena; if this hypothesis were verified according to correct scientific criteria, we could then begin to speak of scientific valencies in the Koran. A concordist can be a valid Muslim scientist if, in practice, he keeps the two levels distinctly separated: the cultural, ideological, religious value and scientific practice. In this way, the two approaches can peacefully coexist.

The contribution of contemporary Muslim religious authorities to the development of biomedical science is generally positive. At times strongly proposing attitudes can be observed and at other times less open ones. On the positive side, we can recall the attitude towards the Human Genome Project openly approved in 1998 by the Islamic Fiqh Academy of Jeddah and by the ISESCO, despite the little interest shown in the topic in the Muslim world.[74]

The example of the debate on the criteria of brain death can be instructive. In some countries, starting with Saudi Arabia, the religious authorities have played a leading role in promoting organ transplants and the criteria of cerebral death. In other countries, e.g. Egypt and Pakistan, the greatest opponents to a law on new criteria of death are to be found in the authorities and religious parties.

Political-religious radicalism does not appear to oppose modernity. Apart from some glaring cases (e.g. Afghanistan of the Taleban), radical governments (especially if wealthy) often seem to be the most uninhibited in driving the country towards biotechnological modernisation (e.g. the case of stem cells), although results are not yet brilliant. In these cases, the myth of the golden age of medieval Islamic science and scientific concordism can provide the ideological grounds for this motivation.

To close problematically the chapter on modernity, which represents the natural premise to the development of bioethics, it is useful to present some

---

[73] Gulshani M., Science and the Muslim Ummah, *Al-Tawhid*, 1404 [1983], 1 (1), 111–131.

[74] See section "The Debate on Genetics" in Chapter 9.

resolutions and recommendations of the IFA-Islamic Fiqh (Muslim Law) Academy of Jeddah (emanating from the OIC) on secularism, modernism and the relationship with modern science. These documents are often harsh and anti-Western in tone (political component); however, it has to be remembered that the same Academy has supported biomedical modernisation in many fields. In addition, many ideals, concepts (but also some prejudices concerning Western materialistic civilisation) in these documents have accompanied us throughout this book.

Resolution no. 99 (2/11) on secularism by the Islamic Fiqh (Muslim Law) Academy (14–19 November 1998, 11th session, Bahrain) states that secularism, i.e. the separation between religion and daily life began as a reaction to the "arbitrary acts" committed by the Christian Church in the Middle Ages (99/1).[75] Secularism subsequently invaded Muslim countries, thanks to the action of colonial powers, sowing doubts on relations between reason and the *Shari'a*. In this way, secularism intends to replace the *Shari'a* by manmade laws (99/2). Secularism diffuses destructive ideologies such as communism, Zionism, racism and freemasonry which have as their aim the destruction of the community of the faithful (*umma*). One of the results is the occupation of Palestine (99/3). Secularism is a system based on the principles of atheism (99/4). Islam is a complete model of life that regulates all areas of life and refuses the separation between religion and life (99/5).

Resolution no. 100 (3/11) on Modernism (14–19 November 1998, 11th session) by the same Academy states that modernism is allied with secularism[76] and expresses complete trust in reason and in the data of empirical sciences. This estranges people from the faith. However, thanks to *ijtihad* (the use of personal reasoning; deduction by analogy, personal interpretation of the Sacred Sources, etc.), Islam is perfectly capable of being updated.

Recommendation 1/c included in Resolution no. 38 (13/4) of the IFA (4th session, Jeddah, 6–11 February 1988) is an invitation to reject any attempt at Westernisation and to fight the cultural invasion that does not respect the principles of Islam. Recommendation 1/d states that: all sciences must be taught from an Islamic perspective in order to enable the *umma* to remain at the vanguard in all sciences.[77] According to Recommendation 3/d, the history of science must be re-formulated to highlight the contributions made by Muslims and freeing it from Eastern and Western theories; the classification of sciences and research curricula must be reviewed from an Islamic perspective. The following point 3/e recommends that sciences which study the universe, man and life are presented as a product of divine creativity.

---

[75] Online available at, www.islamibankbd.com/page/oicres.htm#99 (2/11).
[76] Online available at, www.islamibankbd.com/page/oicres.htm100 (3/11).
[77] Online available at, www.islamibankbd.com/page/oicres.htm#38 (13/4).

This last document of the Islamic Fiqh (Muslim Law) Academy of 1988 invites Muslims to take part in the project of "Islamisation of knowledge" launched by the International Institute of Islamic Thought.

The expression "Islamisation of knowledge" was proposed for the first time by the Malaysian S.M. Naquib al-Attas in his book *Islam and Secularism* (1978) and subsequently used by the Palestinian philosopher Ismail al-Faruqi in 1982. Faruqi thought that at length the constant use of categories, concepts and instruments formulated in the West alienated Muslims from Islam to the extent of inhibiting the very capacity of recognising the violations of the principles of Islam, including from the ethical point of view. As a remedy he proposed recovering the ethical constraints present in Muslim philosophy of the origins. Only in this context it is possible to Islamise modern knowledge as it would remain subordinate to the original ethical values.

However, we should consider the fact that, more often than not, when Islamisation of sciences is discussed, only the Islamicness of the context in which the research is carried out or of the values shared by the researchers is claimed, without this highlighting a new Islamic scientific method. The consequences include the frequent attack of the conception of a pure or neutral science, autonomous from religion; in these cases, there is insistence on reinforcing the Muslim values at the basis of the research and in the operative context of the researchers, in imitation of what allegedly took place in the golden age of Islamic science.

Many authors are in favour of modernisation, but not Westernisation, which is almost always judged as synonymous with materialism, secularism, etc. The return to Islam would avoid these risks.

GENERAL CONCLUSIONS

In theory, at the end of our overview, we can imagine two groups of opposing positions, overlooking the effective importance of each theory in Muslim thought as well as the fact that the relationship of force between the various positions may change as time passes. This theoretical exercise is useful to recall the variety of the positions present.

The first group can include the following positions in favour of: population control and contraception as long as there is the consent of the spouses; abortion before the infusion of the soul on the 120th day (a foetus without a souls may be alleged not to be human) but also afterwards for therapeutic reasons; abortion of handicapped foetuses; research on embryos in the first stages of development or until infusion of the soul; the criteria of brain death; organ transplants; reproductive human cloning if the technique only involves the spouses; research on embryonic stem cells; passive euthanasia; female genital mutilation (in those areas where it is present) if carried out by health care personnel; penal mutilation "assisted" by health care personnel to protect the offender's life and contributing to save the moral level of society; IVFET including with the second wife of the same husband; the donation of ovules (if *Shi'ite*); etc.

The second group can include the following positions opposed to the first group, i.e. opposing contraception and population control; abortion except for therapeutic abortion but only before infusion of the soul around the 40th day; therapeutic abortion; abortion of handicapped foetuses; research on embryos; the brain death criteria; the explantation of organs from a corpse; human cloning; research on embryonic stem cells; every type of euthanasia; female genital mutilation and penal mutilation by doctors, etc.

In fact, we generally find more moderate or articulated positions on each issue in both groups.

Different models of bioethics depend, for example, on the positions regarding animation. When infusion of the soul (on the 40th or 120th day or on another date) is deemed decisive to establish the human nature of the foetus, this can generally lead to permissive positions (and "soft" approaches in criminal proceedings) on abortion before infusion, contraception, research on the embryo, the production of stem cells, the abortion of handicapped foetuses, etc. These positions may oppose those that "overlook" the role of infusion of the soul and intend to protect the embryo from fecundation onwards. These are two anthropological models which are both based on the Sacred Sources.

Some positions on particular issues – although based on different interpretations of the Sacred Sources – are similar to positions claimed in the West by secular lines of thought.

Dariusch Atighetchi, Islamic Bioethics: Problems and Perspectives.
© Springer Science+Business Media B.V. 2009

All this guarantees a strong internal plurality for Islamic bioethics in which the majority position on an individual issue generally has to be identified. These positions (and their relations with one another) may be modified in space and in time. One more example can be given by abortion on which in general the positions expressed by "classic" Muslim law could be rather liberal before the infusion of the soul; only some minority schools of thought (*Malikite*, *Zahirite*, *Jafarite Shi'ites* and the *Ibadite* sect) tended to refuse abortion at any stage. These stricter positions tend to be commoner today than in the past.

In addition, there is the variable represented by national laws and regulations which do not represent a "betrayal" of a mythical and monolithic Islam but, on the contrary, can represent a further elaboration of the very rich classic Islamic background.

Another component of very great importance is represented by the effective attitude to the various issues by individual peoples, social groups and individuals. This covers Muslim bioethics based on practice, much more important and varied than a bioethical approach based on principles which is often abstract, apologetic and, in any case, further from real life.

Today, in bioethical thought in the different religions, the risk of "fundamentalist bioethics" is not to be underestimated. This approach, which is in a minority everywhere, tends to be based exclusively on the Holy Sources of the different religions to provide immediate, simple and undisputable answers to highly complex problems which are in constant evolution and on which it is often difficult for there to be agreement on the very meaning of the terms used. Moreover, the term "fundamentalism" does not concern the positions on the individual issues (moderate or radical is without any influence), but describes the method that leads to these position.

This risk or temptation is perhaps stronger in some Muslim circles as the Koran is the only Sacred Text that contains the direct and omnicomprehensive word of God, as well as the "sayings" of Muhammad, the last of the Prophets sent by God. An arbitrary use of the Holy Texts may be made to limit the multiplication of positions on various issues of bioethics (amongst others). In actual fact, this strategy may produce the opposite effect: different but unquestionable positions as apparently based on the Word of God. Of all bioethics based on religion, Muslim bioethics is perhaps the system which makes the most frequent reference to quotations from its holy texts.

In countries of immigration (but not only there) this attempt often responds to the wishes of the individual Muslim to find clear answers allowing him to oppose the disorienting ethical pluralism present in the West.

Be that as it may, a great wealth of reflection, positions and undertones can be observed in Islamic bioethics; this wealth is not hindered but, on the contrary, stimulated by having its roots in the Sacred Sources, even if the issues discussed concern ethics.

There is not just one Islam but there are many, in ethics and in bioethics as well.

AUTHOR INDEX

A

el-Abbadi, A. S., 110
Abbas, M., 42
Abbas (sheikh), 311
Abd al-Fattah Ali (Minister for Health), 317, 319
Abd al-Fattah (sheikh), 301
Abdalla, A. H., 48
'Abd al-Rahim, 41
'Abd al-Salam al'Izz, 97
Abd el-Dagi, M. E., 52
Abdelaal, M. A., 223
Abdel-Halim, R. E. S., 270
Abdel-Moneim, M., 321
Abdel-Wahab, M., 321
Abdo, G., 45
al-Abdulkareem, A. A., 257–258
Abdullah, A. M., 167, 175
Abdul-Rauf, M., 111, 271, 287
Abid, L., 281–282
Aboulghar, M. A., 152, 180
Abu Aisha, H., 51
Abu el-Magd N., 293
Abu Bakr, 299
Abu Dawud, 3–4, 95, 162–163, 171, 298, 306
Abu Hanifa, 6, 67, 70, 291, 307
Abu Sanah, A. F., 167
Abu Shadi M., 171
Abu Yusuf, 67, 70
Abu Zikri, A., 43
Abu-Sabib, H. A., 312, 324

Abu Sa'id, 65
Abu Sahlieh, S. A. A., 8, 55, 58, 99, 113, 151, 308, 310–312, 316–317, 319, 337
Acharya, T., 238
Adam, 35, 91, 113, 342
al-Adawiya Rabi'a, 33
al-Adawy, A. R., 181
Adib, S. M., 278, 282, 290
Aghajanian, A., 86, 88
Ahmad, F., 46
Ahmed, A. M., 290
Ahmed, S., 251, 262
Aït Zaï, N., 141
Ajlouni, K. M., 47
Ajur, A. W., 295
Akhtar, F., 194
Akrami, S. M., 187, 188
Aksoy, A., 21–22
Aksoy, S., 19, 47, 93, 108
Albar, M. A., 122, 166, 251–252, 263, 339
Ali, Caliph, 94, 149, 308
Alrajhi, A. A., 226
Aluffi Beck-Peccoz, R., 105–106, 138–139, 284
Alami, M., 222
al-Alayli, (sheikh), 58
Alkuraya, F. S., 261–262
Allam, M. F. A., 323
Allpala, J. S., 76
Amaliah, L., 325–326
al-Amili, 308

355

Amr Ibn al-As, M., 67
Anees, M. A., 307–308, 328, 331, 348
Aniba, M., 201
Antes, P., 164
Arda, B., 46, 260, 281, 289
Aroua, A., 118, 252
Aryabakhshahyesh, G., 62
al-Asadiyya, J. B. W., 66
al-Ashqar, M. S., 123
al-Ashqar, O. S., 96, 102, 108
al-Ashqar, S., 108
Asnaghi, L., 168
al-Asqalani Ibn Hajar, 70
Atighetchi, D., 37, 119, 307, 319, 332
'Atiyya Saqr, 41
Averroes, 4, 34, 36, 308
Avicenna, 36, 71, 270, 330, 344
Awad, M. N., 342
al-Awwa, S., 312
Azab, A., 256
Azarmina, P., 45
Azimaraghi, O., 151

B
Babu Sahib, M. M. H., 163
al-Badri, Y., 320–321, 323
Baffioni, C., 267, 329
Baguri, I., 99
Bahar, B., 279
Bahishti, M. H., 86, 113
Bahl, L., 259
al-Bahuti, 71, 308
Bali, S., 48
Ballal, S. G., 257–258
Bälz, K., 322
al-Banna, M., 335
Baqi, S., 218–219
Barsoum, R. S., 189–190
al-Basam, A. A., 167
Bassily, S., 212
Bassiri, A., 186
Bayat, Z., 200–201
Bayoumi Abdel Mo'ti, 146
Bin Baz, (sheikh), 116
Beauchamp, T. L., 21
Belkhodja, M. H., 238, 242
Ben Hamida, F., 108, 240, 252, 288
Benomran, F. A., 106
Berkey, J. P., 308

Bernvil, S. S., 225
Bertini, B., 202, 205
Bhatti M.S., 60
Bhootra, B. L., 302
al-Bijarmi, 96
Black, J., 263
Blanc, F. -P., 139
Blum, J. D., 49
Bökesoy, I., 257, 261
Boland, R., 122
Bonifant, J., 272–273
Borrmans, M., 7, 40, 66, 74, 99, 114, 125,
    128, 142, 144
al-Bostani, A. A., 148, 166, 170, 299
Botiveau, B., 142, 156, 336–338, 348
Boubakeur, H., 174, 286
Bou-Dames, J., 261
Bouhdiba, A., 306
Bouzidi, M., 95, 112
Branca, P., 331, 334
Brockopp, J. E., 109, 291, 297
Brody, B. A., 237, 271
Bronsveld, J., 114
Broumand, B., 186
Bucaille, M., 92, 340–341
Budiharsana, M., 325–326
al-Buti, Muhammad Said Ramadan, 76,
    113, 251
Bukhari, 3, 33-34, 47, 93, 100,
    137, 143, 152, 162, 182, 237,
    269, 291.

C
Carré, O., 348
Castro, F., 1, 8–9, 54, 56, 104, 111, 136,
    149–150
Chaabouni, H., 238
Chafi, M., 135, 137
al-Chahawi, 323
Chams, D., 310, 321–322
el-Chazli, F., 213–314
Chelala, C., 322
Chiffoleau, S., 190
Childress, J. F., 21
Cilardo, A., 104
Colin, J., 106
Compagnoni, F., 37
Court, C., 62
Cullen, R., 259

D

D'Agostino, F., 37
D'Emilia, A., 104, 164
Daar, A. S., 176, 183
Dabu, C., 249
Dada, M. A., 300–301
Dalal, K., 265
Dalla-Vorgia, P., 275
Daoud, R., 24
al-Dardir, 70, 97, 101, 110
Darr, A., 259
Darwich, D., 313
Darwin, 336
Dasilva, E. J., 238
al-Dasuqi, 97, 110
Dawood, N. J., 92
De Caro, A., 19, 238, 242–243
Delavar, B., 87
Demirag, A., 192
Desai, E., 294
al-Dhahabi, 32
Dhami, S., 259
Dhar, J., 324
Dhar, L., 269, 287, 289
Dialmy, A., 84, 201, 219, 222
Dickens, B. M., 147, 260
Digges, A., 322–323
Dimashqi, 101
Dirie, M. A., 314
Dorkenoo, E., 312, 314, 324–325
Dupret, B., 320, 322

E

Ebrahim, A. F. M., 47, 69, 93, 99, 105,
    109, 118–119, 141, 167, 171, 175,
    182, 253, 284–285
Eich, T., 247
Elharti, E., 222
Eligio (Padre), 330
Ellis, M. E., 224
Elmali, A., 19, 21–22, 47
Elworthy, S., 312, 314, 324–325
Emami, A., 45
Enel, C., 233
Engelhardt H.T., 109
Ennaifer, H., 165
Erbayraktar, S., 276
Ezzat, A., 272–273
Ezzat, D., 316–317, 319, 321

F

Fadlallah, S. M. H., 145, 246
Fageeh, W., 180
Fairclough, D., 225
Fakhry, B. M., 223
al-Faqih, S. R., 170
Farag, F., 146
Farahat, M. N., 156
Farid W., N., 242
Farid, S. M., 254, 257
al-Faris, E., 44
Farokhi, M., 249
Farrag, O. A., 83
al-Faruqi I., 351
Fatemi, S. M. G. S., 150–151
Fatima (daughter of the Prophet), 255
Fawzi, M., 320–322
Faye, A., 233
Fletcher, J. C., 257
Foyer, J., 213
Freud, 336
Friend, T., 157
Fujiki, N., 240
Furqan, A., 169

G

Gabr, M., 338
Galbiati, D., 330
Garaudy, R., 338
Gatrad, A. R., 259
el-Gawhary, K., 200, 215–216
Geary, N., 275
Gerin, G., 47, 108, 145, 288
Gezairy H., 211
Ghaffari, M., 150
Ghafour, P. K. A., 264
Ghahramani, N., 187
Ghanem, I., 104, 106, 108, 141, 152
Ghavamzadeh, A., 279
al-Ghazali, 35, 67–69, 74, 82, 95–96, 255,
    267, 306, 333–334
Gholshani M., 346
el-Gibaly, O., 324
Giladi, A., 306–309, 315
Gillon, R., 21, 24, 48
al-Gindi, A. R., 79, 99, 114, 143, 154,
    208, 255
Gobrane N., 191
Göksel, F., 257, 261

Golshani, M., 346
Gumaa, A., 8, 117
Gorinstein, S., 343
Gray, A. J., 272–273
Gregory of Nazianzum, 109
Grønboek, M., 343
Guéguen, R., 343
Gulpayegani (ayatollah), 87, 188
Gulshani M., 349

H
Haberal, M., 192–193
Hafees M. A., 323
Haji-Mahmoodi, M., 46
al-Hajjaj, M. S., 52
al-Hakim Mohsen at-Tabataba'i, 150
al-Hakim Muhammad Sa'id
        at-Tabataba'i, 150, 246
Halawi, J., 295
Halim Abdel R. E. S., 270
Halim, M. A., 224
Hamadeh, G. N., 278, 282
el-Hamamsy, L., 15
Hamarneh, S., 38
al-Hamdan, N., 44
Hamdi, S. A., 225
Hamed, T., 175, 179
Hammani, A., 114, 128
Hamza, A. Ghani., 141
al-Haq, J. (sheikh), 43, 79, 81, 111–112,
        114, 140, 169, 179, 239, 251,
        309–311, 317–318
Harfi, H. A., 223
Harris, J., 146
Harrison, A., 281
Hasan, B. Ali., 149
el-Hashemite, N., 259
Hashmi, K. Z., 59
Hassaballah, A. M., 176
Hathout, H., 95, 103, 107, 110–111, 168,
        237, 271
el-Hazmi, M. A. F., 240, 256–257,
        261, 264
Hedayat, K. M., 164, 188
Helmi, S., 246
Hendriks, H. F. J., 343
Herieka, E., 324
al-Hifnawi, M. H., 338

al-Hilali, 92
al-Hilli J. Ibn al-Hasan, 71, 102, 149,
        298, 308
Himmich, H., 204, 210
Holland, J. C., 275
Holm, S., 146
Honeyman, M. M., 259
Howeidy, A., 317
al-Hudhali, Hamal b. al-Nabigha, 100
Hume D., 267
Hussain, M., 194
Hussain Saddam, 62, 253
Hussain, W., 301
Hussein, K., 347
Hyder, A. A., 217–218

I
Ibn Abbas, 41
Ibn Abdul Aziz Talal (Prince), 264
Ibn Abi Bakr, 97
Ibn Abi Talib, Ali, 308
Ibn Abidin, M. A., 70, 95, 99
Ibn Ahmad, B., 167
Ibn al-Arabi, 66
Ibn al-Humam Kamal, 70, 95
Ibn al-Jallab, 308
Ibn al-Jawzi, 97
Ibn al-Najjar, 70
Ibn al-Sakan, 73
Ibn al-Quff, 270
Ibn Asim, 110
Ibn Babawayh, 308
Ibn Hanbal, Ahmad, 6, 33, 70, 143,
        299, 307
Ibn Hazm, 71, 79–80, 102, 143, 329
Ibn Ishaq, Khalil, 70, 101
Ibn Juzay, 308
Ibn Khuzayma, 32
Ibn Maja, 3, 32, 163, 171, 202, 255,
        298, 340
Ibn Nugaym, 35
Ibn Nujaim, al-Masri, 70, 95
Ibn Qayyim, 31–32, 80, 107, 307, 309
Ibn Qudama, 31, 70, 97, 102, 104–105,
        149, 298, 308
Ibn Sa'd al-L., 256
Ibn Thabit S., 38
Ibn Taymiyya, 33, 69, 102, 307

Ibn Wahban, 95
Ibrahim, A., (sheikh), 78
Ibrahim, B., 324
Ibrahim Ismail, 249
Ibrahim, M. A., 225
Iesa, A. -A., (sheikh), 99
Imane, L., 204, 210
Inhorn M., C., 140, 150, 155–156, 140
al-Iraqi, Abdel Rahman Ibn al Hussein, 70
al-Iraqi, al Hafez (15th century), 70
Irala-Estevez, J., 323
Isbister, W. H., 272–273
Ismail, G., 323
Isobar, A. O., 225
Issef M., 220
Iyilikçi, L., 276

J
al-Jaber, A. M. Y., 145
Jacquemond, R., 336–338, 348
Jafarey, A. M., 17
Jafarzadeh, M., 150
Jahani, F., 120–121
Jarallah, J. S., 44
Jawadi Amuli, A., 328–329
al-Jeilani, M., 32
Jibril, B. Y., 335
Jimenez-David, R., 76
John Paul II, 346
Juynboll, Th. W., 4–5

K
Kagimu, M., 231
Kandela, P., 118, 211, 222–223, 227
Karakayali, H., 192
Karimi, M., 262
Karmi, H. S., 65, 67, 72, 74, 76, 78, 86, 93,
    95, 112–114, 136–137
al-Kasani, 70
al-Kassimi, F. A., 276
Katsouyanni, K., 275
Kawas, S. H., 290
Kazi, M. A., 328, 345–346
Kéchat, L., 202, 205
Kehinde, E. O., 174
Keskin, A., 288
al-Khader, A., 165, 183, 185
Khaïat, L., 213

Khalifa, D., 313
Khalil, A., 212, 213
Khalili, M. A., 158
Khamenei, S. A., 146, 150, 187, 188, 252
Khan, O. A., 217–218
al-Khatib, E., 142, 155
Kheir, M. M., 290
Khomeini (ayatollah), 40, 45, 58, 86–87,
    187, 299
Kilani, R. A., 261–262
Kim, J. Y., 204
Kister, M. J., 308
Koçtürk, T., 255
Kopelman, L. M., 314

L
Labidi, L., 42, 271
Lagarde, E., 233
Lancaster, J., 321
Lane, S. D., 24, 48–49, 127
Lankarani (ayatollah), 188
al-Laqani, 99
Larijani, B., 121, 187
Lawrence, P., 33, 40–41, 278, 296
Lemmen, K., 210, 228–230
Lenton, C., 204, 212
Lickiss, J. N., 274
Lindmark, G., 314
Loo, S. P., 343, 346
Lustig, B. A., 237, 271
Lutfi, S. H., 191
Lykkeberg, J., 59

M
Maamoon (or Mamun) H. (sheikh), 172
MacDonald, D. B., 6
Macer, D. R. J., 240
MacVane Phipps, F., 24, 50
Madani, T. A., 226
Madkur, M. S., 67, 78–79, 95–97,
    99, 102
Madut, Jok. J., 127
el-Magd, A. N., 293
Mageid, A. A., 215
Mahallati, Bahaeddin, 76, 86
al-Mahdi, M., 107, 179
Mahfouz, A. A. R., 211, 216
Mahfouz, N., 246

Mahmood, K., 249
Mahmud, Abd al-Halim, 80, 136, 143,
    300, 337
Mahmud, H., 338
Mahroof, M. M. M., 335
Makarem, S., 164
Makhluf, H. M., 32, 300–301
Malek Hosseini, S. A., 186
Malekzadeh, R., 45
Malik, Ibn. Anas, 6, 49, 101, 106, 143,
    279, 291, 307
Mamun (or Maamoon), H., (sheikh), 72
Manhart, L. E., 219
Manouchehri, M. A., 158
Mansoori, S. -D., 229
Mapes, T., 231
al-Mardawi, 105, 308
al-Marghinani, 70
Marmuch, L., 338
el-Marrouri, 201
Marum, E., 231
al-Marwazi, A. I., 96
Marx, 336
Masoodi, G. S., 269, 287, 289
el-Masry, Y., 307, 310
al-Masruwah Y., 202
Massué, J. -P., 47, 108, 145, 288
el-Matri, A., 195
Maududi, A., 57, 80
al-Mawardi, 105, 308
Mayer, A. E., 157
Mazhar, F., 194
al-Mazkur, K., 47, 95–97, 107, 118, 123,
    152, 163, 167, 179, 269, 296
al-Mazrooa, A. A., 270
al-Mazrou, Y. Y., 226, 254, 257
Mazzetti, M., 307
McKay Duncan, M., 249–250
Mehdi, R., 122
Mehmood, K., 217
Mehryar, A. H., 87
Mekki Naciri M., 73
Merhi, N., 227
Michot, Y., 307
Mikhael, M. N., 212
Minganti, P., 60
Moazam, F., 17, 41, 48, 50, 279
Mobeireek, A. F., 276–277
Modell, B., 259, 263

Mohamed, Y. A., 342
Mohammad, A, M., 320
Mokri, A., 45
Montague, J., 80
Montazeri, A., 46, 278
Moosa, N., 116
Moreau, P., 272–273, 278
Moses, S., 2, 207
Moulin, A. M., 27
Mounir, O., 139, 256
al-Mousawi, M., 175, 189
Mugas, N. V., 75
Mughniyyah, M. G., 136, 149–150
Muhammad, the Prophet, 2–5, 9, 14, 21, 26,
    28, 32, 34–35, 38, 40–41, 48, 55, 57,
    65–67, 72–73, 79–80, 90–93, 95, 106,
    110, 115, 135–136, 143, 148, 152,
    161–163, 168, 170–171, 199, 206, 220,
    236, 254–255, 265, 267–270, 291,
    294, 296, 299, 306, 308–312, 314, 317,
    320–321, 327, 332, 335, 339–340, 347
Mujeeb, S. A., 217
Mumtaz, A. M., 348
Murad, H., 145
Murtaza, M. G., 251, 262
Musallam, B. F., 80, 107
al-Mushidd, Abd Allah, 147
al-Musili, 308
Muslim (d.875), 3, 31–33, 65, 70, 93, 115,
    268, 291
Mussallam, N. S., 265
Mustafa, Mahdi Abd al-Hamid, 337

N
Nab, N., 48
Nabi, N., 218–219
Nabulsi, M., 261
Nacef, T., 42, 271
Nadvi, M. S., 247, 328, 330–332, 335
Nagar, S. E. H., 324
Naguib, M., 334
al-Naggar, A., 311–312
al-Nahyan, S., 265
Najafi Marashi, (ayatollah), 87
Nanji, A. A., 32, 199
Napoleon, 330
Naqui, A., 194
Naquib al-Attas, S.M., 351
al-Naqueeb, N. A., 256

Naqvi, A. R., 119–120
Naqvi, N. H., 59
Naqvi, S. A., 194
al-Nasa'i, 3
Nashabé, H., 145
Nasim, A., 240
Nasr, S. H., 327, 329
Nasserallah, Z., 257, 258
Nassiba, Om Emara, 41
al-Nawawi, 308
Nazer, I., 65, 67, 72, 74, 76, 78, 86, 93, 95,
    112–113, 136–137
Ng, E., 230
Nikbakht, N. A., 45
Nimir, Mahdi Babo, 85
Nolan, H., 63

O
al-Obeid, A. S., 339–340
Oguz N., Y, 289
Omar Ibn al-Khattab (second Caliph), 94,
    162, 254, 307
Omran, A. R., 25, 66, 69–70, 72, 76, 78,
    80–81, 83, 113, 251
Ornek Buken, N., 280–281
Osati, Z., 187–188
Ossoukine, A., 16, 165–166, 201
al-Otaibi, K., 165, 185
Othman A.G., 230
al-Othimin, M. I. S., 245
Othmani, M. T., 251, 262
Ouyahia A., 16

P
Pacini, A., 56
Pahlevi, R., 86
Panter-Brick, C., 83, 250, 254, 258
Pareja, F. M., 6
Pelin, S. S., 46, 260, 281, 289
Pellegrino, E. D., 275
Perrin, P., 63
Peters, R., 61–62
Peto, R., 343
Peyvandi, F., 262
Pirotta, M., 42–43
Pitamber, S., 324
Plummer, F. A., 207
Prader, J., 137
Prystay, C., 231

Q
Qabbani, M. R. (sheikh), 339, 342
Qabil, D., 314, 317, 320
Qadir, C. A., 330
al-Qadhi, A., 339
al-Qalqili, A. A., 72, 112
Qarami, Z. R., 339
Qoéta, M., 191
Querry, A., 71, 103
Quigley, M. A., 207
Qureshi, A. F., 49, 279
al-Qaradawi, Y., 39–40, 55, 58, 94, 98,
    116, 142–143, 164, 182, 243–244,
    247, 251, 292–295, 310, 346
al-Qarafi, 5, 35
al-Qasri Halid b. A., 315
al-Qastallani, 70
al-Qattan, M. B. K., 163, 166
al-Qudat S., 337
al-Qurtubi, M., 97, 166
Qutb Sayyid, 348

R
Rab, M. A., 238
Raffa, H., 180
Rahman, A. Ibn, al Hussein al-Iraqi, 70
Rahman, F., 33, 38, 68, 109, 111, 172,
    200, 267, 270
Rahman, M. S., 83
Rahman, Z., 207–208
al-Rahwani, 97
Rajab, 96
Rashad, A. M., 24, 50
Rasheed, H. Z. A., 172
Raza, M., 164, 188
al-Razi, Abu Bakr, 34, 38, 270, 330, 344
al-Razi Fakhr al-Din, 291
Reich, W., 111, 271
Renaud, S. C., 343
Rida, M. R., 106, 300
Rishmawi, M., 133
Rispler-Chaim, V., 32, 41, 76–77, 106,
    109, 116, 136, 140–141, 143,
    147, 165, 171, 200, 251–253, 298,
    300–301, 307, 344
al-Riyami, A. A., 257
Rizk, D. E. E., 44
Rizk, K., 278
Rizvi, A., 194

Rizvi, S. A. H., 194
Roberts, A., 259
Rosenthal, F., 290–292
Rostagno, L., 7, 74, 114
Rozmus, C., 33, 40–41, 278, 296
Rushd, M. b., 308
Rushwan, H., 312

S
Saad, R., 293–295
al-Saadawi, N., 313, 318
al-Saadi, A. M. H., 281
al-Sabil, A., 294
Sabiq, Sayed, 73
Sabri I., 116
Sachedina, A., 244, 248
Bin Saeed, K. S., 25–26
Saheb, M. S. M., 245
Said, S., 117
al-Saif, A., 47, 95–97, 107, 118, 152, 163,
    167, 179, 269, 296
Sakr, H., 211
Salahi, H., 186–187
Salam, Abdus., 330–332, 346
Salama, M., 191
Saleem, H. M., 335
Saleem, M., 251, 262
Salihu, H. M., 263
Salim, Abdul-Majid (sheikh), 71, 99, 111
Sallam, I., 212
Sallam, S. A., 211, 216
Samavat, A., 263
el-Samerra'i, F., 101, 115
Samhan, M., 189
Sanei, O. Y., 188, 209
Santillana, D., 3–6, 32, 34–35, 101, 110,
    137, 162, 275
Santos, D. J. C., 76
Saoud, T., 220
Saurel, R., 314
al-Sayed Hamdy, 191, 211, 249
Sayeed, S. A., 200
al-Sayyid W. H., 320
Scarcia, Amoretti B., 7
Schacht, J., 4, 55, 276
Schenker, J. G., 157
Schieffelin, O., 71–72, 76, 112
Scholl, J., 343

Seck, A., 232–333
Selim III (Ottoman sultan), 330
Sellami, M. M., 169–170, 177
Semait, S. I., 172
Sen, M., 273, 280
Serour, G. I., 21, 25, 145–147, 152–154,
    180–181, 236, 239, 244–245, 248,
    260, 318
Sgreccia, E., 235
Shafi, M., 167
Shaheb, S., 190–191
Shaheen, F. A. M., 48, 184–185, 197
Shahraz, S., 18, 229
Shalabi, M. M., 147–148
Shaltut, M., (sheikh), 7, 76, 78, 94, 98–99,
    107, 142, 253, 311, 344, 349
Shamsuddin, M. M., (sheikh), 76, 79, 86
al-Shafi'i Muhammad I. Idris (d. 820), 6,
    149, 174, 291, 298, 307–308
al-Sha'rawi, 163, 170–171, 268, 321, 339
el-Shahat Y. I. M., 166, 195
Shahine G., 211, 295
al-Shahri, M. Z., 42–43, 269, 280
al-Shaikh, A., 259
al-Shammari, S., 44, 173
el-Shanti, H., 257
al-Sharabassi, A., 72, 113
al-Sharani, 70
al-Shatibi Abu Ishaq, 347
al-Shaybani, M., 70
al-Sheikh Abdulaziz, 295
el-Sheikh, S., 215
al-Sherbini, I., 47, 269
al-Sheikh, A., 259, 295
Sherafat-Kazemzadeh, R., 229
Sheth, K., 225
Shetty, P., 278
Short, R. V., 207
Shushan, A., 157
Siddiqi, M., 248
Simforoosh, N., 186
Shirazi (ayatollah), 87
al-Shirazi (d. 1083), 70
al-Sistani, U. S. A. H., 151, 159, 167, 299
Soebadi, R. D., 274
Sohaibani, M. O., 302
Souqiyyeh, M. Z., 184–185, 197
Stones-Abbasi, A., 151

Suleiman, A. J., 257
Sullivan, L. E., 33, 109, 164, 267
al-Swailem, A. R., 256–257
al-Sukkari, Abd al-Salam, 170, 309–310, 318
Syed, I. B., 245, 247, 335
Szabo, R., 207

T
al-Tabari Abu Ja'far, 108, 291
Tabrizi (ayatollah), 296
Tadros, M., 117, 190–191
al-Tahawi, 70
al-Taher, A., 312, 324
Talib, N., 49
al-Tamimi, E. A., 117–118
Tantawi Jawhari, 334, 341
Tantawi, M. S., 74–75, 81, 116–117, 146,
        163, 165, 169, 171, 178–179, 190,
        193, 293–294, 297
Tawfik, M. O., 274
al-Tawhidi Abu H., 292
al-Tayyeb, A., 249, 294
Teebi, A. S., 254
Tejawinata, S., 274
Tenik, A., 21
al-Thagafi, M. F., 52
Thanvi A., 80
St. Thomas, 109
Thun, M. J., 343
al-Tirmidhi, 3
Tjønneland, A., 343
La Tour Dupré A., 145
Torki, B., 333, 336, 342
Touré, B., 231, 234
al-Tusi, 149, 298, 308

U
Ukacha, A., 348
Umeh, J. C., 28
Uskul, A. K., 46
Uthman (Caliph), 3

V
Vahdani, M., 278
Vallaro, M., 104

Vawda, A., 182
Veenstra, J., 343

W
Wahba, M., 59
Wahdan, M. H., 204, 212
Wahid Ad-Din, Han, 329, 333
al-Wa'ii, T., 95, 97, 102
Wajid, H., 333
al-Wakeel, D., 264
Walters, L., 249
Wassel N. Farid, 117, 147, 190, 242, 251,
        294, 321
Weiss, H. A., 207
Wertz, D. C., 257
Wiessner, P., 210, 228–230

Y
Yaqubi (sheikh), 165
Yasin, M. N., 99, 107
Yomna Kamel Middle East Times Staff, 181
Younge, D., 272–273, 278
Youssef, R., 245–246
Youssef, M., 191
Youssef, Y., 190, 320
Yusuf, Ali, A., 2, 92, 341

Z
Zadsar, M., 229
Zahed, L., 261
Zahedi, F., 121, 187
Zaki Hasan, K., 21
Zali, M. R., 18
Zarakshi, 105
Zargooshi, J., 187
al-Zarqa, M., 98, 103, 171
Zawawi, T. H., 223
Zeidguy, R., 139
Zemser, M., 343
Ziaei, S., 249
Zindani, A. M., 336–337, 342
Zohra, M. A., 80, 346
Zribi, A., 240
el-Zubeir, M. A., 44
Zurqani, 34–35

SUBJECT INDEX

Page numbers followed by *n* indicates footnotes.

A

Abbasid period, 38
Abortifacient properties, 77
Abortion, 59, 91–133, 262
  acceptance of, 111
  after ensoulment, 98–100
  ahadith on, 100, 107–108, 110
  bioethical problems, 106–111
  birth control, 112
  contraception, 112
  debate and contemporary opinions,
    111–115
  defence of honour, 131–133
  drug, 105
  duration of pregnancy, 105–106
  before ensoulment, 95–97
  for eugenic purposes, 250
  fasting, 105
  Muslim law, 96
  Muslim views, 133
  penal system, 100–105
  prohibition (*haram*) of, 94
  punishments, 131
  rape, adultery and fornication, 115–119
  sources of tradition, 92–95
Abortion, legislation, 119–131
  Algeria, 127
  Bahrain, 127
  Burkina Faso, 129
  Egypt, 126
  Indonesia, 128
  Iran, 119
  Kuwait, 123
  Malaysia, 125
  Mali, 130
  Mauritania, 130
  Morocco, 131
  Nigeria, 129
  Pakistan, 122
  Qatar, 124
  Saudi Arabia, 121
  Senegal, 129
  Sudan, 124
  Tunisia, 125
Abraham, 2
Abrahamic monotheism, 3
Academy of Muslim Law, Human
  Cloning, 238
ACECR, *see* Iranian Academy Centre for
  Education, Culture and Research
ADA, *see* Deficiency of adenosine-
  deaminase
Adoption, 159
Adultery, 138, 201, *See also Zina*
Aga Khan University, Pakistan, 17
Ahadith, 3–5, 14, 79, 93, 162, 199, 292,
  310, 312
  and abortion, 100, 107–108, 110
  as word of Prophet, 10, 57, 65, 306, 327
AIDS, 20, 199–234
  and abortion, 208–209
  and blood donation, 217
  and blood screening, 218
  and blood test, 212–214

AIDS (*Continued*)
  and breastfeeding, 208
  and circumcision, 207
  and condom, 202
  control programme, 229
  health budget, 210
  and marriage, 207–208
  in the Middle East, 204
  in Muslim countries, 204–206
  Muslim customs, 226
  in Muslims, 200
  particular aspects of, 207–210
  prostitution and homosexuality, 217
  screening, 225
  and sin, 201
  spread, 231–232
  transmission, 209
AIDS and countries
  Egypt, 211
  Indonesia, 230
  Iran, 229
  Lebanon, 226
  Libya, 228
  Malaysia, 229
  Mali, 234
  Morocco, 219
  Oman, 228
  Pakistan, 217
  Saudi Arabia, 223
  Senegal, 232
  Syria, 228
  Uganda, 233
AIDS Hotline, 215
Al-azl, 65–66, 70
  Ghazali, 67–68
Alcoholism, 36
Algerian Family Code, 141 *n*
Al-Halal wa-l-Haram fi-l-Islam
    (The lawful and the prohibited
    in Islam), 292
Ali al-Ruhawi, 32
A'mal al-badan, 1
A'mal al-qalb, 1
Amniocentesis, 260
Anencephalic foetus, explantation from, 177
Anthropological models, 353
  and Sacred Sources, 353
Anti-AIDS programme, 233–234
Anti-HIV tests, 228

Apologetics, 42
Archangel Gabriel, 2
ART, *see* Assisted Reproductive
    Technology
Artificial fecundation, 136, 150
Artificial fertilisation, 148
Artificial insemination, 135, 157
  homologous technique of, 144
Artificial procreation, 135, 149, *See also*
    Assisted procreation
  lesbian relationship, 150
Assisted procreation, 135–159
  juridical-religious formulations,
      140–148
  legal adoption, 139–140
  opinions in shi'ite islam, 148–151
  problems relative to embryo, 151–154
  society and legislation, 154–158
Assisted Reproductive Technology
    (ART), 147, 154, 181, 245, 248
Autonomy, 51–54
  in Algeria, 51
  and consent of the patient, 47–54
  in Libya, 51
  in Saudi arabia, 51
  in Tunisia, 51
Azl, 80, *See also Al-azl*

B
Beneficiary of transplant, 165
Beta-thalassaemia, 262
Bid'a (innovation), 13–14
Bioethical issues, 13
  Algeria, 16
  Iran, 16
  Pakistan, 16
  Tunisia, 16
Bioethical issues, opinion on
  biomedical bodies, 13
  doctors, 14
  fatwa committees, 13
  jurists, 14
  Muslim World League, 13
  Organisation of the Islamic
      Conference, 13
  pan-Muslim congresses/conferences,
      13
  political authorities, 14
  religious authorities, 14

Bioethics, 11, 15, *See also* Bioethical
   issues
   and apologetics, 28
   cultural context, 25–28
   dependence on Muslim law, 18–19
   Muslim countries, 28
   principles of, 21–23
   and society, 26–28
Bioethics in Human Reproduction, 239
   in Muslim world, 153
Biomedical ethics, 17
Biomedicine, 5
Blood money, 100, 100 *n*
Blood price, 103
Blood transfusions, 167–168, 223
   and Sacred Sources, 167
Brain death, 189
Burial arrangements, 302

C
Caliph Uthman, 3
Catholic personalism, 23
Causae secundae, 267
Causality, 267
Celibacy, 137
Cerebral death, 175–176, 188
Child prostitution, 222
Christianity, 2
Circumcision, 232
Clitoridectomy, 306
Cloned man and soul, 245
Cloning, act of creation, techniques of, 242
CNEM, *see* National Committee of
   Medical Ethics
Code de Statut Personnel (Moudawana),
   Morocco, 138
Code of Ethics, of PMDC, 17
Coitus interruptus, 65–67, 94
Commercial sex workers (CSW), 218
Committee of *Fatwas* of the University
   of Al-Azhar, 8
Compensation/testamentary adoption
   (*jaza*), 139
Conception, artificial means for, 140
Concordism, 332
   radical, 334–336, 341, 345
   and *Shari'a*, 333
Condom, 231
   fornication, 220

Congenital malformations
   beta-thalassaemia, 250
   haemophilia, 250
   mucoviscidosis, 250
   sickle cell anaemia, 250
Congenital syphilis (*bijel*), 258
Consanguineous marriage
   congenital malformations, 254
   genetic pathologies, 254
   neural tube defects, 254
   neuro-metabolic disorders, 254
   precocious mortality, 254
*Conseil National de l'Éthique des Sciences
   de la Santé*, Algeria, 16
Consent, free and informed, 152
Contagion, 199–200
Contraception, 65, 87, 231
   coitus interruptus, 77
   condom, 77
   diaphragm, 77
   hormonal devices, 77
   mechanical interceptive devices, 77
   opponents of, 79–82
   and pro-contraception jurists, 71–78
   progestogen pill, 77
   socio-political context, 82–86
   and socio-political context, 82–86
   spermicides, 77
   survey, 83 *n*
Contraceptive pills, 220
Contraceptive techniques, 78
Contragestives, 77
Cornea transplant, 172, 195
Corporal punishments, 57, 61
Criteria of death, debate on, 174–178
Cryopreservation of ovocytes, 158
Cryopreserved embryos, 146, 152, 159
Cryopreserved sperm, 147–148
CSW, *see* Commercial sex workers
Cultural sensibilities, medical ethics
   and Islamic bioethics, 23–26
Cystic fibrosis, 236

D
Dead foetus, 177
Death
   end of life and, 296–297
   ensoulment, 108
   penalty, 57, 59–60

Deceased or Brain Dead Patients Organ
    Transplantation Act, 188
Deceased to be buried, 299
Deficiency of adenosine-deaminase
    (ADA), 236
Deontological theory, 21
Development of foetus, 107
Dhimmi, 164
Dhimmis (protected; Christians
    and Jews), 20
Diyya, 71, 94, 101–102, 104
DNR, see Do-not-resuscitate
Doctor, Muslim civilisation, 36
Doctor–patient relationship, 48
    and aspects of medical ethics, 37–39
    and khalwa, 37–39
Donor's card, 173
Do-not-resuscitate (DNR), 276
Down syndrome, 250
Drug addicts, 218
Duchenne type muscular dystrophy, 236
Duration of pregnancy, 106

E

Embryo, problems, 151–154
Embryonic development, 92, 107
    and Sacred Sources, 108
Embryonic stem cells, 248
Embryos, experimentation on, 152
Embryo-transfer, juridical implications, 144
End of life, 267–304
    death, 296–297
    and Euthanasia, 285–290
    incurably and terminally ill patient,
        271–272
    and living will, 283–285
    palliative care, 272–274
    and post-mortems, 297–303
    seriously and terminally ill patient, infor-
        mation and consent of, 274–283
    suffering and illness, 267–270
    suicide and martyrdom, 290–295
Ensoulment, abortion, 101, 103–104
    before, 95–97, 113, 153
    after, 98–99, 112–113, 153
    and sources of tradition, 92–94
    and three bioethical problems, 106–111
Ethical Guidelines in Human Research, in
    Islamic World, 25

Ethical Implications of ART, 147
Ethical-juridical principles, organ
    transplants, 161–163
Ethical pluralism, problems of
    in Egypt, 15
    In Europe, 16
    and interreligious pluralism, 16
    and Islamic bioethics, 14–16
    Muslim communities, 14
    and Sacred Sources, 15
    and second hadith, 14
Ethics, obsolete conception of, 16
Ethics Committee for Research in Human
    Reproduction, 25
Euthanasia, 273, 285
    active and passive, 286
    end of life and, 285–290
    Islam, 286–290
Expiation (kaffara), 55
Eye transplant, 172

F

Fard kifaya, 32
Fatawa (sing. fatwa), 7–8, 13
    and sheikh of Al-Azhar, 8
Fecundation, 107
Female circumcision, 305
Female genital mutilation (FGM), 20
    debate, in some countries, 316–326
    historical-juridical elements of,
        305–309
    opinions, against of, 311–316
    opinions, in favour of, 309–311
    and Shari'a, 322–323
    in specific Muslim areas, 305–326
    types of, 306
Fertilised ovules, 151–152
Fertility, 135
FGM, see Female genital mutilation
Flagellation, 60
Foetal sex selection, 153
Foetus, 93
    adultery, 115
    animation, 107
    autonomy of, 111
    ensoulment, 93, 106–107
    miscarriage of, 102
    mother, 110
Foreign workers, 224–225

Frozen embryo, 158
Fundamental principle of necessity to
    save human lives (*al-dururat tubth
    al-mahzurat*), 40
Fundamentals of Islam, 327

G
Gene therapy, 235
    on germinal cells, 236
    on somatic cells, 236
Genetically identical human, 244
Genetic counselling, 264
Genetic engineering, reasons
    diagnostic, 235
    manipulating, 235
    manufacturing, 235
    regarding therapeutic, 235
Genetic fingerprints, 240
Genetics
    and abortion of handicapped foetuses,
        250–254
    and consanguineous marriage, 254–259
    debate on, 235–245
    human cloning, positions tolerating,
        245–247
    Islamic Perspective, 263
    and Muslims law, 239
    and Muslims world, 238–240
    opinions on, 235–265
    and pre-natal diagnosis, 259–265
    principles and values, 235–237
    and research on stem cells, 248–250
Genital mutilation
    debate in countries, 316–326
    jurist opinion, 306–309
    opinions against, 311–316
    opinions in favour, 309–311
Ghurra, 94, 101–104
Gonorrhoea, 36
Gynaecological clinic, survey at, 44

H
HAART, *see* Highly active antiretroviral
    therapy
Hadith (pl. *ahadith*), 3–5, 14
    and authenticity, 3
    good or acceptable (*hasan*), 3
    healthy (*sahih*), 3
    Muslim, 3

Sunna (Tradition), 3
    weak (*da'if*), 3
Haemophilia, 236, 262
Hanafites, 5
Handicapped foetuses, and genetics
    abortion of, 250–254
        Muslim countries positions, 253
        rigid positions, 252
        tolerant positions, 251
Hereditary hearing impairment, 257
Heterologous artificial fertilisation, 135,
    145–146
Heterologous artificial insemination,
    141, 148
Heterologous artificial procreation, 138
Heterologous procreation, 150
    techniques, 159
Heterosexual transmission, 230
HGP, *see* Human Genome Project
Highly active antiretroviral therapy
    (HAART), 230
Hiraba, 209
HIV, mother–foetus, 225
Homologous artificial insemination,
    135, 151
Homologous *in vitro* foeundation,
    148, 151
Homologous IVF, 159
Hormonal contraceptives, 78
HUGO, *see* Human Genome
    Organisation
Human clone, resurrection, 247
Human cloning, 241–245
    act of creation, 242
    immoral, 242
    married couples, 245–250
    motives in favour, 242
    negative opinion, 243–245
    new procreative technique, 246
    positions in Islam, 241–245
    positive opinions, 245–250
    therapeutic instrument, 249
    tolerating positions, 245–250
Human embryos, therapeutic cloning of,
    249
Human Genome Organisation (HUGO),
    240
Human Genome Project (HGP), 240
Human life, start of, 108–109 *n*

Human Reproduction in Islam, 79
Human Rights in Islam, 56
Hymenorraphy, 117

I
ICSI, *see* Intracytoplasmic sperm
　　injection
ICSI technique, 158
Igma' (unanimous consensus), 2–6, 9
　　and elementary duties, 4
　　and the Koran, 4
　　worship and law, rules of, 4
Igtihad, 6–7
Illegitimate child (*walad az-zina*), 138
Illegitimate relationship, 119
Illness, islamic view, 268
Imitation (*taqlid*), 6
Infertility
　　divorce, 137
　　in Muslim, 136
　　polygamy, 137
Infidels
　　idol-worshippers, 164
　　people of the Book, 164
International Seminar on Organ
　　Transplantation, 162
Interreligious pluralism, 16
Intracytoplasmic sperm injection (ICSI),
　　156
Intravenous drug abuse, 230
In vitro fertilisation (IVF), 136, 140, 143,
　　157–158
In vitro fertilisation with embryo transfer
　　(IVFET), 142, 148, 154–155
Iranian Academy Centre for Education,
　　Culture and Research (ACECR), 18
Iranian National Commission for
　　UNESCO, 18
Islam, 2
　　defence of, 9
　　and family planning, 78
　　and the Koran, 15
　　and Population Policy, 83
　　and Prophet Muhammad, 2
　　and the *Shari'a*, 15
Islamic bioethics, features of, 13–29
　　and apologetics, 28
　　cultural sensibilities and, 23–26

dependence on Muslim law and, 18–19
ethical pluralism, problems of, 14–16
and modernisation of Muslim law, 14
monolithic image, 19
and Muslim countries, 28
political dimension of, 19–21
and society, 26–28
and value of different positions, 13–14
Islamic Code for Medical and Health
　　Ethics, 39
Islamic Code of Medical Ethics, 32, 36
Islamic Manual of Family Planning, 75
Islamic medicine, 270–271
Islamise modernity, 333
IVF, *see In vitro* fertilisation
IVFET, *see In vitro* fertilisation with
　　embryo transfer

J
Jesus Christ, 2
Jihad, 294–295
Juridical approach, 39
Juridical capacity, limitation of, 276
Juridical principles, 163
Juridical-religious authority, 5
Juridical-religious formulations, assisted
　　procreation and, 140–148
Jurists, 7
　　fuqaha, 1
　　muftis, 1, 8
　　pro-contraception, 71–78
　　ulama, 1

K
Kafala, 139, 141 *n*
Kaffara, 100
Khafd or *khifad*, 306
Khalwa, 39–41
　　doctor–patient relationship, 37–39
Kidney donors, 187
Kidney transplant, 172, 174, 184, 189,
　　192, 194, 1186
The Koran, 2, 5, 14, 21–22, 33–35, 39,
　　54–56
　　as authentic word of God, 4
　　as divine miracle, 3
　　and *igma'* (unanimous consensus), 4
　　and Islam, 15

The Koran (*Continued*)
and modern science, 27, 327–354
positions in, 20
Sacred Sources, 28
scientific exegesis, 332–345
the *Sunna*, supplement to, 3

L
Lantiretroviral drugs, 211
Laws to control/expel foreigners infected
with HIV/AIDS, 210
Lebanese Committee of Bioethics, 18
Legal adoption, and assisted procreation,
139–140
Legally dead, 176
Lesbian relationship, 149
Life, end of, 267–304
death, 296–297
and Euthanasia, 285–290
incurably and terminally ill patient,
271–272
and living will, 283–285
palliative care, 272–274
and post-mortems, 297–303
seriously and terminally ill patient,
information and consent of,
274–283
suffering and illness, 267–270
suicide and martyrdom, 290–295
Life, ensoulment, 108
Living non-related donors (LNRD), 168
Living related donors (LRD), 168
Living unrelated donors (LURD), 168
Living will (*al-wasiya*), 283
LNRD, *see* Living non-related donors
LRD, *see* Living related donors
LURD, *see* Living unrelated donors

M
Malik Ibn Anas, 6
Malikites, 5
Manslaughter, 100
Mansuri hospital, 38
Maple syrup urine disease
(MSUD), 257
Mapping of genes, 240
Marriage and family in Muslims, 136
Marriage prohibits, 255–256

Martyrdom, 292
and suicide, 290–296
Martyrs' attacks, 294
Maslaha, 37
Masturbation, 143
Medical ethics, aspects of
autonomy and consent of patient,
47–50
doctor–patient relationship and,
37–39
doctors and penal mutilation, 57–63
men and women, 39–45
principles and characteristics, 33–37
problem of penal mutilation, 54–57
Medical ethics (*akhlaq tibbiyyah*), 19
cultural sensibilities in, 23–26
Medical ethics and Muslim law
principles and guiding values, 236–237
Medical paternalism, 48, 283
Medical practice
in Algeria, 51
in Libya, 51
in Saudi arabia, 51–54
in Tunisia, 51
Medicine for the spirit, 270
Medicine of the Prophet, 32
Menopause, 146
Mental health, 38
Microinjection, 236
Milk banks, 255
Milk children, 256
Ministry for Justice, 16
Moderate concordism, 345–347
Modern science, and the Koran
concordism, 345–347
scientific exegesis, 332–345
scientific exegesis, opponents of,
347–349
Monolithic image, of Islamic bioethics, 19
Monotheistic faiths, 2
Morphine, 273
Mother–foetus transmission, 223
MSUD, *see* Maple syrup urine disease
Muftis, 8, 10
Muhammad al-Shafi 'i, 6
Muslim bioethics, 19, *See also* Biomedical
ethics
Muslim Code of Medical Ethics, 58

Muslim corpse, guidelines to be followed, 296
Muslim law
  and genetics, 239
  and Islamic bioethics, dependence of, 13–14
  modernisation of, 14
  origin of, 1–6
  in the penal context, 54–55
  present, 7–9
  and universal declaration of human rights, 56
Muslim law, categories of
  compulsory (*fard, wagib*), 1
  forbidden (*haram, mahzur*), 1
  free (*ja'iz, mubah*), 1
  recommended (*mandub, mustahabb*), 1
  unadvised (*makruh*), 1
Muslim law, present, 7–9
  and analogical method (*qiyas*), 7
  and issue *fatawa* (sing. *fatwa*), 7–8
  muftis and, 8
  and personal interpretation (*igtihad*), 7
Muslim law, roots of, 2
  igma' (unanimous consensus), 2
  the Koran, 2
  qiyas (reasoning by analogy), 2
  Sunna (Tradition), 2–3
Muslim Law and Universal Declaration of Human Rights, 56
Muslim Law Council, UK, 177
Muslim medical anthropology, 270
Muslim medical ethics, 33, 203
Muslim medical-juridical literature, 13
Muslim monotheism, 91
Muslim patrimonial law, 162 *n*
Muslim penal law, 57
Muslim territory (*dar al-Islam*), 164
Muslim world
  awakening of, 7
  and genetics, 238–240
Muslim World League, 13

N
National Bioethics Committee, 15
  Egypt, 18
  Iran, 18
  Pakistan, 17

National Committee of Medical Ethics (CNEM), Tunisia, 16–17
Neonatal screening
  hypothyroidism, 265
  phenylketonuria, 265
New reproductive techniques (NRT), 154–155
Non-reproductive cloning, 245, 248
NRT, *see* New reproductive techniques
Nutfa, 341

O
Organ donation, 163, 163 *n*, 169
Organs of animals, 182
Organ trade, 178
  and organ transplantation, 178–180
Organ traffic, 178
Organ transplants, 161
  from corpses, 170–174
  corpses of sentenced, 165
  and debate on the criteria of death, 174–178
  development of, 161–197
  ethical-juridical principles, 161–163
  features of the debate, 163–168
  from living donors, 168–170
  Muslims and non-Muslims, 166–168
  organ trade, 178–180
  and uterine transplantation, 180–181
  xenotransplantation and, 181–183
Organ transplants, national legislations, 183–196
  Egypt, 189
  Iran, 186
  Jordan, 196
  Kuwait, 188
  Morocco, 193
  Pakistan, 194
  Saudi Arabia, 183
  Tunisia, 195
  Turkey, 192
  United Arab Emirates, 195
Ottoman Empire, territories of, 6

P
Pain management, 274
Pakistan Medical and Dental Council (PMDC), 17

Palliative care
  end of life and, 272–275
    main problems, 273
    socio-cultural obstacles, 273
    use of morphine, 273
Patient, incurably and terminally ill, 271
Patient, seriously and terminally ill
  information and consent, 274–283
  Muslim law, 276
Patients, egalitarian trend for, 38
Penal mutilation, 58, 60–61
  and aspects of medical ethics, 57–63
Penal mutilation, and doctors
  and aspects of medical ethics, 57–63
Penal mutilation, problem of, 54
  Muslim Law, 39–46, 54–55
Pessary (diaphragm), 69 n
PGD, see Preimplantation genetic
  diagnosis
Phenylketonuria, 258
Physical penalties, 57
Plague in Tunisia, 199
Pleasure trips, 223
PMDC, see Pakistan Medical and Dental
  Council
Political dimension, of Islamic bioethics,
  19–21
Polygamy, 135
Population control, contraception and
  classic formulations of, 65–71
  Iran, 86–89
  opponents of contraception and,
    79–82
  pro-contraception jurists, 71–78
  socio-political context, 82–86
  sterilisation and, 78–79
Post-menopausal pregnancy, 147
Post-mortem pro-donation, 188 n
Post-mortems, end of life and,
  297–303
Pre-embryos, 152
Pregnant wife, burial, 99
Preimplantation genetic diagnosis (PGD),
  154, 260
Premarital blood test, 264
Premarital HIV test, 230
Premarital medical certificate, 226
Premarital relations, 219

Pre-natal diagnosis, and genetics,
  259–265
  abortion, 261–262
  anti-ethical purposes, 260–261
  ultrasonography, 260
Presumed consent, 173
Principalism, 21
Principle of medical ethics (maslaha), 37
Principles
  of Islamic bioethics, 21–23
  of public good/usefulness (maslaha), 35
Procreative union (al-istibdaa), 135
Procured abortion, 94
  animated foetus, 101
Professional Codes of Muslim countries, 48
Prophet Muhammad, 2
Prophet (nabi), 2
Protection of patient's rights, 227
Psycho-physical inferiority, 47–48
Psychosomatic medicine, 38, 270
Public benefit (maslaha), 22
Punishments
  diyya, 55
  hudud, 54
  qisas, 54
  ta'zir, 54
Purchase of organs, 178

Q
Qiyas
  logic and deduction, method of, 4

R
Recessive autosomal genes, 254
Religions of the Book, 2
Reproductive cloning, 245
Reproductive human cloning, 154
Reproductive medicine, 158
Restrictive juridical reflection, 7

S
Sacred Sources (the Koran and the
  Sunna), 28, 148, 306, 312
  anthropological models, based on, 353
  and blood transfusion, 167
  and embryonic development, 108
  and problems of ethical pluralism, 28
Salvific religion, 31

Science
  Islamic interpretation of, 327
  Islam's role in, 331
  of law, 5
Science of canonical prescriptions
    (*ilm al-fi qh*), 5
Scientific anticipations, 335
Scientific exegesis, 332–345
  opponents of, 347–349
Scientific Signs in the Koran and in the
    Sunna, 339–340
Second mother, juridical condition of, 144
Secret/minor infanticide, 66
Self-transplantation, 165, 183
Sexual discrimination, 41
Sexual ethics, 118
Sexuality, 36
Sexually transmitted diseases (STDs),
    203, 219
Sexual revolution, effect of, 45
Shafi'ites, 5
Shari'a, religious Law of divine,
    1–2, 6–7, 10, 14–15, 305, 310–312,
    331, 344, 350
  and concordism, 333
  and FGM, 322–323
  orthodoxy, 1
  roots of the law (*usul al-fi qh*) and, 2
  sexual relations, 115
Shi'ite Islam, 6
  opinions in, 148–151
Sick
  fasting, 37 *n*
  prayer, 37 *n*, 38
Silence-assent, 21
Sleeping foetus, 106
Smoking, 36
Social security, 43
Sodomy, 201
Sperm banks, 158
Spread of AIDS, 221–223
State of necessity (*darura*), 34 *n*
STD, *see* Sexually transmitted diseases
Stem cells, genetics and, 245
  from miscarriages, 248
  research on, 248–250
  from surplus ova, 248
Sterilisation, 78–79

Sterility, 135, 159
Stillborn, funeral to, 99
Subsidiary rules of law
  criterion of equity (*istihsan*), 5
  criterion of utility (*istislah*), 5
  the principle of utility (*maslaha*), 5
Suffering and illness, end of life,
    267–270
Suicide, 290
  attacks, 295
  and martyrdom, 290–296
Suicide–martyrdom relationship, 293
Sunna (Tradition), 2–3, 5, 10, 306
  in *ahadith*, 3
Sunni Islam, schools of
  and effect of *igma'*, 6
  Hanafi te, 6
  Hanbalite, 6
  Malikite, 6
  and Muslim orthodoxy, 6
  Shafi'ite, 6
Supreme juridical-religious authority, 9
Surrogate maternity, 146
Surrogate maternity/mother, 145
Syphilis, 36

T
Tawafuq (matchmaking) project, 265
Ta'zir, 209
Temporary marriage (*mut'a*), 151
Territory of war (*dar al-harb*), 164
Thalassaemia, 258
The Higher Muslim Council, 16
The Organisation of the Islamic
    Conference, 13
Therapeutic abortion, 98
Trade in organs, prohibition of, 179
Transplants, 164–165, *See also* Organ
    transplants
  at molecular level, 240
  of nerve tissues and brain cells, 177
Tribadism-lesbianism, 149 *n*
Trisomy 21, 250
Typhoid and cholera in Tunisia, 199

U
Umma (Muslim community), 9, 20
Unborn child, inheritance, 110

Universal Declaration of Human
    Rights, and Muslim Law, 56
Uterine transplantation, 180
    and organ transplants,
        180–181

V
Voluntary abortion, 101

W
Woman's rights, abuse of the, 43

X
Xenotransplantation, 181–183

Z
Zina (fornication), 115, 137–138, 141

CPSIA information can be obtained
at www.ICGtesting.com
Printed in the USA
LVHW010803200722
723842LV00004B/76

9 781402 096150